SEPARATION PROCESSES IN WASTE MINIMIZATION

Environmental Science and Pollution Control Series

1. Toxic Metal Chemistry in Marine Environments, *Muhammad Sadiq*
2. Handbook of Polymer Degradation, *edited by S. Halim Hamid, Mohamed B. Amin, and Ali G. Maadhah*
3. Unit Processes in Drinking Water Treatment, *Willy J. Masschelein*
4. Groundwater Contamination and Analysis at Hazardous Waste Sites, *edited by Suzanne Lesage and Richard E. Jackson*
5. Plastics Waste Management: Disposal, Recycling, and Reuse, *edited by Nabil Mustafa*
6. Hazardous Waste Site Soil Remediation: Theory and Application of Innovative Technologies, *edited by David J. Wilson and Ann N. Clarke*
7. Process Engineering for Pollution Control and Waste Minimization, *edited by Donald L. Wise and Debra J. Trantolo*
8. Remediation of Hazardous Waste Contaminated Soils, *edited by Donald L. Wise and Debra J. Trantolo*
9. Water Contamination and Health: Integration of Exposure Assessment, Toxicology, and Risk Assessment, *edited by Rhoda G. M. Wang*
10. Pollution Control in Fertilizer Production, *edited by Charles A. Hodge and Neculai N. Popovici*
11. Groundwater Contamination and Control, *edited by Uri Zoller*
12. Toxic Properties of Pesticides, *Nicholas P. Cheremisinoff and John A. King*
13. Combustion and Incineration Processes: Applications in Environmental Engineering, Second Edition, Revised and Expanded, *Walter R. Niessen*
14. Hazardous Chemicals in the Polymer Industry, *Nicholas P. Cheremisinoff*
15. Handbook of Highly Toxic Materials Handling and Management, *edited by Stanley S. Grossel and Daniel A. Crowl*
16. Separation Processes in Waste Minimization, *Robert B. Long*

Additional Volumes in Preparation

SEPARATION PROCESSES IN WASTE MINIMIZATION

ROBERT B. LONG

Long Consulting, Inc.
Austin, Texas

CRC Press is an imprint of the
Taylor & Francis Group, an **informa** business

CRC Press
Taylor & Francis Group
6000 Broken Sound Parkway NW, Suite 300
Boca Raton, FL 33487-2742

First issued in paperback 2019

ISBN-13: 978-0-8247-9634-1 (hbk)
ISBN-13: 978-0-367-40171-9 (pbk)

Visit the Taylor & Francis Web site at
http://www.taylorandfrancis.com

and the CRC Press Web site at
http://www.crcpress.com

Preface

The U.S. Environmental Protection Agency has made the minimization of hazardous waste production its top priority in hazardous waste management. The principal ways to achieve this minimization are (1) reduction of the amounts of waste generated at the source, (2) recovery and reuse through recycling of specific hazardous materials recovered from the waste streams, and (3) removal of the hazardous materials from the waste stream to both reduce the volume of hazardous material and render the waste stream nonhazardous. Separation processes play a major role in all three of these methods.

Chemical engineering explores the science and technology of separation processes and provides the technical basis for design, construction, and operation of equipment used for waste minimization through separations. However, the operators of separation equipment and their immediate supervisors in the waste management industry usually are not chemical engineers familiar with the theory of separation processes. Therefore, this book was written to provide a simplified discussion of the separation processes used in waste minimization. It describes for laymen how separation processes work in practice and provides guidelines for troubleshooting the operation of separations equipment.

The book begins with a background section, which is followed by three sections devoted to describing individual separation processes in some detail. These latter three sections group the processes roughly according to the principles on which they are based, that is, by mechanical forces, by differences in rates of motion, or by differences in equilibrium concentrations between different states of matter in contact. Related processes are discussed together in the same chapter.

The three chapters in the first section provide an overview of what is and what is not waste minimization, a simplified summary of separation science, and a discussion of the types of separation processes used in waste minimization, including where they are described in this book. Each chapter in the remaining sections explains the fundamentals of the separation processes discussed, the applications of these processes to waste minimization, the types of equipment that are available along with the vendors, and the solutions to some of the common problems encountered operating the equipment.

The separation processes included in the book are all commonly used industrially but are at various stages of growth in their application to waste minimization. The economic incentive for their use changes as the cost of alternatives for waste disposal changes. At present, the cost of waste disposal continues to increase. On the other hand, many valuable materials can be recovered from streams that formerly went to waste. Thus, many separation processes once thought too expensive for processing wastes are finding their way into broad application due to the value of the products recovered for recycling.

This book is a cross between an operating manual and a text; it is definitely not a design manual. There are many good texts on separation processes written for design engineers. Four of them were used as primary references in writing this book. There are many books written on waste management, treatment, and disposal, also used as references. This book is confined to the separation processes either currently used in waste minimization or that have good potential for use in the future. It excludes both separations not expected to be applied to waste minimization and waste treatments that are not separation processes. The book attempts to give enough understanding of the separation processes covered to permit intelligent operation of such equipment after a little practical experience. Including only enough theory to grasp the basic concepts underlying the various separations discussed, it is truly aimed at the operators of separation equipment for waste minimization. However, it may also be useful as a text for introductory courses in environmental engineering and chemical engineering.

U.S. hazardous waste regulations are written in both the metric (centimeter, gram, second) and English (foot, pound, hour) systems of units. Both of these systems are used by chemical engineers, whose training is in a transition from the historically used English system to the metric system. To make this book more easily understandable to more people, the English system of units is used as the primary nomenclature. The exceptions to this practice are when the common unit in the trade for a given measurement is not in the English system. For example, with very small dimensions, as with particle sizes, the unit is the micron. At times, for the sake of clarity, units for both systems are given. The International System of Units (SI), which is commonly used only by scientists, is not used in this book.

Robert B. Long

Contents

1 Waste Minimization

1.1 INTRODUCTION

The United States is in the midst of a very exciting period of change in the way hazardous wastes are managed. This has been brought about by a change in the philosophy of regulating hazardous wastes. This change has resulted in the passing of two statutes by the Congress in the mid 1980s which removed ambiguities and closed loopholes in the basic waste management laws. These statutes also showed a much needed strong support from Congress for the management of hazardous wastes. They provided the Environmental Protection Agency (EPA) with clear and specific direction from the legislation itself. However, complying with the regulations is shooting at a moving target as the laws are changed to make sure they are effective at protecting the people and the environment.

The EPA had been established as the environmental regulatory agency of the federal government by the President in 1970. It is an independent agency in the executive branch of the U.S. Government. It has regulatory responsibility for administering the laws governing air pollution, water pollution, solid wastes, pesticides, and environmental radiation. The laws governing these areas of concern are the Clean Air Act (CAA), the Clean Water Act (CWA), the Toxic Substances Control Act (TSCA), the Resource Conservation and Recovery Act (RCRA), the Federal Insecticide, Fungicide, and Rodenticide Act (FIFRA), and the Atomic Energy Act of 1954. The administration of the federal waste management laws thus fell under EPA jurisdiction. EPA experience in trying to operate under the existing waste management laws had demonstrated to the EPA and to Congress by the early 1980s the need for major amendments to these laws.

The first of the amending statutes was the Hazardous and Solid Waste Amendments (HSWA) passed in 1984. It tightened up the Resource Conservation and Recovery Act (RCRA) of 1976 and changed the philosophy of hazardous waste management from "bury it" to "don't produce it." RCRA was, and with its amendments still is, the basic hazardous waste management law. It was designed as a preventive regulatory policy to cover hazardous waste management and provide "cradle-to-grave" control over hazardous wastes. However, this law as it was originally written relied heavily on unrestricted land disposal as the primary method of managing hazardous wastes. This discouraged industrial firms from investing capital in more effective, but more expensive, waste management technologies. Such more expensive technologies would often put them at a cost disadvantage when competing with companies relying solely on low cost land disposal. The changes made by the HSWA statute began a phasing out of landfills as the primary disposal method for hazardous wastes. It thus encouraged development and use of innovative alternate treatment technologies.

The second important amending statute, passed in 1986, was the Superfund Amendments and Re-authorization Act (SARA), also known as the CERCLA amendments, or Superfund amendments. This law amended the Comprehensive Environmental Response and Compensation Liability Act (CERCLA) of 1980, widely known as Superfund. In contrast to RCRA, which is a prevention law, Superfund is a remediation law to control cleanup of past mistakes. It provides for cleanup, compensation, and assignment of liability for release of hazardous substances into the air, land, or water. Thus, it is the primary law controlling cleanup of abandoned landfill sites which contain hazardous wastes. In its original form it relied heavily on sealing, flushing, treating, or excavating and moving hazardous wastes found in old landfills. The hazardous waste problems that were the reason for the remediation in the first place usually returned after a period of time to the dismay of all. The Superfund amendments changed the emphasis in cleaning up hazardous waste sites from (1) sealing in place or (2) excavating and transporting the hazard to a new disposal site to the much more desirable alternative of actual destruction of the hazardous waste. This led to the development of mobile equipment for on-site waste treatment to remove, concentrate, or destroy the hazardous material. It also led to the formation of commercial companies for the destruction of hazardous wastes. Destruction solves the problem once and for all, preventing later recurrence.

These two amending statutes, the RCRA amendments and SARA, have provided a strong regulatory framework to guide hazardous waste management in the United States. They provide a sense of certainty to waste management regulation and encourage industry to make the necessary investments to comply with hazardous waste minimization and treatment. Together, they are the main reason for the sense of excitement in hazardous waste management in the United States. They represent a national commitment to minimizing the hazardous component of waste management. They will be discussed in more detail in the following sections, particularly as they influence waste minimization.

1.2 WASTE MINIMIZATION

The Congress of the United States made hazardous waste minimization the policy of the country by the following statement in the 1984 amendments to RCRA:

> The Congress hereby declares it to be the national policy of the United States that, wherever feasible, the generation of hazardous wastes is to be reduced or eliminated as expeditiously as possible. Waste that is nevertheless generated should be treated, stored, or disposed of so as to minimize the present and future threat to human health and the environment (RCRA Amendments, 1984).

In its 1986 Report To Congress (EPA, 1986), EPA felt that specific quota regulations for waste minimization at that time would not be helpful but promised to update its viewpoint by 1990. This viewpoint still prevails at EPA in 1995, although strong support exists for voluntary waste minimization. Thus, mandatory quota regulations for hazardous waste minimization are not yet the law of the land.

1.2.1 Definition of Waste Minimization

EPA has adopted the following priority order for the management of hazardous wastes:

1. Source reduction
2. Recycling and recovery
3. Treatment
4. Disposal

Waste minimization is defined as the first two terms of this heirarchy, which are further defined as follows:

> *Source reduction*: The reduction or elimination of hazardous waste at the source, usually within a process. Source reduction measures include

process modifications, feedstock substitutions, improvements in feedstock purity, housekeeping and management practice changes, increases in the efficiency of equipment, and recycling within a process.

Recycling and recovery: The use or reuse of hazardous waste as an effective substitute for a commercial product or as an ingredient or feedstock in an industrial process. It includes the reclamation of useful constituent fractions within a waste material or the removal of contaminants from a waste to allow it to be reused.

Waste minimization does not include the various waste treatment techniques such as incineration, chemical oxidation, encapsulation, etc., which are highly effective in minimizing or eliminating the hazards of disposal of hazardous wastes. Furthermore, it deals with newly generated wastes and not with the aged materials found in Superfund sites.

1.2.2 EPA Policy on Waste Minimization

EPA (1987) summarized its policy on waste minimization in a report issued in 1987 entitled "The Hazardous Waste System." This summary is directly quoted below.

> EPA strongly favors preventing the generation of waste rather than controlling waste after it is generated. It is a national policy that the generation of hazardous waste be reduced as expeditiously as possible.

Within the private sector, strong incentives already exist to promote waste minimization. These incentives include:

- Large increases in the price of treating and disposing of hazardous wastes
- Difficulties in siting and permitting new hazardous waste units
- Concern with liability associated with managing hazardous waste
- Public pressure on industry to reduce waste generation

According to a recent EPA study, "Waste Minimization Issues and Options," a 20–30% reduction in waste volume may be possible through process changes, product substitution, and good housekeeping practices. Many firms have already markedly reduced and are continuing to reduce the amount of hazardous waste they produce. This is done through a variety of waste minimization techniques including:

- Source reduction
- Waste separation and concentration
- Waste exchange
- Reuse and recycling of waste

At present, there are three statutory requirements relating to waste minimization, all of them enacted in the 1984 Hazardous and Solid Waste Amendments. The requirements are summarized in the following:

- Generators must certify on their manifests that they have a program in place to reduce the volume and toxicity of waste [Section 3002(b)].
- Any new treatment, storage, or disposal permit must include a waste minimization certification statement [Section 3005(h)].
- As part of a generator's biennial report, generators must describe the efforts undertaken during the year to reduce the volume and toxicity of wastes generated [Section 3002(a)(b)] and document actual reduction achieved.

EPA's waste minimization program has as its main objective to foster the use of waste minimization through dissemination of technology and information. As mentioned earlier, EPA does not recommend mandatory quotas for hazardous waste reduction. However, EPA has launched a voluntary air pollution reduction program with targets of 33% reduction by 1992 and 50% reduction by 1995 for 17 chemicals which are produced in high volume, are considered of serious health and environmental concern, and can be reduced by pollution prevention. This program has been well received by industry and is being rapidly implemented.

In conclusion, concern over economic and liability issues are driving generators to reduce the volume and toxicity of hazardous waste produced. Waste minimization can alleviate the capacity problem by reducing the volume of waste requiring treatment and disposal.

1.2.3 Where Separation Processes Fit

The principal applications of separation and purification processes to hazardous waste minimization are (1) the removal of small amounts of hazardous material from large volumes of non-hazardous diluent to permit easier disposal of both the purified diluent and the much smaller volume of recovered hazardous waste and (2) the recovery for reuse of high quality components which have become contaminated in use and are no longer suitable for use unless the contaminants are removed. Typical of the first type of application are pretreatments of various wastewaters to remove toxic materials which prevent their treatment in public water treating plants. After the pretreatment, the purified water can usually be disposed of in the same way as more conventional wastewaters. The separation processes usually involved are adsorption, membrane processes, ion exchange, gas stripping, filtration, settling, and flotation, among others.

The second type of application is represented by the recovery and recycling of waste solvents, waste oils, waste paper, waste plastic, and scrap metals, to name a few. Recovery of useful materials for recycling is greatly simplified if the mixture of materials in the waste is not overly complex. This means that careful segregation of wastes by the generator can make recovery and reuse much less complex and more economically attractive. For example, if a generator pours all waste solvents and waste oils into the same container, it becomes very difficult to recover any of the specific solvents or oils in sufficient purity for reuse. On the contrary, if individual waste solvents are segregated as they are generated, their recovery in sufficient purity for reuse can be achieved using relatively simple separation processes. Economical small recovery units are available for these applications. Thus, effective recycling starts with segregation of wastes by the generator. The separation processes involved in these applications are sorting, magnetic and electrostatic separations, distillation, solvent extraction, evaporation, absorption, stripping, scrubbing, etc.

1.2.4 Segregation, Labeling, and Manifesting

Segregation of wastes is of paramount importance to the success of any recycling or reclamation program. The requirements for segregation depend upon the quality and quantities of waste available at each location, plus a knowledge of the viable options for recycling and disposal.

Labeling and proper manifesting of the segregated product is essential to maintaining the integrity of individual waste types. Each container should be labeled with the common name of the contents to permit decisions on recycling versus disposal, and to minimize cross contamination. Also, a large easy-to-read sign should be placed on top of or over the collection container.

Labeling of containers for collection should be done by the generator. The labels should include the waste accumulation start date, the name or type of waste, and the designation "Hazardous Waste," if appropriate. When the containers are ready for transportation to either an on-site or off-site recovery and recycling facility, it is good practice to use a manifesting system for record keeping, even though the universal manifest system is required only for off-site transportation.

1.2.5 On-Site Recovery

The decision between on-site and off-site recovery of recyclable materials depends upon the volumes and complexity of the wastes generated, the potential liability of moving them off-site, the degree of sophistication of the generator at operating recovery equipment, and the organizational philosophy of the generating entity. For large corporations having major capital resources, technically sophisticated personnel, and large volumes of a variety of hazardous wastes, it is probably preferable to organize a waste managemment group. This group would operate a waste minimization program companywide, including source reduction, waste segregation, waste movements and record keeping, and recovery and recycling. This group could be part of a larger group whose responsibilities also include hazardous waste treatment and disposal.

At the other extreme there are the smaller, less sophisticated generators who might produce only one or two kinds of waste in relatively small quantities. Examples of such generators might be dry cleaning establishments, automotive shops, and printing shops. These shops frequently have choices among (1) installing their own solvent recovery equipment, (2) dealing with a solvent supplier who provides fresh solvent and picks up waste solvent for off-site recovery and recycling, and (3) sending waste solvent to a commercial recycler. The choice depends on the volume and value of the solvent being recovered, the availability and costs of solvent suppliers who pick up waste solvent, and the transportation distance to commercial recyclers. In general, high volume or high solvent cost favor on-site recycling. Off-site recycling is favored by small operations and others with readily availabile commercial recyclers or solvent suppliers who recycle purified solvents recovered from waste solvents.

1.2.6 Off-Site Recovery

In most cases the best place to recover hazardous wastes is within the generator's facility because of the additional potential liability of moving the wastes off-site. However, when the volumes generated are small and the recovered material cannot be used on-site, it is probably best to recover the wastes at an off-site facility. Preference should usually be given to a raw material supplier who picks up the waste, recovers the raw material, and returns it for reuse, thus satisfying the requirements for recycling and waste minimization.

If the recovered material is not of sufficient quality to be recycled to its original use, the material can enter commerce for less demanding uses. Commercial off-site recyclers exist who will take waste solvents, recover and purify them to the extent that they can reenter commerce, and thus qualify the waste solvent as being recycled for waste minimization purposes. If the original generator does not have alternate uses for the recovered solvent, it is sold in the open market by the commercial recycler.

1.3 SUMMARY

The revisions to the hazardous waste management laws passed by the U.S. Congress in the 1980s and administered by the EPA have provided a firm basis for cradle-to-grave management of hazardous materials. This is to protect both the public and the environment. The emphasis of the law has changed the waste management priority from disposal of hazardous wastes in landfills to the minimization of the amount of hazardous waste generated in the first place. This minimization is achieved by waste reduction at the source, waste separation and concentration for volume reduction, waste exchange for reuse, or waste recovery and recycling for reuse. By definition, waste minimization does not include treatment and disposal techniques such as incineration, oxidation, and combustion, which also can protect the public and the environment from the undesirable effects of hazardous wastes.

This shift in emphasis has caused a corresponding shift in required expertise toward chemical engineering technology with particular emphasis on separation processes, reaction kinetics, heat and mass transfer, mixing, diffusion, and the like. Separation processes are of particular utility in waste concentration and in waste recovery for recycling. Their use in these applications is the subject of the rest of this book.

REFERENCES

Clean Air Act Amendments (1990). Clean Air Act Law and Explanation, Commerce Clearing House, Inc., Chicago.

Environmental Protection Agency (1986). Report to Congress: Waste Minimization, Vols I and II, EPA/530-SW-86-033 and EPA/530-SW-86-034.

Environmental Protection Agency (1987). The Hazardous Waste System, EPA Office of Solid Waste and Emergency Response, June, 1987, p. 3-30.

Environmental Protection Agency (1990). The Nation's Hazardous Waste Management Program at a Crossroads, EPA/530-SW-90-069.

Pollution Prevention Act (1990). Public Law 101-508, Nov. 5, 1990.

RCRA Amendments (1984). Hazardous and Solid Waste Amendments of 1984 (HSWA), Public Law 48-616, 1984, included in 40 CFR 260-272, July 1, 1991.

RCRA (1991). Protection of Environment, 40 CFR 260-272, July 1, 1991.

2

Basic Principles of Separation Science

2.1 INTRODUCTION

This book is not a design manual for separation processes and equipment. Such books already exist in Schweitzer (1979), King (1981a), Perry and Green (1984), and Rousseau (1987) and are primary references for designers of separations equipment. Rather, this book is intended for the operator of proposed or already existing equipment to help him or her to understand how the equipment works. It should also help in troubleshooting unsatisfactory performance during operation. The book is focused primarily on the separation processes encountered in hazardous waste minimization. It relies heavily on more easily understood graphical treatments rather than computer-based calculations.

In this chapter, the fundamentals that are generic to separation science are discussed. Extra details specific to the individual processes are detailed in Chapters 4–15. These chapters discuss applications of specific separation processes to waste minimization.

2.1.1 Definition of a Separation Process

It has been known for over a century that the natural tendency of all substances is to mix together spontaneously. This is a consequence of the second law of thermodynamics. Thermodynamics is the field of science which describes the relationship between heat and other forms of energy. There are three primary laws in thermodynamics. The first law states that although energy exists in many forms, the total quantity in an isolated system remains constant. That is, energy is neither created nor destroyed and if it disappears in one form it appears simultaneously in other forms. The third law states that the disorder of a substance, as defined by a term called entropy, may become zero at absolute zero in temperature. It actually does so for perfect crystalline substances in which all molecular motion has stopped. However, all substances at temperatures above absolute zero have a positive entropy and an increase in randomness.

The second law of thermodynamics states that all natural processes occur in a way that tends to increase the randomness (entropy), or homogeneity, of the universe. In other words, all naturally occurring processes are inherently mixing processes. The result of this thermodynamic relationship is that to separate mixtures into two or more products of different composition, energy must be supplied. This energy must be sufficient to overcome the entropy of mixing and thus cause the separation to occur. Thus, separation processes are in reality demixing processes and the theoretical minimum amount of energy required to make the separation is that equivalent to the entropy of mixing.

It takes work (energy) to bring order out of chaos (randomness), while the production of chaos from order can occur spontaneously. For separation processes, chaos is represented by homogeneous mixtures and order is represented by pure compounds. Small differences between product compositions represent small amounts of demixing and small energy requirements. Conversely, high product purities represent a large amount of demixing and thus a much larger energy requirement. The theoretical and practical energy requirements of separation processes are discussed later in Section 2.4.

In the second edition of his book on separation processes, King (1981b) defined separation processes as "those operations which transform a mixture of substances into two or more products which differ from each other in composition." This describes exactly what is meant by separation processes in this book.

2.1.2 Characteristics of Separation Processes

A large number of different separation processes exist which are based on different techniques to generate a composition difference between products. King (1981c) neatly categorized these different processes into types defined by (1) the nature of the feed, (2) the method of generating concentration differences, and (3) whether or not the separation depended on the addition of special separating agents as well as energy. These types of processes are now discussed.

Heterogeneous Feeds

Heterogeneous feeds contain more than one phase of matter (i.e., solid, liquid, or gas) and can contain almost any combination of solid, liquid, and gas in a solid, liquid, or gas matrix. The exception is gas in gas systems, which are always homogeneous except for some very special mixtures which separate into two gas phases at very high pressure. Many of the feed stocks encountered in waste management are of the heterogeneous type. Typical examples are slurries of solids, sludges, and oils in wastewaters; dusts and mists in exit gases from manufacturing processes or incinerators; mixtures of solid phases encountered in ore processing or solid waste handling; and emulsions from metalworking plants, to name a few. These types of feedstocks are separated by *mechanical separation processes* such as settling, centrifugation, flotation, filtration, magnetic separation, and electrostatic precipitation.

Homogeneous Feeds

Relatively simple mechanical equipment is needed to separate heterogeneous feeds into their individual phases. However, homogeneous feeds usually require a diffusional transfer of matter to one of the product streams. These processes work by generating two phases of different composition from the feed and then separating the phases to give two products of composition different from both the feed and each other. The processes are called *equilibrium processes* if the two product phases are in equilibrium when they are generated, as in simple distillation. Alternatively, the phases may be thoroughly mixed with one another to achieve equilibrium prior to phase separation. Examples of such equilibrium processes are distillation, extraction, adsorption, ion exchange, crystallization, evaporation, etc.

Unlike the equilibrium processes, some separation processes get their driving force from a gradient of temperature, pressure, composition, electric potential, or the like. They achieve a change in composition of the product from differences in diffusion rate of feed components through a barrier across which the gradient is applied. Processes of this type are called *rate-controlled processes*. They are illustrated by (1) membrane processes such as gaseous diffusion, ultrafiltration, reverse osmosis, and permeation and (2) processes depending on slow chemical reactions to make the separation, as in some carbon dioxide scrubbing processes.

2.2 EQUILIBRIUM AND THE THEORETICAL STAGE

Equilibrium processes are in essence phase-mixing/phase-separation processes. Equilibrium processes which can be used to separate homogeneous feeds require

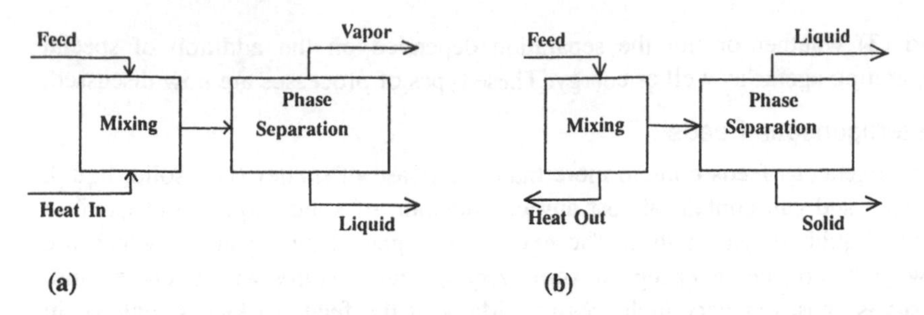

Figure 2.1 Schematic flow plans for (a) continuous distillation and (b) crystallization processes.

some way of generating a second phase of different composition from the bulk feed. The two phases must be thoroughly mixed and then separated to get products. With distillation and crystallization from a melt, this is done by either adding or removing heat, respectively, as shown in Figure 2.1.

The new phases generated by the addition or removal of heat from the liquid feeds have a composition in equilibrium with the residual liquid from the separation step. The composition difference between the phases leaving the separator may be either quite small, as is the case for distillation of close boiling liquid mixtures, or quite large, as is often encountered in crystallization.

For many homogeneous feeds containing small amounts of high boiling or high melting impurities, it is impractical or expensive to use energy alone to generate the second phase. For example, with small amounts of heavy oils or chemicals in water, it is usually unwise to distil off all the water to remove and

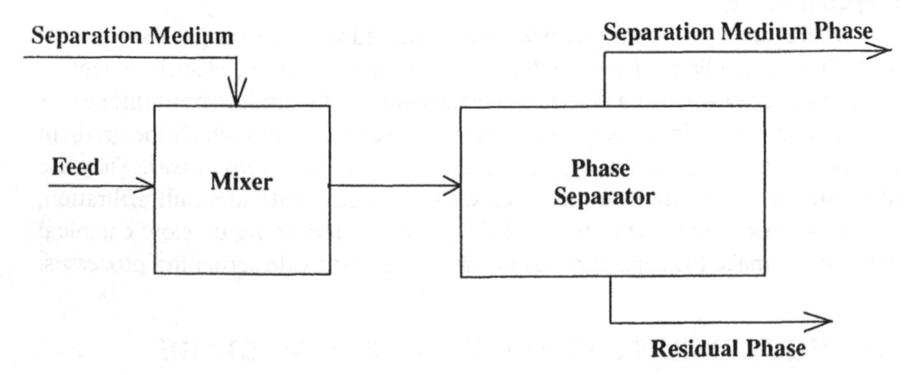

Figure 2.2 Use of a separation medium to generate two phases.

recover the small amount of impurity. Similarly, freezing all the water to remove it as pure water is also expensive. However, freezing has been used in seawater desalting where the clean water was the desired product.

To provide more economical separation processes, the required second phase can be obtained by the addition of a special material insoluble in the homogeneous feed. This material will selectively interact with one of the components in the feed giving a second phase differing in composition from the feed. This is illustrated in Figure 2.2. The separation medium can be a solid as in the case of adsorbents, a liquid as in the case of extraction solvents and absorber fluids, or a gas as in the case of stripping agents. These materials are known in each case as the separation medium and (along with heat) are the agents that make most separations possible.

2.2.1 The Separation Medium

The separation medium can be defined as an additional stream of matter or energy which must be supplied to the separation process. It generates the two phases of different composition which are necessary to achieve a separation. In the case of distillation, the separating medium is the heat which generates the vapor phase. This provides the necessary two phases (i.e., vapor and liquid) to get a separation. In the case of liquid-liquid extraction, the separating medium is the solvent or solvents necessary to provide the two liquid phases. Finally, for crystallization, the separating agent is the refrigeration or precipitation agent necessary to make the system supersaturated. The solid phase then comes out of solution as an insoluble precipitate. Obviously, combinations of media can be used to generate two phases. For example, in extractive distillation both a solvent and heat are used to generate a two-phase system of vapor and liquid.

It may sometimes appear that the use of a separation medium other than heat to cause two phases to be formed gets around the need for energy to achieve demixing. For example, in solvent extraction energy is usually not required to generate two phases of different composition. However, the energy of demixing is provided when the components of the feed are recovered from the solvent, usually by distillation. There isn't any free lunch, just a minimum of waste energy. This is generally true of all separation processes which use a medium other than energy to generate two phases. Some other examples of such processes are (1) adsorption, where a solid is used to generate the adsorbed phase in equilibrium with either a bulk liquid or a bulk gas, (2) scrubbing, where a liquid is used to generate a solution phase in equilibrium with the gas phase, and (3) precipitation from solution by the addition of a second solvent to generate supersaturation. In all these cases the energy required for demixing is provided during the step which regenerates the separation medium for recycling and continued use.

2.2.2 Vapor-Liquid Equilibria

Vapor-liquid, liquid-liquid, and solid-liquid equilibria are the underlying bases for most of the separation processes applicable to the minization of homogeneous hazardous wastes. The major exceptions are the diffusional rate-controlled membrane processes and certain gas scrubbing processes which are also rate controlled. The equilibrium relationships at constant temperature for vapor-liquid, liquid-liquid, and solid-liquid phase equilibria are illustrated schematically for binary feedstocks in Figures 2.3 through 2.6.

Figure 2.3 shows the *vapor-liquid equilibrium* versus temperature at constant total pressure for an ideal system. Typical ideal systems are mixtures of saturated hydrocarbons or mixtures of aromatic hydrocarbon. The composition of the mixture is given in terms of the mole fraction, which is the ratio of the moles of one component divided by the total moles in the mixture. It can vary from zero to one. A mole is an amount of material equal to one molecular weight of that material. For example, a mole of water which has a molecular weight of 18 would be 18 units of weight in any consistent units such as pounds, grams, tons, etc. Pure water would have a mole fraction of one because it would contain 18 pounds of water in a total of 18 pounds, i.e., one pound-mole of water in a total of one pound-mole. A 50:50 mixture on a weight basis of ethyl alcohol and water would contain 50/46 = 1.09 pound-moles of alcohol and 50/18 = 2.78 pound-moles of water. This would be equivalent to a mole fraction of 1.09/(1.09 + 2.77) = 0.28 for alcohol and 2.77/(1.09 + 2.77) = 0.72 for water

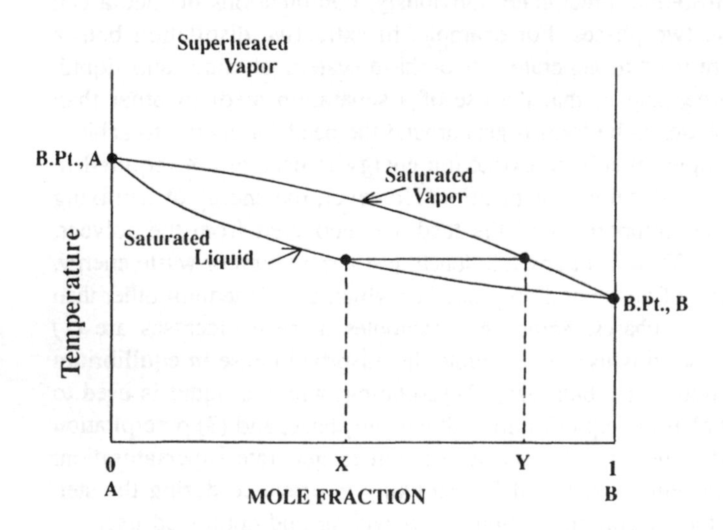

Figure 2.3 Vapor-liquid equilibrium for ideal binary liquid mixtures.

in a 50:50 mixture. If the weights were in grams, the moles would be gram-moles and if in tons, ton-moles. Furthermore, if the mixtures contained more than two components, the mole fractions would still be the moles of an individual component divided by the total moles of all components.

Returning to Figure 2.3, the left-hand intercept is the temperature at which the vapor pressure of the higher boiling (or lower vapor pressure) component, A, of the feed is equal to the total pressure on the system. Similarly, the right-hand intercept is the temperature at which the vapor pressure of the more volatile or lower boiling component, B, of the feed is equal to the total pressure. Changing the total pressure will change both of these intercepts in the same direction, but not necessarily by the same amount. The horizontal line at a given temperature is called a "tie line" because it ties together the two equilibrium compositions of vapor and liquid at a given temperature and pressure. These two compositions define the vapor-liquid equilibrium for a given temperature and pressure. By drawing horizontal lines at various temperatures between the left and right intercepts, the equilibrium relationship for the entire composition range can be defined.

Special equilibrium modifying agents are frequently added to the above systems to increase the composition differences at equilibrium. For example, a vapor-liquid system which would normally be the basis for a distillation process could have a liquid selective solvent added to improve the vapor-liquid equilibrium. This would convert the distillation process into an extractive distillation process. Added materials can also be used to generate a two phase system from a feed mixture that would otherwise be homogeneous at the desired operating conditions. For example, a solvent immiscible with the feed could be added to generate two phases from a homogeneous feed and thus provide the equilibrium basis for an extraction process. As an extension of this, two immiscible solvents can be used to distribute the feed between them to provide the basis for a double solvent extraction. Examples of this are the Duo Sol process for heavy oils, which uses liquid propane and phenol as solvents, or the analytical technique which uses countercurrent distribution of feed between aqueous and non-aqueous solvents.

2.2.3 Liquid-Liquid Equilibria

There are many variations in *liquid-liquid extraction equilibria*, among which are the three types shown in Figure 2.4. These three are (1) partial miscibility of a feed with a selective solvent as illustrated in Figures 2.4a and b, (2) distribution of a single compound between two immiscible phases as illustrated in Figure 2.4c, and (3) the distribution of a feed mixture between two immiscible solvents as illustrated in Figure 2.4d. In the first case, the relative concentration of feed components dissolved in the solvent is different from their relative concentration in the non-solvent phase. This is typical of solvent extraction of

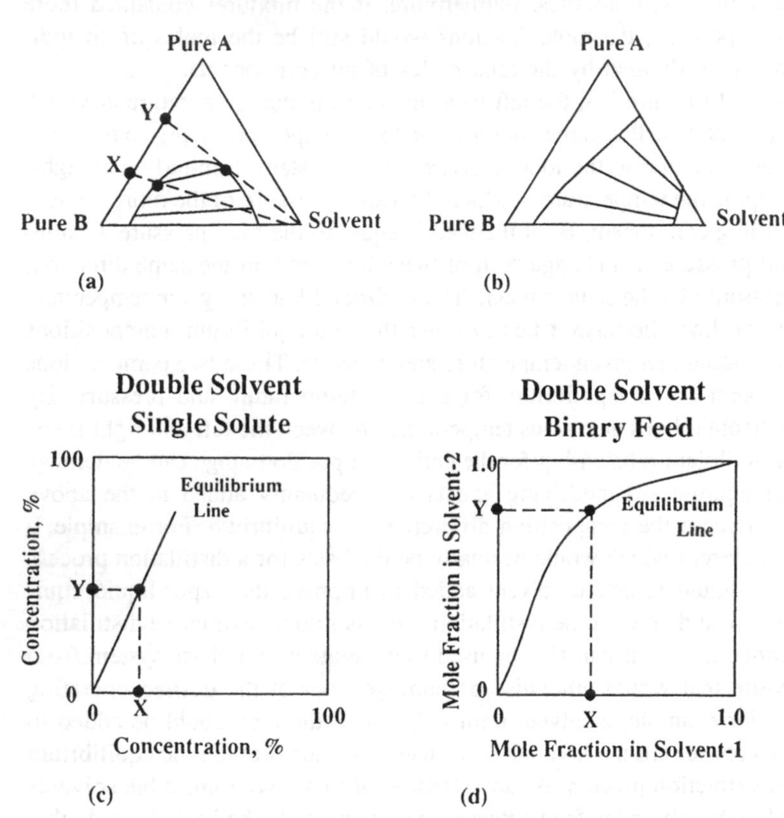

Figure 2.4 Representation of various types of liquid-liquid extraction equilibria. (a) Closed two-phase area, (b) open two-phase area, (c) distribution of solute, (d) solvent-free equilibrium.

aromatics from lube oils or kerosine. The second type is exemplified by extraction of phenol from a wastewater stream with kerosine. The third type is often used where the feed would be soluble in either solvent in the absence of the second solvent. A typical system of this type is the distribution of low molecular weight alcohols and ketones between water and naphtha.

Single solvent, mixed feed systems are much like vapor-liquid systems except that instead of using heat to generate a second phase, the second phase is generated by the addition of a solvent selective for some feed components over others. The equilibrium relationships for these systems are often plotted in a triangular diagram, as illustrated in the two graphs in Figures 2.4a and b. The triangular diagram in Figure 2.4a shows an equilibrium of the type encountered in the extraction of aromatics from kerosine using liquid sulfur dioxide as solvent. Here the apex A of the triangle represents pure aromatics, the left-hand corner B represents pure non-aromatics, and the right-hand corner S represents the pure solvent, sulfur dioxide. Two liquid phases exist inside the curved envelope and a single phase exists outside the two-phase envelope. The maximum aromatic concentration that can be obtained in this extraction is determined by the intecept on the left side of the triangle of a tangent to the two-phase region drawn from the solvent vertex. This is the upper limit of aromatic concentration which will still permit two phases to exist. At higher aromatic concentrations, the entire hydrocarbon phase would be soluble in the solvent, the second phase would disappear, and no further enrichment of aromatics could occur.

The intercepts on the base of the triangle for the two-phase region show the solubility of oil in the solvent phase on the right and the solubility of solvent in the oil phase on the left. The nearly horizontal tie lines inside the two-phase region connect the equilibrium compositions of the two phases. By projecting lines from the solvent vertex through the ends of the tie lines to the solvent-free left side of the triangle, the solvent-free equilibrium compositions X and Y can be obtained for use in two-dimensional graphic analysis methods. The value of Y is the mole fraction of aromatics in the extract phase and the value of X is the mole fraction of aromatics in the raffinate phase. The two-dimensional graphic analysis will be discussed later in Section 2.3.

The triangular diagram shown in Figure 2.4b has a two-phase region which intercepts both the base and the right-hand side of the triangle. This occurs when the solvent is only partially miscible with both pure components of the feed and sometimes occurs naturally by the proper choice of solvent and operating temperature. It can sometimes be forced to happen by using as solvent a mixture of totally miscible solvents which have widely different solubilities for the feed components. By varying their proportions as the more soluble component in the feed is enriched, the solubility can be maintained relatively constant. A typical natural system showing this behavior is the *n*-heptane, methylcyclohexane, aniline system studied by Varteressian and Fenske (1937). It is typical of systems of this type that the tie lines do not give great concentration differences between the solvent-free equilibrium phases. Thus, separations based on this type of equilibrium are inherently difficult.

The equilibrium between the compositions of feed components in the solvent and oil phases is affected by the solubility of oil in the solvent phase. Larger differences in solvent-free composition occur at lower oil solubilities. This is because for a given slope of tie line, the longer it is, the farther apart are the solvent-free compositions of the phases. Since tie line slope does not change much with oil solubility for a given system, better separation equilibria are obtained at lower solubilities at the expense of increased solvent requirement.

The double solvent, single solute systems are frequently encountered in management of wastewaters containing low concentrations of toxic materials. These are produced in many industrial operations. An example is the phenolic water that was produced in the past by lube oil plants, which used phenol as an extraction solvent. This undesirable material in the wastewater can be removed by extracting the wastewater with an immiscible solvent, in this case kerosine. This permits simpler disposal of the phenol-free wastewater. In such cases the equilibrium between the concentration of solute in the water phase and the concentration of the solute in the solvent phase is relatively constant. This constant is known as the distribution constant. It is usually written as the ratio of concentrations rather than mole fractions for the solute, phenol, between the two phases. This type of equilibrium is illustrated in Figure 2.4c.

The double solvent, mixed feed systems are usually used to provide a second phase where the feed would be completely miscible with the selective solvent chosen. For example, if it is desired to use water as a solvent to separate a mixture of low molecular weight oxygenated compounds such as alcohols, esters, and ketones, a second phase can be generated by adding a second solvent having low solubility in water. This second phase can, like water, have a high mutual solubility with the feed mixture. Such materials can be hydrocarbon mixtures such as naphtha or pure hydrocarbons. In such a case, the feed components in the water phase would be enriched in the more water-soluble compounds, and the less water-soluble feed components would tend to concentrate in the naphtha phase. To get a clear picture of what is happening to the feed components, such equilibria are usually plotted on a solvent-free basis at constant temperature, as is shown in Figure 2.4d.

2.2.4 Solid-Liquid Equilibria

The equilibrium relationships for *solid-liquid systems* fall into two general types, eutectic systems and solid solution systems. A typical eutectic system is shown in Figure 2.5a.

The eutectic equilibrium relationship between solids and liquids is frequently encountered in either crystallization from a melt or crystallization from a solution. It is extremely useful because it permits recovery of very pure materials from mixtures. For example, in Figure 2.5a the temperature is plotted against the composition of a mixture of components A and B. The area named "Solution" is the temperature-composition region where all the compositions are liquid. If we start at a composition C in the solution region above the area labeled "A + Solution" and start lowering the temperature, component A starts to crystallize out of solution when the temperature reaches the upper boundary of "A + Solution." This precipitation of essentially pure A continues as temperature is lowered. The composition of the residual solution in equilibrium with

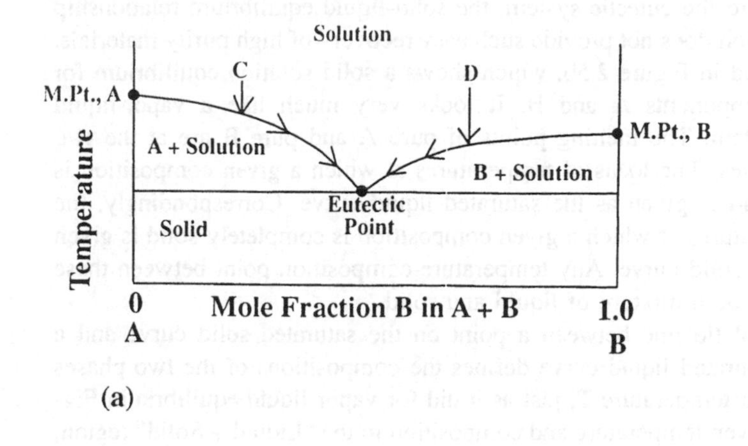

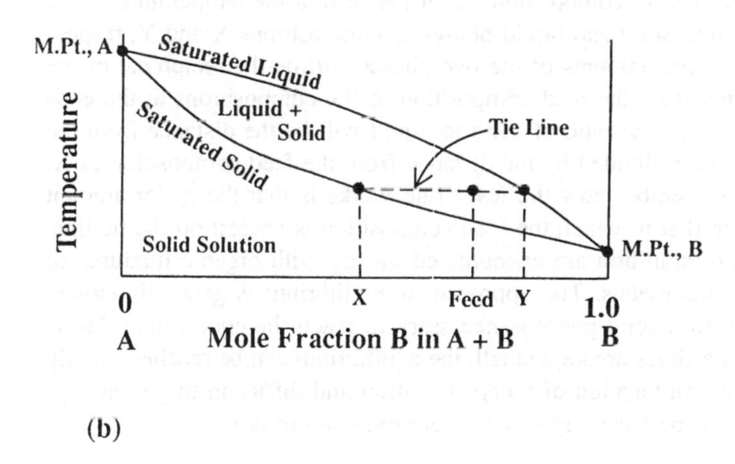

Figure 2.5 Solid-liquid equilibria for (a) eutectic mixtures and (b) solid solutions.

the crystals follows the path of the arrows until it reaches the eutectic point. At this point the entire system solidifies and no further change takes place. However, if the pure A crystals were mechanically removed from the slurry with solution before everything turned solid, a good yield of pure A crystals would be obtained. Similarly if one started with an overall composition at D and cooled the solution, one could get pure B crystals to precipitate out of the solution in the temperature-composition region labeled "B + Solution." This illustrates the tremendous separating power of crystallization from a eutectic system.

In contrast to the eutectic system, the solid-liquid equilibrium relationship for a solid solution does not provide such easy recovery of high purity materials. This is illustrated in Figure 2.5b, which shows a solid solution equilibrium for mixtures of components A and B. It looks very much like a vapor-liquid equilibrium system. The melting points of pure A and pure B are at the two ends of the curves. The locus of temperatures at which a given composition is completely liquid is given as the saturated liquid curve. Correspondingly, the locus of temperatures at which a given composition is completely solid is given as the saturated solid curve. Any temperature-composition point between these two curves will be a mixture of liquid and solid.

A horizontal tie line between a point on the saturated solid curve and a point on the saturated liquid curve defines the compositions of the two phases in equilibrium at temperature T, just as it did for vapor-liquid equilibria in Figure 2.3. For a given temperature and composition in the "Liquid + Solid" region, the quantities of the two phases are determined by the so-called lever rule. According to this rule an overall composition, X, of the feed at the temperature of the tie line will split into solid and liquid phases of compositions X and Y, respectively. The ratio of the amounts of the two phases will be the reciprocal of the ratio of the lengths from the feed composition to the compositions at the ends of the tie line. That is, the ratio of solid to liquid will be the distance from the feed composition to Y divided by the distance from the feed composition to X. An easy way to remember how the lever rule works is that the major amount of material will be that to which the feed composition is nearest on the tie line.

Solid solution equilibria are encountered mostly with organic mixtures or mixtures of precious metals. The approach to equilibrium is generally slower because diffusion in a solid phase is necessary to reach the equilibrium. However, if the solid particles are kept small, the equilibrium can be reached rapidly enough through a combination of recrystallization and diffusion to permit separation processes to be based upon solid solution equilibrium.

2.2.5 Solid-Vapor Equilibria

The equilibrium between solids and vapors is of importance mainly in sublimation processes in which a heated solid is vaporized directly without passing through the liquid state. Such processes need to be carried out at or below the vapor pressure of the solid at the temperature of sublimation. Thus, relatively high vacuum is frequently required. Alternatively a carrier gas can be used to sweep the vapors out of the vaporization chamber. The vapors released from the solid are condensed and deposited on a cold surface as a separate solid. The deposited solids are the sublimed product. The equilibrium involved is that between the solid feed and the vapors derived from it. It is much like vapor-liquid equilibrium and is illustrated in Figure 2.6.

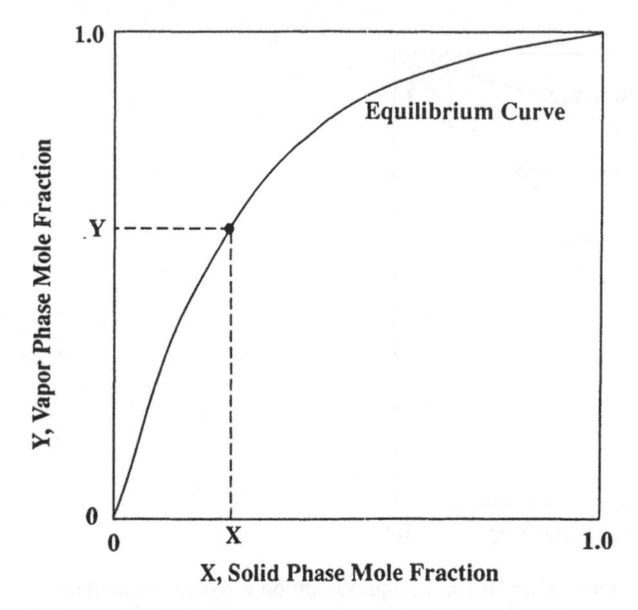

Figure 2.6 Schematic representation of solid-vapor equilibria.

Solid-vapor equilibrium does not play a significant role in waste minimization because sublimations are usually limited to a single contacting stage and therefore provide limited separation. They also require expensive vacuum equipment and in some cases refrigeration equipment. However, in cases where only one component of the feed is volatile, such as in freeze-drying for removal of water from heat sensitive materials, solid-vapor equilibrium can be exploited effectively.

2.2.6 Generalized Graphic Treatment of Equilibrium

The preceding types of equilibria can be more conveniently displayed by plotting the equilibrium composition, Y, of one phase against the equilibrium composition, X, of the other phase on a separation medium-free basis. This is shown in Figure 2.7 for vapor-liquid equilibria. For solvent extraction, this is a solvent-free basis, for adsorption it is an adsorbent-free basis; for extractive distillation and absorption, it is a solvent-free basis; and so on. The convention is to make Y the composition of the phase enriched by the equilibration. For example, in Figure 2.7, Y is the composition of the more volatile component in the vapor phase. For solvent extraction, Y would be the composition of the more soluble component in the solvent phase, i.e., the more soluble component in the solute in the solvent phase. Similarly, Y would refer to the adsorbed material in an

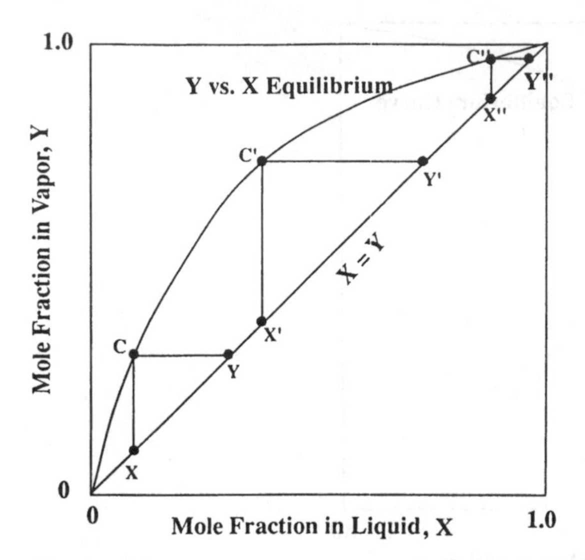

Figure 2.7 Generalized graphical treatment of equilibrium on a separation medium-free basis. Illustration is for vapor-liquid equilibria.

adsorption, the vapor phase in absorption and extractive distillation, and the solid phase in crystallization. Correspondingly, X would refer to the composition of the alternate phase on a separation-medium-free basis.

The discussion in the rest of this chapter will use distillation and vapor-liquid equilibria to illustrate various fundamental phenomena in separation processes. However, the reader should understand that corresponding phenomena exist by analogy in processes using other phase relationships. They can be readily visualized using the X-Y method of plotting equilibria data.

In Figure 2.7, each point on the equilibrium curve is the value of Y in the vapor phase that is in equilibrium with X for the liquid phase. By drawing a horizontal line from a point C on the equilibrium curve to where it intersects with the the X = Y line at Y, we derive the composition of the liquid which is obtained by condensing the equilibrium vapor. The composition difference between the liquid and vapor products from a point on the equilibrium curve defines one equilibrium stage or theoretical plate. The amount of the difference changes depending on the composition, X, X′, or X″ of the equilibrium liquid. A theoretical plate, i.e., the step from X to Y, X′ to Y′, or X″ to Y″, corresponds to larger composition differences near the middle of the equilibrium curve than it does at the ends. This shows that it is easier to achieve large composition differences for nearly equimolar mixtures than it is to achieve products of high purity. Figure 2.7 displays the vapor-liquid equilibrium for an ideal system, i.e., one that obeys

Raoult's law. Raoult's law states that the vapor pressure, or partial pressure, of each component in the vapor phase is equal to its mole fraction in the liquid phase times the vapor pressure of the pure liquid component at the same temperature. This is illustrated for components A and B of an ideal system in Eq. 1.

$$p_A = P_A^\circ X_A$$
$$p_B = P_B^\circ X_B \qquad (1)$$

From Dalton's law of partial pressures, the total pressure (Π) is the sum of the partial pressures of the individual components, as illustrated in Eq. 2.

$$\Pi = p_A + p_B \qquad (2)$$

Therefore, the mole fraction, Y, of each component in the vapor phase is the partial pressure of each individual component divided by the total pressure, as illustrated by Eq. 3.

$$Y_A = \frac{p_A}{\Pi} = \frac{p_A^\circ X_A}{\pi}$$
$$Y_B = \frac{p_B}{\Pi} = \frac{P_B^\circ X_B}{\Pi} \qquad (3)$$

$$\frac{Y_A}{Y_B} = \frac{P_A^\circ X_A}{P_B^\circ X_B} = \alpha \frac{X_A}{X_B} \qquad (4)$$

The term $P_A^\circ / P_B^\circ = \alpha$ in Eq. 4 is a constant at constant temperature for an ideal system. The constant, α, is called the relative volatility between components A and B of the binary mixture. The relative volatility varies with temperature, and thus with pressure, generally increasing as temperature and pressure are decreased. However, for close boiling ideal systems, changes in relative volatility with temperature are usually negligible because the vapor pressure-temperature curves of the two components are so similar.

2.2.7 Azeotropes

In ideal systems the intermolecular attractive forces are very similar for both of the pure components of the feed mixture. The molecules cannot tell the difference in attractions between like molecules and between unlike molecules. However, there are many systems where the differences are more pronounced, leading to non-ideality. These systems usually contain at least one component made up of polar molecules which interact strongly with other polar molecules through hydrogen bonding. The polarity is usually caused by the presence of an oxygen, nitrogen, or sulfur atom in the molecule which strongly attracts

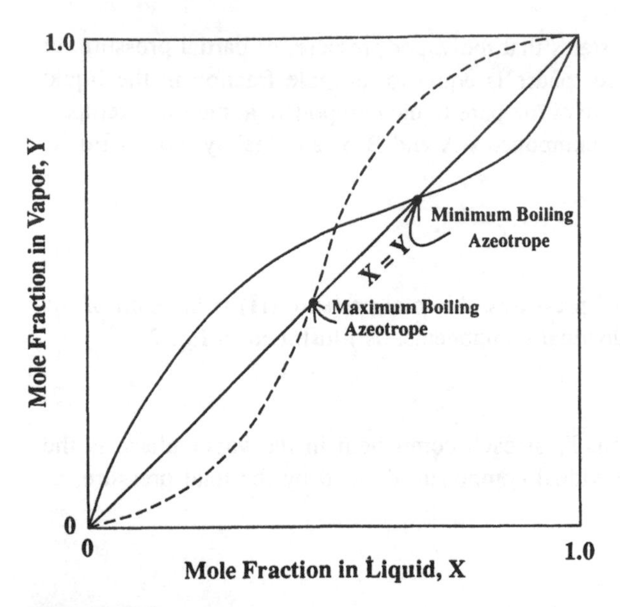

Figure 2.8 Graphical illustration of equilibrium for maximum boiling and minimum boiling azeotropes.

hydrogen atoms from other molecules. Mixing a highly polar material with a non-polar material can reduce the intermolecular bonding in the polar material and raise its vapor pressure and lower its boiling point. Similarly, mixing a weakly polar material with a strongly polar material can increase the strength of intermolecular bonding, decrease the vapor pressure, and thus raise the boiling point. In the first case the boiling point goes through a minimum with changing composition, and in the second case the boiling point goes through a maximum with changing composition. The equilibrium curves for such systems actually cross the X = Y line as shown in Figure 2.8. They cross from above for a minimum boiling mixture or from below for a maximum boiling mixture.

The composition at which the equilibrium curve crosses the X = Y line is called the azeotropic composition and the mixture having this composition is called an azeotrope. Because the equilibrium point at the azeotropic composition has the coordinates X = Y, no composition change is obtainable by equilibration of the two phases at this point. If the equilibrium curve crosses the X = Y line from below as concentration of the more volatile component in the liquid is increased, the azeotrope is called a maximum boiling azeotrope. In this case the two components in the feed interact with each other more strongly than they do with themselves and the vapor pressure of the mixture is less than that of the pure components. This gives a higher boiling point at constant pressure for the azeotrope than for either of its pure components and produces a maximum boiling azeotrope.

Similarly, if the equilibrium curve crosses the X = Y line from above, the azeotrope is called a minimum boiling azeotrope. Here the two components of the feed interact more strongly with themselves than they do with each other, leading to higher vapor pressures for mixtures than for the pure components. This gives a lower boiling point at constant pressure for the mixtures and for the azeotrope. Such an azeotrope has a lower boiling point than either of its pure components and is called a minimum boiling azeotrope.

In order for an azeotrope to form, the boiling points of the pure components must usually be within about 30°C of each other. The likelihood of formation increases as the boiling point difference decreases. Azeotropy is also enhanced when the components of the feed mixture differ greatly in chemical type, and thus in the slopes of their vapor pressure versus temperature curves. For example, mixtures of highly polar compounds such as alcohols and organic acids with relatively non-polar hydrocarbons such as paraffins and aromatics lead to minimum boiling azeotropes. This is because the non-polar component dilutes the strong intermolecular attraction of the polar component and leads to increased vapor pressure for the mixtures over what would be expected for an ideal mixture. This gives rise to a minimum boiling point azeotrope.

Azeotrope formation can frequently be avoided by changing the operating pressure and thus the operating temperature of the separation. A technique for determining the temperature and pressure range over which azeotropes can exist for a given pair of components is given by Nutting and Horsley (1952). This technique predicts that binary azeotropic systems can become non-azeotropic at both low and high pressures while exhibiting azeotropy at intermediate pressures. However, only a few actual systems have been found that show signs of this behavior, principal among which is the methanol-acetone system.

Another way of eliminating azeotrope formation consists of changing the vapor-liquid equilibrium for the system by adding materials which change the relative volatility of the components. This can be done by adding a higher boiling selective solvent to increase the relative volatility sufficiently that no azeotrope forms. This technique is called extractive distillation. It works by suppressing the vapor pressure of one of the components of the azeotrope, preferably the higher boiling one, by strong interaction with the solvent.

While the formation of azeotropes is frequently a nuisance in attempts to purify materials by distillation, it can also be a great benefit. For example, ethyl alcohol is usually produced as the binary azeotrope with water which contains 95% ethanol. To make pure ethanol, benzene or cyclohexane is frequently added to the feed to form a minimum boiling ternary azeotrope of alcohol, water, and hydrocarbon containing more water than the binary ethanol azeotrope. This removes the water overhead and produces pure alcohol as the bottom product

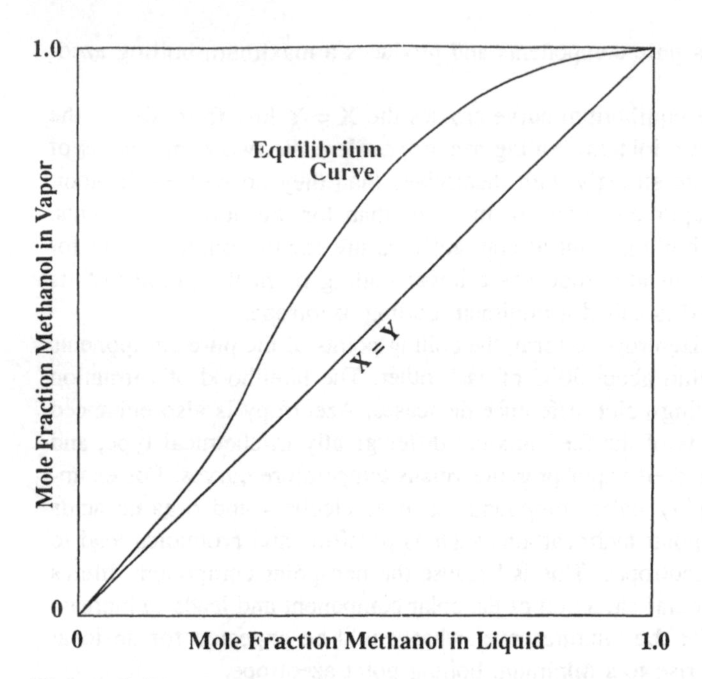

Figure 2.9 Illustration of non-ideal behavior of methanol-acetone equilibrium.

from the distillation. Such added materials are called entrainers because they usually form a minimum boiling azeotrope and carry the desired component overhead.

Azeotrope formation is caused by the non-ideality in the physical chemistry of solutions. However, it is possible to have non-ideality produce equilibrium curves of strange shapes without actually forming azeotropes. That is, the equilibrium curve does not cross the X = Y line. This is illustrated in Figure 2.9 for methanol-acetone equilibrium at 0.27 atmospheres (200 mm Hg) pressure, which was calculated from data given in Britton, Nutting, and Horsley (1952). The methanol/acetone system does not form an azeotrope at this pressure but the system is clearly non-ideal. At atmospheric pressure this same system does form an azeotrope showing that increasing pressure sometimes eliminates azeotropes and sometimes causes them. In this particular case it does both because the azeotrope disappears again at 20 atmospheres (300 psig) pressure.

The equivalent of azeotrope formation in distillation can be encountered in other separation processes where the relative composition of the feed components in the two phases becomes equal. This can happen in solvent extraction or extractive distillation, for example.

The method of displaying the vapor-liquid equilibrium shown in Figure 2.7 is the basis for the McCabe-Thiele (1925) graphical method for analyzing the design and performance of distillation columns. This will be discussed further in Section 2.3 for the separation of binary mixtures. However, by making some simplifying assumptions which do not greatly depart from reality, this type of equilibrium curve is widely used to understand many different types of separation processes. For example, solvent extraction equilibria are often plotted on triangular diagrams to show the ternary systems in detail, as was described for Figure 2.4. Plotting the equilibrium data on a solvent-free basis as Y versus X for the components of the feed stock, gives an equilibrium curve very similar to that shown in Figure 2.7. This is true for both solvent extraction and extractive distillation.

While the above equilibrium diagrams use phase composition in mole fractions as the basis for the equilibrium relationships, any additive property, such as density, refractive index, etc., can be used with multicomponent systems to simplify the equilibrium relationships and permit analysis by the McCabe-Thiele method.

2.3 GRAPHICAL TREATMENT OF MULTISTAGE SEPARATIONS

The McCabe-Thiele plot provides a useful way of using basic equilibrium data to predict the outcome of various types of separation processes. In this graph the equilibrium composition of one phase is plotted against the corresponding equilibrium composition of the other phase. This is illustrated in Figure 2.10.

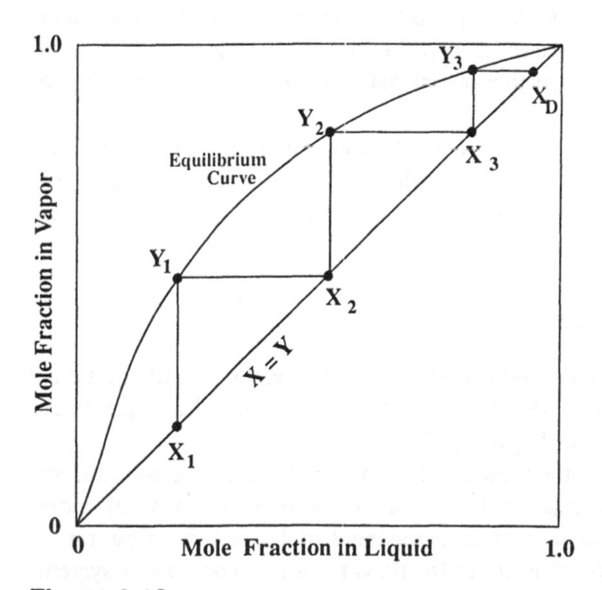

Figure 2.10 McCabe-Thiele plot for a three-stage distillation.

For distillation the values of Y are the mole fractions of the more volatile component in the vapor phase. The X values are the liquid phase mole fractions of the more volatile component. The points on the equilibrium curve are the liquid and vapor phase equilibrium points over the entire range of liquid compositions. This type of equilibrium plot can also be used for other types of separation processes, such as extraction, extractive distillation, countercurrent distribution, etc., by plotting the equilibrium on a separation-medium-free basis. For example, liquid extraction equilibria are easier to understand on a solvent-free basis than when plotted on a triangular diagram. Furthermore, if extraction is done with mixed solvents, as is often the case, the triangular diagram loses some of its rigor and also becomes an approximation.

2.3.1 The Concept of the Theoretical Stage

Extending the vapor phase composition from a point on the equilibrium curve in Figure 2.10 until it intersects the X = Y line is the same as condensing the vapor phase. Thus the vapor phase in equilibrium with liquid X_1 is condensed to get a second liquid composition X_2. The change in composition achieved by one equilibrium contact defines the equilibrium stage in separation processes. This is often called the theoretical plate in vapor-liquid contacting processes. The vapor in equilibrium with liquid X_2 can also be condensed to form a liquid of composition X_3 and so on for as many stages as are required to get the desired composition of the final distillate. The number of steps needed to get from liquid composition X_1 to the desired product composition X_n is the number of theoretical stages or plates required to make the desired separation. This is illustrated in Figure 2.10 as three theoretical plates required to get from X_1 to X_D.

The greater the distance between the equilibrium curve and the X = Y line, the greater the concentration difference achieved in one stage or in a given number of stages. Thus, the ease of separating a given mixture is related to the "fatness" of the equilibrium curve.

2.3.2 Separation Factors

The fundamental factor characterizing a separation process is called the separation factor, which relates the two product compositions from the single-stage separation process illustrated in Figure 2.11.

For equilibrium processes the separation factor is defined by the equilibrium between the two product compositions. It can be written in terms of mole fractions, weight fractions, or individual component molar or mass flow rates. The separation factor, α, written in mole fractions for a two-component system is related to product composition by Eq. 5:

$$\alpha = \frac{X_{A,1}/X_{B,1}}{X_{A,2}/X_{B,2}} \tag{5}$$

where:

$X_{A,1}$ = Mole fraction A in phase 1
$X_{B,1}$ = Mole fraction B in phase 1
$X_{A,2}$ = Mole fraction A in phase 2
$X_{B,2}$ = Mole fraction B in phase 2

For an ideal binary system, it can be seen from the discussion in Section 2.2 that the separation factor, α, for distillation is identically the relative volatility of the components in the feed.

In distillation it has become customary to call the vapor phase mole fractions by the letter Y and the liquid phase mole fractions by the letter X. Furthermore, for binary systems, if one component in the liquid phase has mole fraction X the other component has the mole fraction $(1 - X)$. Similarly, for the vapor phase the values of the two mole fractions are Y and $(1 - Y)$. Substituting these values in Eq. 5 gives an equation for equilibrium concentration of:

$$\alpha = \frac{Y/(1-Y)}{X/(1-X)} \tag{6}$$

Rearranging Eq. 6 we get:

$$\frac{Y}{(1-Y)} = \alpha \frac{X}{(1-X)} \tag{7}$$

It has become a convention in equilibrium separations to use Y for the mole fraction of the components in the vapor phase for vapor-liquid equilibria, in the solvent phase for extraction, and in the solid phase for adsorption and

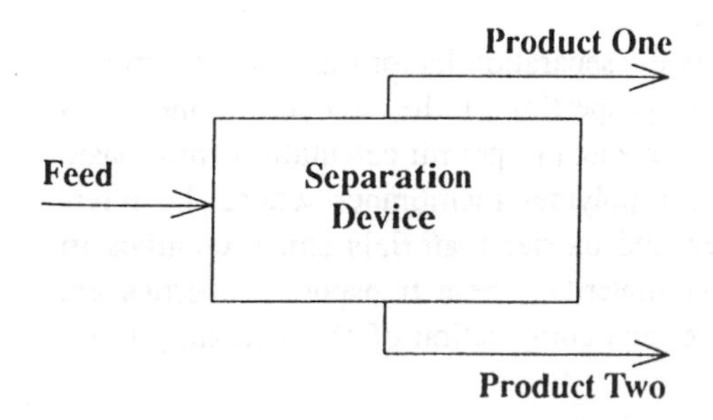

Figure 2.11 Schematic of a single-stage separation process.

ion exchange. Furthermore, the X and Y in Eq. 7 refer to the more volatile, more soluble, or more strongly adsorbed component of the binary pair being separated.

Equation 7 can be rearranged to relate vapor phase composition directly to liquid phase composition through the separation factor. This permits the direct calculation of the equilibrium curve from the liquid phase composition and the separation factor. This is shown in Eq. 8:

$$Y = \frac{\alpha X}{1 + (\alpha - 1) X} \tag{8}$$

This equation is used to calculate the equilibrium curve for binary systems or systems that can be approximated as binary systems, such as solvent extractions and extractive distillations, for example.

For *rate-controlled processes*, which often involve different diffusion rates of feed components through a barrier, the separation factor is determined by the relative diffusion rates through the barrier of the individual components of the feed. The barrier may be porous, as with gas diffusion and adsorption in porous solids, or solid but very thin, as with reverse osmosis and ultrafiltration. The flow rate of each component is proportional to the driving force divided by the resistance to flow. The driving force can be any of a number of gradients, such as pressure, concentration, and various kinds of fields. The resistance to flow depends on the thickness, chemical composition, and physical properties of the barrier. It also depends on the chemical and physical properties of the diffusing molecules.

Another type of rate-controlled process involves slow reactions of reactive gases with the chemical solutions which are used in scrubbing of exhaust gases. An example is the recovery of acid gases such as hydrogen sulfide and carbon dioxide with carbonate or amine aqueous solutions. In these cases the separation process is controlled by the reaction rates of the complex chemical reactions which tie up the carbon dioxide or hydrogen sulfide. It is also influenced by the rates of the various physical processes (solubilization and diffusion) taking place prior to the chemical reactions.

For most rate-controlled processes the separation factor must be determined experimentally because the transport properties of the feed components in the barrier material are not well known enough to permit calculation from basic principles. This is particularly true for polymer membranes where the interactions between diffusing molecules and barrier materials cause changes in the transport properties of the barrier material. These transport properties are also sensitive to temperature, pressure, and composition of the diffusing molecules.

2.3.3 Batch Versus Continuous Processes

Batch processes are those in which a separation device is filled with the feed to be separated and the separation is carried out without any further addition of feed during the separation process. Product of constantly changing composition is recovered from the device and the final residue is removed from the device at the end of the separation. For distillation this is called a Rayleigh distillation after Lord Rayleigh (1902), who first adequately described it. The equipment and equilibrium curve for a Rayleigh distillation are shown in Figure 2.12. The compositions of the distillate and liquid in the still are shown as D and L for the beginning of the distillation. At the end of the distillation they are shown as D′ and L′ on the equilibrium diagram.

Continuous processes feed the mixture to be separated continuously to the separation device and produce two products of constant composition throughout the separation. For distillation this is illustrated by the simple flash drum which equilibrates liquid and vapor derived from heating a continuous feed mixture. Both of the products are collected continuously. This device is particularly effective when the products are widely different in boiling point. Examples are the removal of dissolved low-boiling solvents from waste lubricating oils or of dissolved gases from liquids. The equipment and equilibrium relationship for a continuous single-stage distillation are sketched in Figure 2.13.

Although the Rayleigh distillation suffers the operating disadvantage that the still must be emptied and refilled between distillations, it gives a purer overhead product than flash distillation. This is because the average liquid

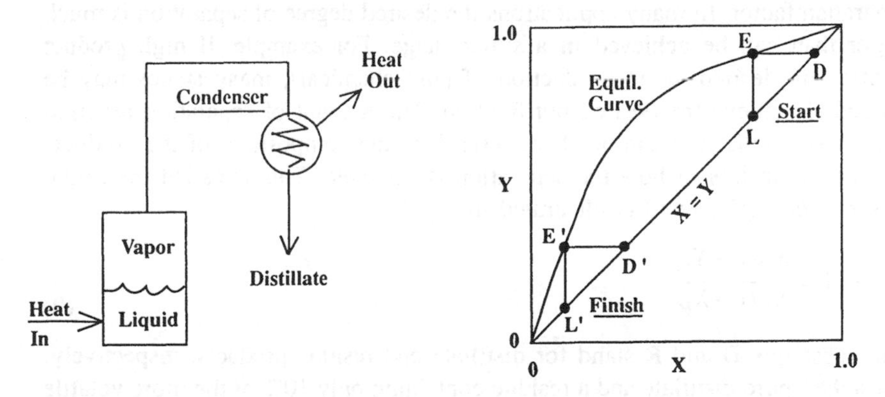

Figure 2.12 Representation of equipment and graphical equilibrium treatment for a Rayleigh distillation. Left: apparatus, right: equilibrium diagram.

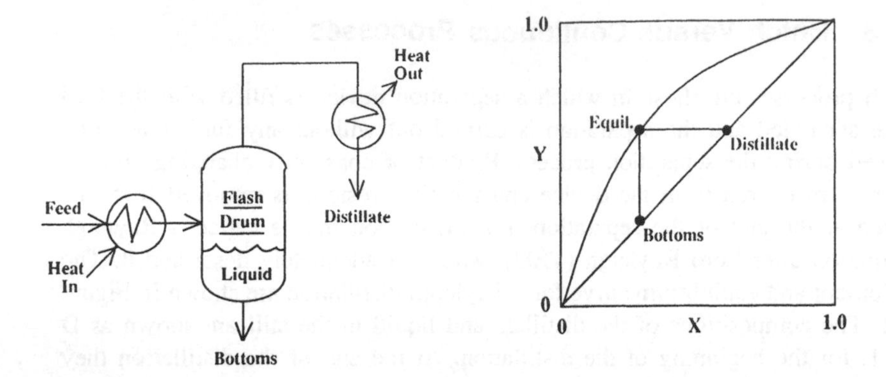

Figure 2.13 Representation of equipment and graphical equilibrium treatment for a continuous flash distillation. Left: apparatus, right: equilibrium diagram.

composition from which the distillate is derived is higher in concentration of overhead product than the final bottoms composition. For equilibrium flash distillation, the entire distillate is in equilibrium with the final bottoms composition. Thus it contains less of the more volatile component than the product from the Rayleigh distillation.

2.3.4 Single-Stage Versus Multistage Operation

The degree of separation achievable in a single-stage separation process is limited by the equilibrium between the two product phases and thus by the separation factor. In many applications the desired degree of separation is much larger than can be achieved in a single stage. For example, if high product purities are desired, as in production of pure chemicals, many stages may be needed to achieve the desired purification. The amount of separation required in a given application can be characterized by the composition of the products desired, regardless of how the separation is achieved. This is called the *extent of separation* (E.S.) and is illustrated in Eq. 9.

$$\text{E.S.} = \frac{X_D/1 - X_D}{X_R/1 - X_R} \tag{9}$$

The subscripts D and R stand for distillate and residue products, respectively. For a 99% pure distillate and a residue containing only 10% of the more volatile component, the extent of separation is 99/1 divided by 10/90, giving a value of 891. For high purity chemicals, the extent of separation can range up to 100,000. Most separations encountered in hazardous waste management are much less demanding than this and are more on the order of 100. The methods for

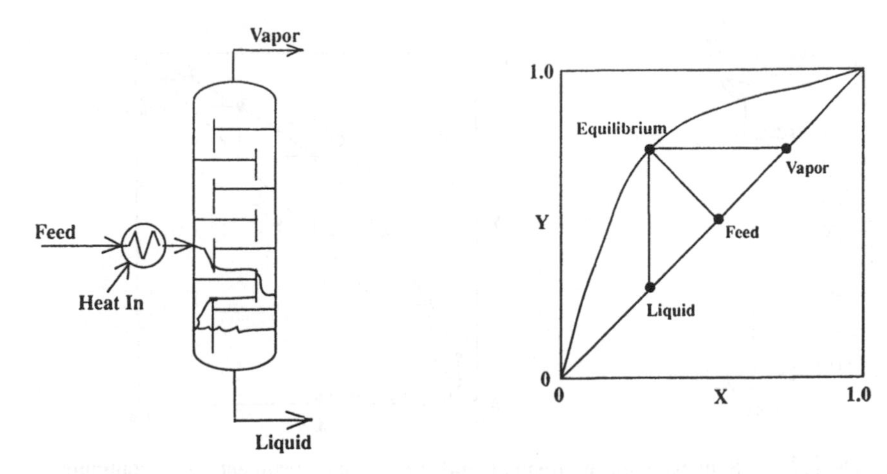

Figure 2.14 Illustration of ineffectual use of a multiplate distillation tower in the absence of two-phase counterflow. Left: apparatus, right: equilibrium diagram.

achieving high extents of separation for multistage processes are discussed below.

For multiple stages to be effective, there must be a counterflow of the two phases being contacted. In distillation this counterflow is usually generated by having liquid flow downward through the separation apparatus against a rising stream of vapor. The liquid and vapor are equilibrated at each equilibrium stage or theoretical plate achieved by the apparatus. In the absence of a counterflowing liquid phase above the point of vapor generation, the entire upper part of the contacting apparatus is the equivalent of only the one theoretical plate achieved by vaporizing part of the feed. Similarly, if no further vapor generation is carried out below the point of feed vaporization, no further two-phase contacting is achieved in the lower part of the apparatus. If both of these limitations apply, a large multistage separation apparatus is reduced to a single-stage equilibrium flash unit with long useless contacting units both above and below the equilibrium flash stage. This is illustrated in Figure 2.14.

A liquid phase can be obtained for the upper part of the unit by condensing some of the final vapor from the top of the multistage contacting device and feeding the resulting liquid back to the top stage. This liquid then overflows countercurrently to the rising vapor phase, giving effective re-equilibration of the phases many times between the feed vaporization point and the final vapor condensation point. The stream of condensate that is fed back is called the reflux stream. Its flow rate determines the effectiveness of the contacting stages above the feed point for making products of high purity. It does this by deter-

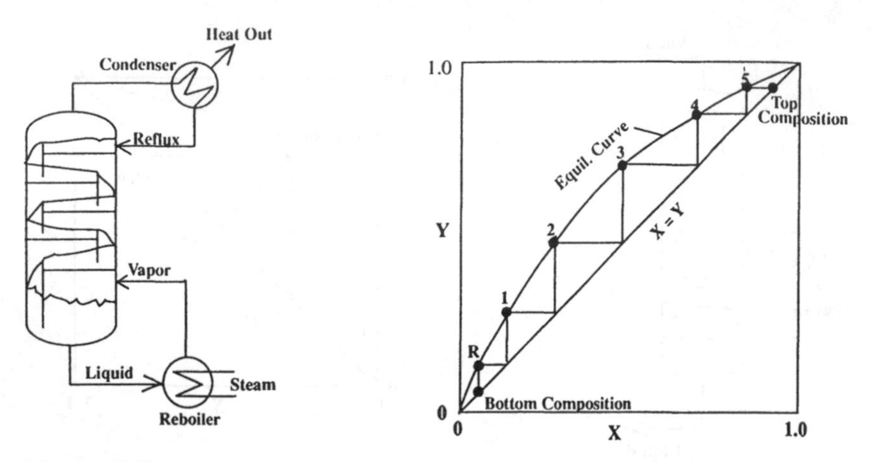

Figure 2.15 Schematic of equipment and equilibrium treatment of a continuous enriching column at total reflux. Left: apparatus, right: equilibrium diagram.

mining the slope of the operating line in the enriching section of the separation apparatus. The operating line is a line through the final composition of the distillate having a slope equal to the ratio of liquid flow to vapor flow inside the column. The enriching section is the region of the separation apparatus between the feed point and the end of the apparatus where the more volatile component of the feed is concentrated, i.e., it is enriched. The enriching section of a distillation column is illustrated in Figure 2.15 along with the McCabe-Thiele diagram for its operation at total reflux. At total reflux the slope of the operating line is 1.0 and X = Y is the actual operating line. At total reflux all of the condensate is returned to the column and no product is taken.

Reflux can be generated in a variety of ways but its effect is always determined by the effect it has on the liquid-to-vapor ratio, and thus on the slope of the operating line. The most common method of generating reflux in distillation is by totally condensing the overhead vapor product and pumping or gravity feeding part of the resulting liquid back to the top of the column. The rest of the condensate is taken as product. However, reflux can also be generated by intentional partial condensation of vapors inside the column by means of a cooling coil. Partial condensation can also occur from unintentional heat leakage from the distillation column to the surroundings. Unintentional heat leakage is usually more important in small laboratory columns.

Operating lines for the enriching section are illustrated for distillation in Figure 2.16. They pass through the final distillate composition and have slopes equal to the ratio on a mass basis of liquid overflow to vapor flow between

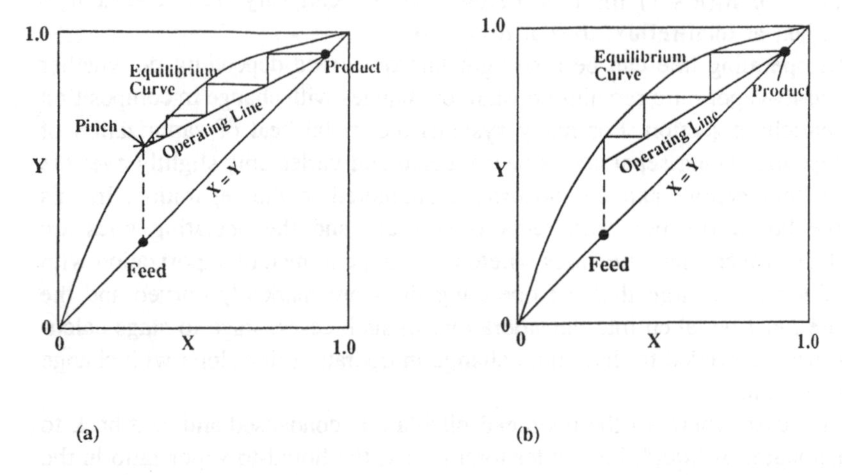

Figure 2.16 The effect of increasing reflux ratio on the required number of theoretical plates for a separation by distillation in an enriching column. (a) Minimum reflux ratio; (b) 1.5 times minimum reflux.

stages above the feed vaporization point. They are often called the locus of compositions of passing phases. They are determined by material balances on a mass basis around the enriching section, as shown in Eq. 10.

$$VY = LX + DX_D \tag{10}$$

where:

VY	=	Upward flow of the more volatile component
LX	=	Downward flow of the more volatile component
DX_D	=	Product flow of the more volatile component as distillate, also called net flow of the more volatile component.

Dividing Eq. 10 by V gives the usual form of the operating line shown in Eq. 11.

$$Y = \frac{L}{V}X + \frac{D}{V}X_D \tag{11}$$

The slope L/V of the operating line is called the internal reflux ratio and is related to the external reflux ratio, R, by Eq. 12 when the reflux stream is liquid at its boiling point.

$$\frac{L}{V} = \frac{R}{R+1} \tag{12}$$

The values of $R/(R+1)$ must be below one and can vary from one at R = infinity, i.e., at total reflux, to 0 at R = zero.

The operating line can be a straight line or curved depending on whether the liquid-to-vapor ratio remains constant or changes with change in composition in the enriching section. For many systems the molal heat of vaporization of the compounds being separated is nearly equal and varies only slightly over the range of temperature and composition encountered in the separation. In this case, the liquid-to-vapor ratio remains constant and the operating lines are straight. However, there are cases where the change in heat of vaporization with composition is so large that the operating lines are markedly curved and the curvature must be taken into consideration. In such cases, stage to stage calculations may be needed to determine change in operating line slope with change in composition.

In the case where all the overhead distillate is condensed and sent back to the fractionator as liquid, i.e., under total reflux, the liquid-to-vapor ratio in the enriching section is 1.0 and the slope of the operating line is 1.0. Under these conditions the extent of separation desired can be achieved using the smallest number of theoretical plates. That is, total reflux represents the minimum theoretical plates for a given extent of separation or the maximum extent of separation achievable with a given number of theoretical plates. For total reflux, a constant separation factor, and a binary system, Fenske (1932) derived the equation below, which has been named after him. It relates the extent of separation to the separation factor and the minimum number of plates.

$$\frac{X_D}{1-X_D} = \alpha^n \frac{X_R}{1-X_R} \tag{13}$$

The minimum number of theoretical plates capable of making the desired separation of feed into specific distillate and residue product compositions is given as n. For separation of a mixture of A and B into a distillate of 90% A and a residue of 90% B, the extent of separation, E.S., is:

$$\text{E.S.} = \frac{90/10}{10/90} = 81 = \alpha^n \tag{14}$$

For $\alpha = 9$, only two stages would be required to reach the 90/10 concentrations at total reflux. For $\alpha = 3$, the separation would require a minimum of four theoretical stages at total reflux. For $\alpha = 1.1$ the separation would require a minimum of 47 theoretical stages. Finally, at $\alpha = 1.00$, no separation would occur even with an infinite number of stages. Thus, the value of the separation factor clearly tells the degree of difficulty in making a given separation. For 99% pure products the extent of separation would approach 10,000 and much larger numbers of stages would be needed to achieve these purities.

Unfortunately, under total reflux no product is obtained. Therefore, to obtain the maximum amount of product for a given expenditure of energy, it is desirable to operate at the lowest reflux ratio which can achieve the desired separation with the number of theoretical plates available in the separation apparatus. This requires the lowest practical amount of separation medium, i.e., heat for distillation processes and solvent for extraction processes. The minimum reflux ratio is the lowest reflux ratio which will achieve the desired separation with an infinite number of theoretical plates. This is illustrated in Figure 2.16a for distillation in a continuous enriching still making a 93% pure product from a 30:70 two-component feed. Almost all of the infinite number of plates required at the minimum reflux ratio of 1.0 are wasted in the "pinch" where the operating line and the equilibrium curve intersect. By increasing the reflux ratio from the minimum of 1.0 to a practical value of 1.5, as illustrated in Figure 2.16b, the required number of theoretical plates for the separation is decreased from infinity to about 3. Thus, a practical compromise exists using a reflux ratio of about 1.5 times the minimum. This reflux ratio requires a number of theoretical plates only about 1.5 times the minimum of 2.1 plates required to achieve the separation at total reflux. This compromise permits a balance to be made between the cost of additional plates and the cost of additional vapor generation to provide higher amounts of reflux.

Clearly, it is undesirable for economic reasons to build a distillation unit with an infinite number of plates or to operate a process which produces no product. Therefore, distillation practice avoids these conditions by keeping the operating line far enough away from the equilibrium curve to avoid wasting plates in "pinch points." It also uses reflux ratios which are a compromise between the economic costs of equipment and energy for the desired separation.

Practical reflux ratios for multistage distillation operations can run from less than one to as high as 75 depending on the difficulty of the separation and the extent of separation desired. Reflux ratios of zero are used in the batch stills for the recovery of solvents from used oils. This is possible because the used oil is essentially non-volatile under the conditions of distillation. Values less than one are used in distillation of crude petroleum which has a wide range of volatility for the components in the feed. The high values are used in the manufacture of high purity chemicals where the equilibrium curve is very close to the $X = Y$ line. The usual levels of practical reflux ratios are from 1.1 to 1.5 times the minimum. Correspondingly, the usual levels of practical numbers of plates or stages are close to 2.0 times the minimum.

The operation of a stripping still is entirely analogous to the above description of an enriching still. The slope of the operating line is now greater than one because the product is taken off as a liquid at the bottom of the column. The amount of feed recovered from the bottom of the still determines how much vapor must be generated in the reboiler to achieve this bottoms yield, i.e., to

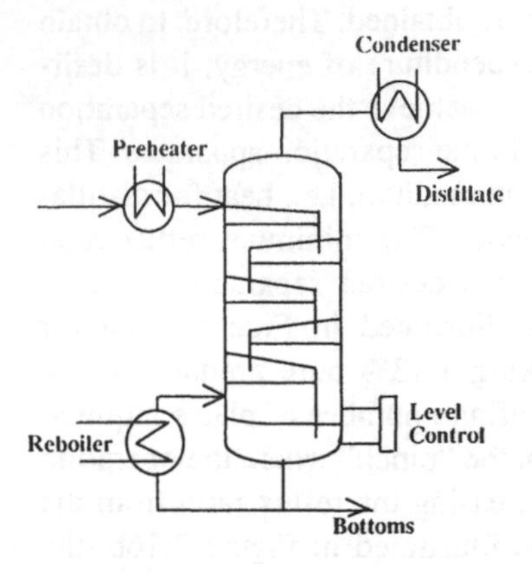

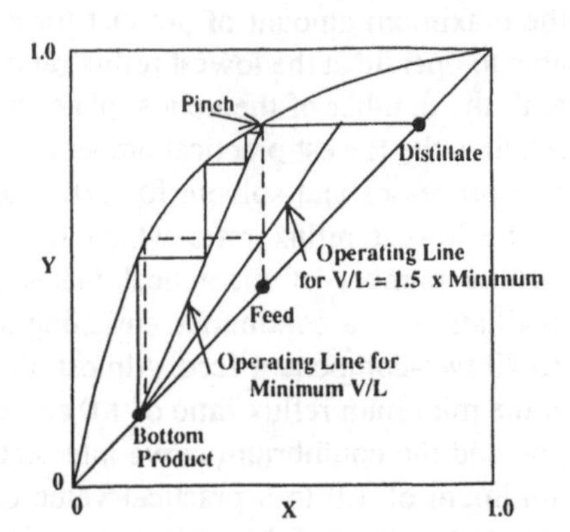

Figure 2.17 Sketch of a stripping column and the graphical effect of increasing gas-to-liquid flow rate ratio on the required number of theoretical plates for a separation by distillation in a stripping column. Left: equipment schematic, right: equilibrium diagram.

carry the rest of the feed overhead as vapor. The minimum amount of vapor generation that can achieve this result with an infinite number of theoretical plates is fixed by an operating line drawn from the feed composition on the equilibrium curve to the bottoms product composition on the X = Y line. This operating line is determined by a material balance around the stripping column. It is also a locus of compositions of passing phases between plates in the stripping column. The stripping operation is illustrated in Figure 2.17 with the resulting operating line given in Eq. 15.

$$Y = \frac{L}{V}X - \frac{B}{V}X_{\mathrm{B}} \tag{15}$$

As with the enriching section, the slope of the operating line in the stripping section is also equal to the molal liquid-to-vapor ratio. However, this slope is now greater than one because the net flow of product is now out the bottom of the stripping section rather than out the top as for an enriching section. The minimum vapor-to-liquid ratio which will achieve the desired separation is the reciprocal of the slope of the operating line shown in Figure 2.17. The practical vapor-to-liquid ratio needed for a finite number of plates will be higher than the minimum, just as the liquid-to-vapor ratio must be higher than the minimum in an enriching column to achieve a reflux ratio above the minimum.

As a general rule, enriching columns are used to produce high distillate purity for the more volatile component and stripping columns are used to produce

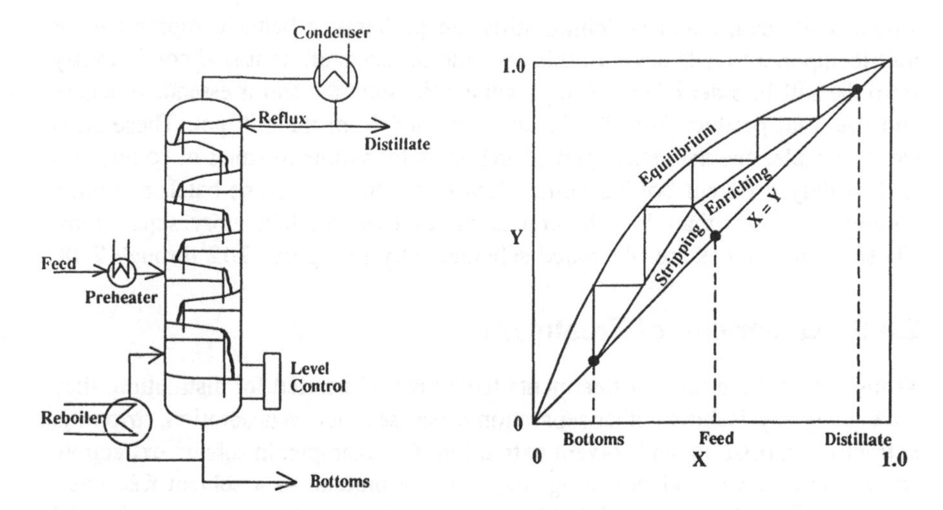

Figure 2.18 Sketch of the (left) apparatus and (right) equilibrium treatment for distillation in a compound column.

high bottoms purity for the less volatile component. When they are combined, as in a compound column, both objectives are achieved. This gives both high yield and high purity of products with a corresponding high extent of separation. The compound column is illustrated for distillation in Figure 2.18 and represents the maximum separating power achievable by continuous distillation.

In a compound column, the intersection of the operating lines between the enriching and stripping sections is determined by the simultaneous solution of the equations for the enriching and stripping operating lines. Since both of the operating lines are material balances, depending on the net flow of products from the enriching and stripping sections of the column, their intersection depends on the condition of the feed to the column. That is, if the feed is all liquid at its boiling point, the operating line intersection falls on a vertical line through the liquid feed composition. For a saturated vapor feed to the column, the intersection is on a horizontal line through the vapor feed composition. For mixtures of liquid and vapor, the intersection is on a line through the overall feed composition on the X = Y line having a slope of $-L_F/V_F$, where L_F is the moles of liquid in the feed and V_F is the moles of vapor in the feed. This slope can vary from zero for all vapor feed to minus infinity for all liquid feed.

The compound column is limited to providing only one pure product per distillation column for a multicomponent feed and requires $n - 1$ columns to produce n pure compounds from an n-component feed. This is expensive and is useful only where large amounts of feedstock are to be processed. For smaller amounts of feed, batch enriching stills are probably a better compromise for multicomponent feedstocks. In this case the products are removed continuously from the still in order of increasing boiling point until the still is essentially empty and the final product from the batch is drained from the still pot. These stills can be simple, one theoretical plate, batch stills for simple mixtures of compounds with widely different boiling points. However, there are some batch enriching columns of more than 100 theoretical plates used for laboratory separations. These column types are illustrated schematically in Figures 2.12 through 2.18.

2.3.5 Generality of Treatment

While the above graphical treatments have been illustrated for distillation, they are equally applicable to other separation processes such as absorption, stripping, extractive distillation, and solvent extraction. For example, in solvent extraction, equilibrium curves and operating lines can be plotted on a solvent-free basis and behave much as those of distillation. However, to approximate equal molal overflow, the solute content of the extraction solvent must be kept approximately constant. This is done by changing the operating temperature with distance through the column, thus increasing or decreasing solubility. The solubility can also be reduced by adding small amounts of antisolvents such as water to a polar solvent as it passes through the extraction unit.

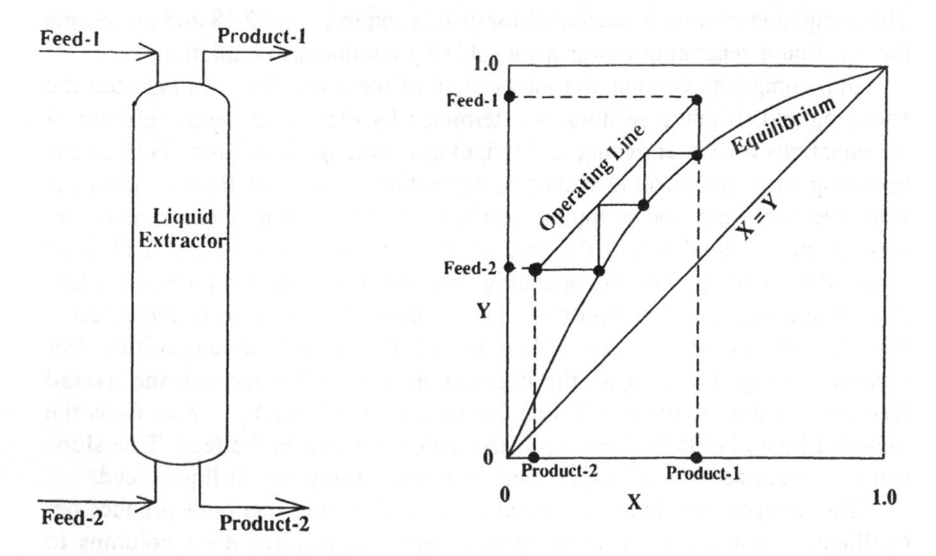

Figure 2.19 Graphical treatment of solvent extraction with the operating line outside the equilibrium curve. Left: equipment schematic, right: solvent-free equilibrium diagram.

In the above graphical treatments the operating line was always in between the equilibrium curve and the X = Y line. This does not always have to be the case as is illustrated for solvent extraction in Figure 2.19.

In this case the operating line, which is still derived by material balance, is outside the equilibrium curve. However, the theoretical plates required for a given separation are still stepped off between the operating line and the equilibrium curve.

2.4 FLUID DYNAMICS

As discussed earlier in this chapter, most separation processes consist of the equilibration of two phases followed by their separation. The separated phases are then collected as products in single-stage separations. In the case of multistage operation, they are introduced separately into additional stages to increase the extent of separation. The effectiveness of the multistage devices depends very strongly on the establishment of suitable flow patterns both within a given stage and between consecutive stages. This section discusses the flow regimes encountered in separation devices, the types of phase contacting used to achieve equilibrium between phases, and the effects of entrainment, channeling, and bypassing.

2.4.1 Flow Regimes

The flow through a single-stage separation device can be either cocurrent, countercurrent, or crossflow. This is illustrated in Figures 2.20a for extraction in a mixer-settler, 2.20b for distillation in a packed bed, and 2.20c for stripping in a sieve plate, respectively. In the mixer-settler, the feed and solvent are mixed in a batch mixer. The resulting emulsion passes to a settler where the settling phases flow cocurrently while the settling takes place. The settled phases then exit from the downstream end of the settler.

In the packed bed, the vapor phase passes upward through the open space in the packing while the liquid flows downward on the surface of the packing. The packing spreads out the liquid into thin films, which give efficient mass transfer between the liquid and vapor phases.

In the crossflow sieve plate in Figure 2.20c the liquid flows across the perforated plate from right to left. The vapors pass through the holes in the plate and form gas bubbles which rise through the liquid phase. The high surface provided by the bubbles is where most of the mass transfer and phase equilibration occur.

In a multistage separation apparatus the flow between stages should be countercurrent to get the maximum extent of separation for a given number of theoretical stages. This is relatively easy to do in distillation by using gravity to pass liquid down from plate to plate through a multiplate vapor-liquid contacting tower. The large density difference between the liquid and vapor phases

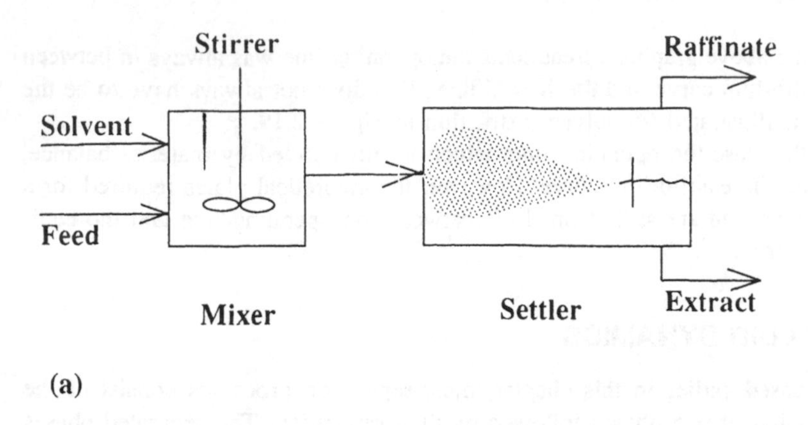

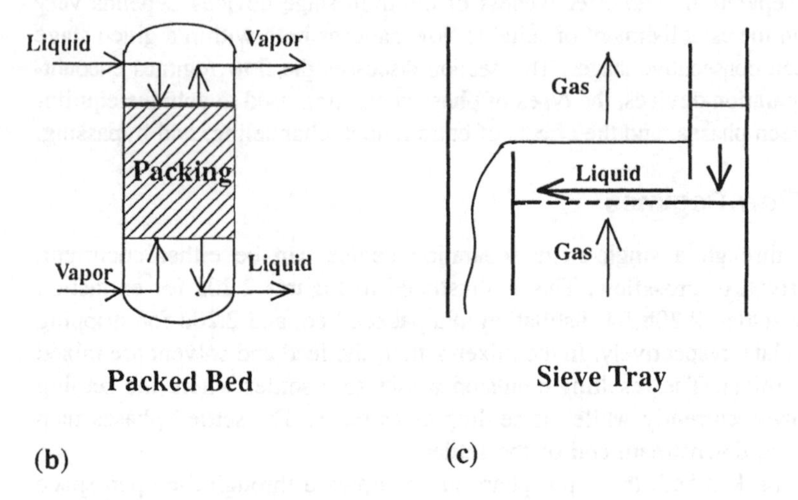

Figure 2.20 Representation of various types of flow through single-stage contacting devices. (a) Cocurrent flow in a mixer-settler, (b) countercurrent flow in a packed bed, and (c) crossflow in a sieve tray.

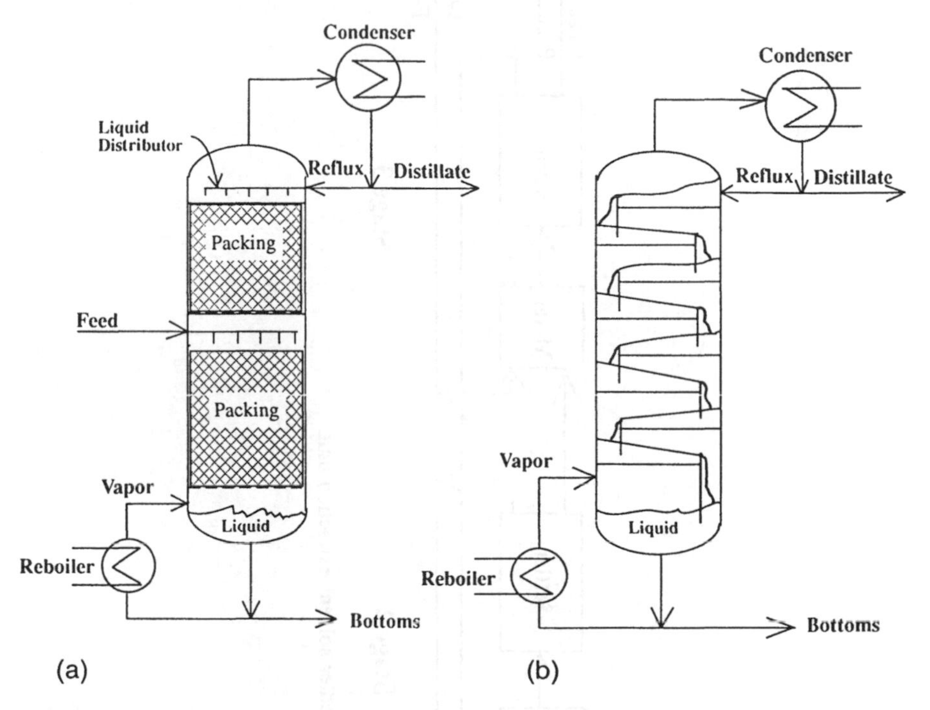

Figure 2.21 Representation of countercurrent vapor-liquid contacting in (a) a compound packed tower and (b) a multiplate sieve tray tower.

makes high countercurrent flow rates possible. The high density difference also causes the counterflow in packed towers and accounts for the popularity of plate towers and packed towers for vapor-liquid contacting. These operations are illustrated in Figure 2.21.

Multistage solvent extraction can also be carried out in plate columns and packed columns. However, the density difference between the two liquid phases is much lower and thus the driving force for counterflow is much lower. This limits flow rates of the two liquids. To overcome this limitation, multistage solvent extraction often employs a series of mixer-settler devices. This is illustrated in Figure 2.22 for a three-stage extraction of an organic feed with water. The pressure to move the phases through the apparatus is provided by pumps (not shown) on the two feed streams. The interface between the settled phases in each settler is controlled by an interface level controller operating a control valve in the exit line for one of the settled phases. The flow rate of the other phase is controlled by a back pressure regulator. These control devices are also not shown in the illustration.

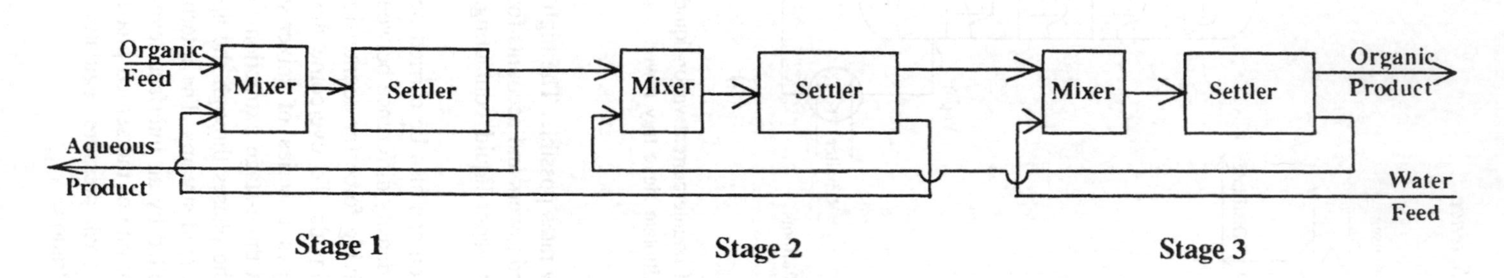

Figure 2.22 Schematic of a three-stage countercurrent mixer-settler solvent extraction unit.

2.4.2 Phase Contacting

The methods used to get intimate contact between two phases vary widely. They can be based on the influence of various fields such as gravity, electrostatic, and magnetic fields or through mechanical energy supplied by pressure drop or mechanical mixers. The density difference between vapor and liquid can be used both to achieve counterflow of the phases and to provide mixing energy for equilibrating the phases. This is done in plate type and packed-bed distillation units. The available energy for flow and mixing is the difference in density between the phases, multiplied by the height of the contacting column. This is true of both distillation and extraction columns.

Mixer-settler devices are frequently used in solvent extraction separations, particularly where the density difference between the liquid phases is large, the interfacial tension between the phases is high, and mixing is therefore difficult. However, where interfacial tension and density difference between phases are low and mixing is consequently easy, packed towers and plate towers can be used as liquid-liquid contacting devices. These milder forms of mixing make the subsequent settling and separation of the two phases much easier. Thus, they help avoid entrainment in the counterflow of the two phases. The best of both worlds has been achieved at the expense of capital investment by centrifugal extractors. In these units the heavy (denser) phase is fed to the center of the centrifuge and the lighter phase (less dense) is fed to the periphery of the centrifuge. The two phases counterflow under the influence of centrifugal force generated by the centrifuge. The centrifuge is equipped with a spiral perforated plate between the inlets and outlets to provide the mixing during the counterflow. The centrifuge multiplies the density difference between the phases, thus enhancing both the countercurrent mixing and the settling of the two phases.

Another form of mechanical mixing is achieved by the line mixer which uses cocurrent flow of the two phases through devices which convert pressure drop into turbulence and thus mix the two phases. The pressure drop generating devices may be orifice plates, packings, baffles, etc. These mixers are limited as compared to countercurrent devices in that cocurrent flow is capable of achieving only one theoretical stage, no matter how long the mixing device. However, they can serve as pump-around mixers in a mixer-settler arrangement as shown in Figure 2.23. They are capable of mixing any fluid phases and are also useful for blending completely miscible mixtures of liquids and gases or vapors.

When using contacting devices which do not achieve a theoretical stage for each actual stage, the concept of stage efficiency is used. This allows us to determine the number of actual stages equivalent to a given number of theoretical stages. For example, if a given contacting device gives a 60% approach to the theoretical equilibrium, the actual number of stages needed for a given separation is the number of theoretical stages needed divided by 0.6, the numerical stage efficiency.

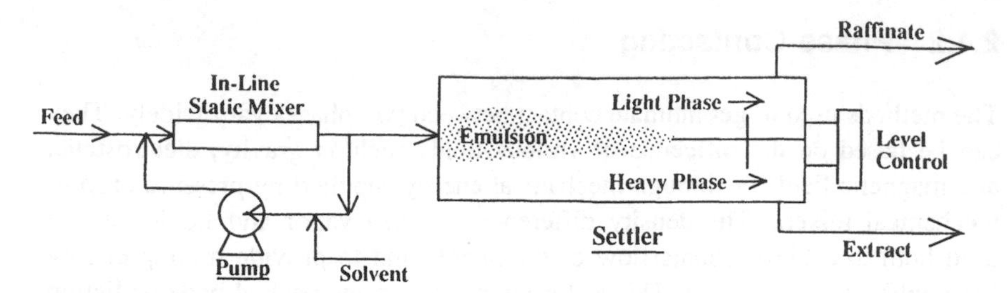

Figure 2.23 Sketch of the use of a pump-around circuit as a mixer in a mixer-settler extraction unit.

2.4.3 Entrainment, Channeling, and Bypassing

Entrainment is the term used for material of one phase carried along in the other phase due to incomplete phase separation after contacting. It can be a result of (1) liquid mists or foams in vapor streams from distillation, absorption and stripping, (2) emulsions and fine droplets in either phase from extraction, and (3) incomplete washing of mother liquor from crystals in crystallization. The mother liquor in crystallization is the remaining solution from which the crystals have been precipitated. The net result of entrainment is to decrease the efficiency of each stage in a separation process and degrade the final product. It can be eliminated by providing improved phase separation capabilities in the separation device.

Channeling and bypassing are similar to each other in that they both involve inadequate phase contacting. They often occur in packed beds when the counterflowing phases find paths through the packing that avoid each other. They are frequently encountered in vapor-liquid contacting, liquid-liquid contacting, solid-liquid contacting, and solid-vapor contacting. Channeling usually involves a flowing stream finding a low resistance path through the packed bed which keeps it from adequately contacting the counterflowing or solid phase. This leads to inadequate equilibration of the phases and loss of effectiveness for the separation device. Bypassing is usually used to describe flow along the interior walls of the vessel containing the packed bed, thus completely avoiding contact with the bed or the counterflowing phase and making the separation device ineffective.

The deleterious effects of entrainment, channeling, and bypassing are a matter of degree. They can vary from relatively minor effects on product purity from small amounts to almost complete loss of effectiveness of the separation process for large amounts. When a separation unit does not perform as expected, these three possibilities are the first that should be investigated.

2.5 ENERGY REQUIREMENTS

The theoretical maximum entropy of mixing and the resulting minimum energy of demixing are relatively small quantities, on the order of 400 calories per gram mole or 720 Btu per pound mole. For water this is only about 4% of the heat of vaporization. The actual amounts of energy needed to run a separation process can often be very large, i.e., many times the minimum. Economy is achieved by matching the separation process to the materials to be separated in such a way that the excess energy needed over the theoretical minimum is kept as small as possible. In a rather obvious example, if we need to separate a slurry of sand and water, it makes no sense to distil off the water when permitting the sand to settle out of the water is so readily achievable and consumes so much less energy and capital.

2.5.1 Theoretical Minimum Work of Separation

The theoretical minimum work or heat equivalent of separation is a thermodynamic state property. That means it depends only on the temperature, pressure, and composition of the feed mixture to, and the products from, the separation. It does not depend on the path of the separation or the process used to make the separation. All separations of the same feed into the same products have the same minimum work of separation, i.e., the demixing energy. Equation 16 gives the minimum work of separation for a binary ideal system.

$$W_{Min} = -RT\,[X_A \ln X_A + (1 - X_A) \ln (1 - X_A)] \tag{16}$$

where:

W_{Min} = minimum work per mole of feed
R = the gas constant
T = absolute temperature
X = mole fraction of component in feed

Substituting values for a 50:50 feed mixture at 25°C (298°K), we get Eq. 17.

$$W_{Min} = -RT(0.5 \ln 0.5 + 0.5 \ln 0.5)$$

$$W_{Min} = +RT \ln 2 = 411 \text{ cal/gram–mole} \tag{17}$$

The minimum work of 411 calories per gram-mole amounts to 4.11 calories per gram or 7.4 Btu per pound for feed components of molecular weight equal to 100. This is about 5% of the heat of vaporization for such components. The actual energy required for separation is nearer to the heat of vaporization than to the theoretical minimum. For feed compositions other than 50:50 and products less than 100% pure, the minimum theoretical work decreases further.

The minimum energy requirement represents the lowest theoretically possible energy requirement to achieve a given separation. All actual separation processes have an energy (work/heat) requirement greater than the theoretical minimum, frequently many times the minimum. Choosing the actual separation process which has inherently the lowest energy consumption for a given separation can often make the difference between a successful project and a major waste of money. However, it must be remembered that increasing energy efficiency may require extra equipment. This can increase both the capital costs and the complexity of the equipment required to achieve a given separation. The optimum separation process will be one that provides the minimum total operating cost, including both capital and energy costs. It should also permit sufficient flexibility to operate the equipment on a stand alone basis, if necessary. It is thus the result of an economic balance and can change with changes in cost of energy versus capital equipment.

2.5.2 Ways to Reduce Energy Consumption

A number of different methods for minimizing energy consumption have been developed. They are used in petroleum refineries, large scale chemical manufacturing facilities, and other large scale processing plants which operate around the clock. These techniques for improving efficiency rely on the effective integration of heat sources and heat sinks available in the entire processing complex. This can be done through (1) heat exchange, (2) prevention of heat loss by insulation, (3) cascading of heat from high temperature sources through a number of operations before disposal, (4) use of mechanical energy to pump heat from lower temperature to a higher temperature for reuse as in vapor compression, and (5) reuse of heat as in multi-effect evaporation. This can lead to such interdependency of processing units that if a major energy-using or energy-producing unit has to be shut down, the entire plant must be shut down. This is caused by loss of energy sources and sinks for the rest of the units. Therefore, the level of heat integration desired in a separation process used in waste minimization depends on the on-stream compatibility of the integrated units. For stand alone waste recovery units the energy integration should usually not go beyond the limits of the recovery unit to permit flexibility of operation. However, for large integrated distillation plants for commercial recovery of a variety of waste solvents, heat integration of the various stills can produce major savings in energy costs.

Heat Exchange

Heat exchange can be used to transfer either sensible heat, as in heating or cooling without phase change, or latent heat as in the condensation or vaporization of either phase. Sensible heat exchange is frequently used to transfer heat from products to preheat the feed to high temperature processes. It can also be used to transfer heat from feed to the cold products in prechilling the feed to low temperature processes. Latent heat exchange is commonly encountered in the condensation of vapors and the boiling of liquids in distillation and evaporation. It is also involved in the formation of solid phases in the crystallization of solids from melts. In most processes encountered in waste management, the latent heats are much larger than the sensible heats and the heats of vaporization are also larger than heats of melting (or fusion). For example, for water the sensible heat for warming one pound of water by 100°F is 100 Btu, while the heat of vaporization for one pound of water is 973 Btu and the heat of fusion is 144 Btu. The British thermal unit represents a quantity of heat and is the amount needed to heat one pound of water one degree Fahrenheit.

Typical examples of heat exchange are shown in Figure 2.24. The left side of the figure shows the transfer of sensible heat between two liquids, two gases, and a liquid and a gas. The top right shows latent heat of vaporization being removed by a liquid in a condenser to both preheat the liquid and condense the vapors. The bottom right shows heat of vaporization being applied to an evaporator by condensing steam. All these types of heat exchange are in common use and there are many suppliers of various types of heat exchange equipment.

Cascaded Heat

Heat can be cascaded through a number of processes in sequence of reducing temperature between a high temperature source and a low temperature final sink before it is discarded. The heat can be transferred (1) partly as sensible heat and partly as latent heat from the heat-carrying medium, and partly as sensible heat and partly as latent heat to the heat receiving medium, (2) totally as sensible heat or totally as latent heat from the heat source to the heat sink, and (3) almost any other combination of latent and sensible heat. An example of such cascading is given in Figure 2.25.

In Figure 2.25 the cold feed to the still is warmed up partially by using it remove heat of condensation in the distillate condenser. It then gets more heat from the tower bottoms by a second heat exchanger. This also performs the desired cooling of the tower bottoms prior to storage. The heat provided to the reboiler of the still also gets used effectively to preheat the feed to the still.

A second type of cascade can be generated by reducing pressure, and thus liquid boiling point, as in the multi-effect evaporator shown in Figure 2.26. This triple-effect steam-heated evaporator is being used to recover distilled water from seawater. The vapors from the first effect of the evaporator, which is steam heated, are fed to the heating coil of the second effect. The second effect is held at a reduced pressure and thus a lower liquid boiling point than the first effect. The vapors from the first effect condense in the heating coil of the second effect, giving up their latent heat of condensation to the boiling liquid in the

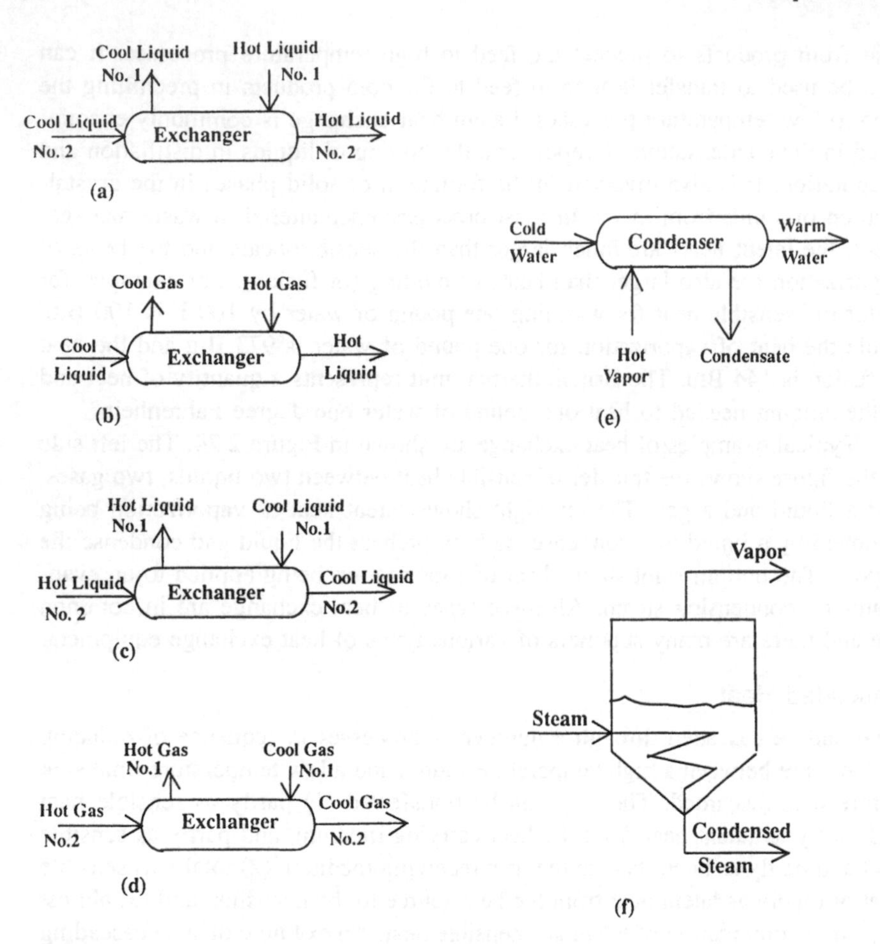

Figure 2.24 Sketches of various types of heat exchange including examples of transfer of (a–d) sensible heat and of (e,f) latent heat. (a) Liquid-liquid exchange heating, (b) liquid-gas exchange, (c) liquid-liquid exchange cooling, (d) gas-gas exchange, (e) condensation, (f) boiling.

second effect and generating a new vapor stream carrying the latent heat. The vapor from the second effect can be passed on to the heating coil of the third effect at further reduced pressure, thus using the latent heat again. In this illustration the heating value of the steam is used three times before it is discarded. The number of effects that can be used depends on (1) the temperature difference between the original steam source and the final sink temperature available to condense the final vapor stream, (2) the boiling point rise in each effect due to concentrating the liquid solution, and (3) the amount of vacuum

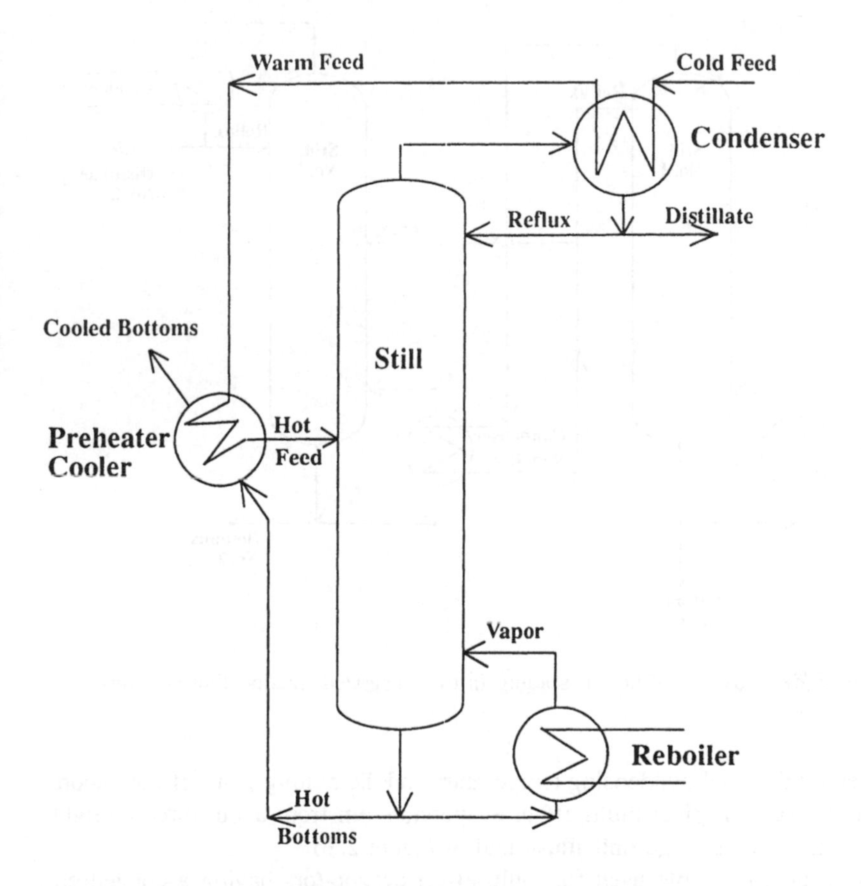

Figure 2.25 Illustration of cascading heat through several exchangers in a single processing unit.

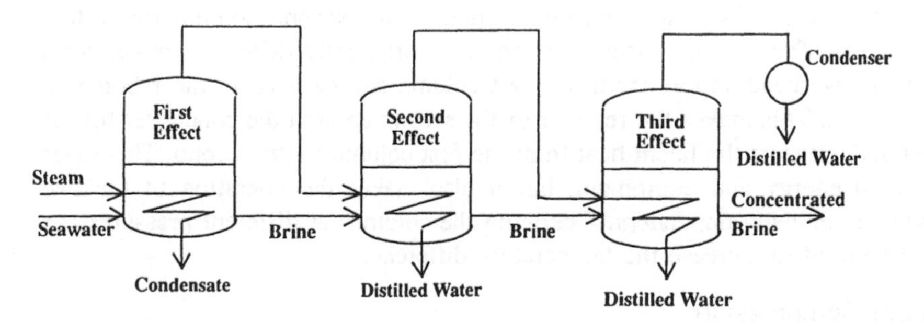

Figure 2.26 Sketch of a triple-effect steam-heated evaporator for recovery of distilled water from seawater.

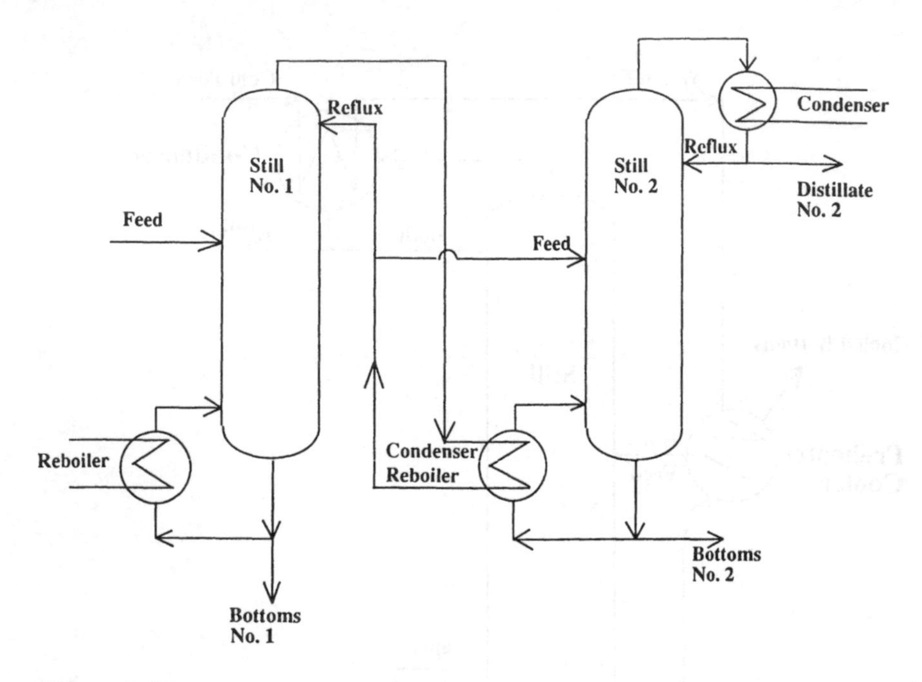

Figure 2.27 Sketch of heat cascading in two integrated fractionating columns.

allowed by the final condensing temperature and the boiling point of the vapors at that pressure. Typical multi-effect evaporators run from about three to eight stages with a three-stage unit illustrated in Figure 2.26.

The same principle used for multi-effect evaporators having a condenser/ boiler for each stage can also be used with fractionating columns. Figure 2.27 shows an example of this type of energy-saving operation. The two columns can operate on separate feedstocks or can be integrated as shown in the illustration to feed the distillate from one column to the second column for further separation. The columns need to operate at sufficiently different temperatures that the overhead vapors from the first column can be used in the reboiler of the second. This makes the reboiler of the second column the condenser for the first and transfers the latent heat from the first column to the second. This saves both on energy and equipment, but it also makes the operation of the two columns mutually dependent. Operating the columns at different pressures can also be used to increase the temperature difference.

Vapor Compression

Another way to use latent heat repeatedly is to use the heat pump principle to pump latent heat from a lower temperature to a higher temperature. This permits

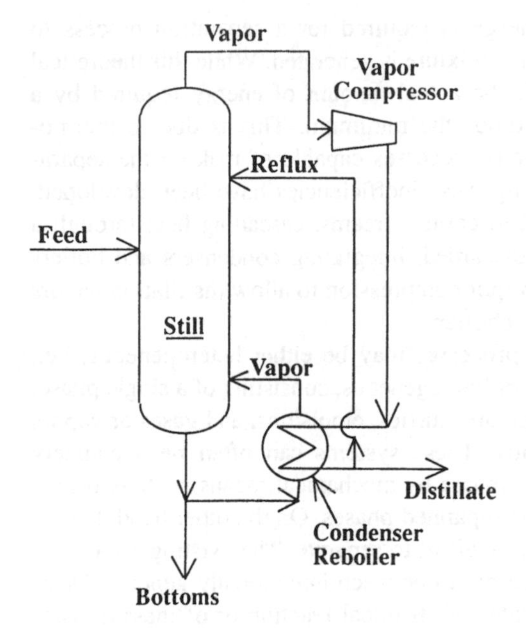

Figure 2.28 Sketch of the use of vapor compression to reuse latent heat of condensation in a reboiler.

the use of the latent heat of condensation of the distillate from a still to provide the heat of vaporization to the reboiler of the same still. This is illustrated in Figure 2.28. The vapors coming overhead from the still are mechanically compressed to a higher pressure and thus a higher condensing temperature. They are then fed to the reboiler of the same still. The vapors condense in the reboiler and give up their latent heat to the boiling liquid in the still.

Vapor compression is most effective where it is needed most, that is, for very difficult distillations of close-boiling components. These distillations require a lot of plates and thus relatively high reflux ratios. Such high reflux ratios demand a lot of latent heat for the formation of all the vapor needed for reflux. Fortunately, such distillations usually have a relatively small temperature difference in boiling point between the distillate and the liquid in the reboiler. They are good candidates for vapor compression because they don't require a very high compression ratio to be effective.

2.6 SUMMARY

Separation processes are the opposite of mixing processes and are often called de-mixing processes. Unlike mixing processes which can occur spontaneously, a certain minimum amount of energy is required for a separation process to overcome the energy released when a mixture is generated. While this theoretical minimum mixing energy is small, the actual amount of energy required by a separation process can be many times the minimum. This is due to thermodynamic inefficiencies in the various processes capable of making the separation. A variety of ways for reducing these inefficiencies have been developed, such as exchanging heat from hot to cooler streams, cascading heat through a number of processes before it is discarded, integrating condensers and boilers to reuse latent heat, and applying vapor compression to allow distillation vapors to be used as a source of heat in reboilers.

The feedstocks to separation processes may be either heterogeneous, i.e., consisting of more than one phase, or homogeneous, consisting of a single phase. Examples of the heterogeneous type are slurries, emulsions, and gases or vapors carrying entrained solids or liquids. These systems can often be adequately purified by separating the discrete phases by mechanical means such as filters, centrifuges, etc., and recovering the separated phases. On the other hand, homogeneous systems usually require more effort to separate. The existing separation techniques are based on either equilibrium between intentionally generated separate phases or on differences in rates of chemical reaction or of mass transfer for the different components of the feed. The first are called equilibrium processes and the latter are called rate-controlled processes. Many practical and effective separation processes are based on each of these approaches.

The equipment used in separation processes is designed to equilibrate two phases and then separate them. If the two phases don't already exist in the feedstock, it is necessary to generate the two phases by means of a separating medium such as heat, solvent, adsorbent, refrigeration, etc. The separation apparatus must efficiently generate the two phases, bring them to equilibrium, separate them, and if multistage operation is used, conduct each phase to the next appropriate contacting stage. Furthermore, suitable equipment must be provided to regenerate the separating medium for recycling if it is an added material such as a solvent or adsorbent. Many very ingenious designs have been invented to permit multistage operation of separation processes when there is a need to achieve high yields of high purity materials.

Graphical treatments can be used to visualize and understand the fundamentals of a wide variety of separation processes. They are often not exactly correct in every detail because they make use of a number of inexact approximations for simplicity's sake. However, they do give a better intuitive feel for what is happening in many separation processes and are therefore of great help in understanding and troubleshooting these processes.

REFERENCES

Britton, E. C., Nutting, H. S., and Horsley, L. H. (1952). Graphical method for predicting effect of pressure on azeotropic systems, *Azeotropic Data, Advances in Chemistry Series No. 6*, p. 317.

Fenske, M. R. (1932). *Ind. Eng. Chem.*, *24*:482.

King, C. (1981a). *Separation Processes*, 2nd ed, McGraw-Hill, New York.

King, C. (1981b). *Separation Processes*, 2nd ed, McGraw-Hill, New York. p. 1.

King, C. (1981c). *Separation Processes*, 2nd ed, McGraw-Hill, New York. p. 18.

Nutting, H. S., Horsley, L. H. (1952). Graphical method for predicting effect of pressure on azeotropic systems, *Azeotropic Data, Advances in Chemistry Series No. 6*, p. 318.

McCabe, W. L., Thiele, E. W. (1925). *Ind. Eng. Chem.*, *17*:605.

Perry, R. H., Green, D. W. (eds.) (1984). *Perry's Chemical Engineers' Handbook*, 6th ed., McGraw-Hill, New York.

Rayleigh, Lord (1902). *Phil. Mag.* [vi], *4(23)*:521.

Rousseau, R. W. (ed.) (1987). *Handbook of Separation Process Technology*, Wiley-Interscience, New York.

Schweitzer, P. A. (ed.) (1979). *Handbook of Separation Techniques for Chemical Engineers*, McGraw-Hill, New York.

Varteressian, K. A., Fenske, M. R., (1937). *Ind. Eng. Chem.*, *29*:270.

3

Types of Separation Processes Used in Waste Minimization

3.1 INTRODUCTION

Separation processes are widely used both in the minimization of hazardous wastes and in the treatment of hazardous wastes before disposal. The treatment before disposal is done to recover valuable components and at the same time remove hazardous materials from the disposed stream. In addition to separation processes, many types of conversion processes are also used to treat waste streams before or during disposal. Some examples are incineration, wet oxidation, pyrolysis, hydrolysis, dehalogenation, or photolysis. The purpose of these conversion processes is to destroy hazardous components in a waste either prior to or during disposal. Freeman (1988) gives a good discussion of these conversion processes in his *Standard Handbook of Hazardous Waste Treatment and Disposal.*

In contrast to the conversion processes, which destroy hazardous materials once and for all, the separation processes are closely linked to waste minimization practices because they recover valuable materials. These materials can be reused or recycled to give both a saving in raw material costs and a saving in waste disposal cost. This book is primarily concerned with separation processes and waste minimization, and therefore conversion processes will not be discussed except incidentally, as they provide the basis for a separation process.

3.1.1 EPA Definition of Waste Minimization

Waste minimization has been defined by the EPA, as discussed in Chapter 1, as source reduction, waste separation and concentration, reuse and recycling of components recovered from waste streams, and waste exchange. *Source reduction* includes recycling of materials within a process, thus extending the life of the materials before disposal and as a result decreasing the amount of waste material generated. Typical operations falling into this category are (1) recycling of cutting oils in metal fabrication with chip removal by gravity or mechanical means, (2) recycling of metal degreasing solvents with grease removal by distillation, and (3) recycling of dry cleaning solvents by distillation. Manufacturers of equipment for these operations often include the accessory equipment needed for recycling right in the process equipment package, thereby greatly extending the life of the necessary process fluids. When the sludge level in or the quality of the spent process fluids becomes too bad for "in process" recycling, the spent fluid can be removed from the process equipment and recovered for recycling in external on-site or off-site equipment. Thus, the only materials that must be finally discarded are the metal chips, sludges, and impurities recovered from the separation processes in the recycling circuits.

Waste separation and concentration describes the techniques used to recover valuable components from a waste stream in concentrations sufficiently high to make them useful as feedstocks. They could be sent back to the process from which they came or used as raw materials to other processes. In this latter case they would be good candidates for a waste exchange where people operating different processes could trade raw materials recovered from wastes.

In many cases waste separation and concentration has a dual beneficial effect in that the removal of toxic materials from wastewater streams can make the wastewaters non-toxic and suitable for recycling or simple disposal. Thus, both the impurity and the wastewater may be recycled and removed from the hazardous waste definition.

Reuse and recycling refers to the recovery of useful material from a waste stream for reuse. This reuse can be either in recycling to the process it came from, if it is recovered in sufficient purity, or in using it for a less demanding purpose. For example, solvents used in the formulation of paints can be recovered by distillation and used for cleaning painting equipment. Another example is that spent high purity solvents used to clean circuit boards can be used in degreasers or for general metal cleaning even without a purification process if they are properly segregated by the generator. With a suitable purification process they may even be suitable for recycling to the original use.

Waste exchange takes advantage in an organized way of the concept that one man's waste may be another man's valuable resource. In many cases valuable materials can be recovered from waste streams, but the generator has no on-site opportunities to recycle or reuse this material. On the other hand, other divisions or units of the same company or outside companies may be purchasing this same material as a feedstock to their manufacturing operations. By selling the recovered material from the generating company to a user company at a price reflecting the savings to both companies then (1) the seller doesn't have the expense of disposing of the material and (2) the buyer gets some economic incentive to use the recovered material rather than new material purchased in the open market.

Waste exchanges can be very effective for laboratory chemicals that are no longer needed for the purpose for which they were originally purchased. They may be leftovers from completed or terminated projects or merely excess inventory. If they are effectively collected, labeled, and stored in the original containers in a convenient place accessible to potential users, they can be used by personnel other than the original purchasers in preference to making new purchases. This system works well and saves money when used inside a single corporation having large laboratories. This is largely because there is no question about what company receives the financial benefits of the operation. However, it does require a control system to provide assurance that identity and purity of the chemicals are correct and that unstable materials are not included in the chemical storage inventory.

Waste exchanges between companies are usually worked out on single items which pose no threat to either company in terms of disclosure of trade secrets or proprietary process technology, or exposure to sanctions under antitrust laws against cooperation. These factors along with the problems of negotiating equitable prices for recovered material as feedstocks have prevented waste exchange between companies from achieving very much of its potential. However, Du Pont (1991) has finessed this problem by offering for sale in the open market chemicals recovered as by-products from waste streams generated in chemical manufacturing. This arm's length approach may be the direction that inter-company waste recycling takes in the future.

In spite of the institutional obstacles to waste exchange, its great potential for waste minimization and financial return should lead to increased application of this technique.

3.2 TYPES OF HAZARDOUS WASTE STREAMS

The types of hazardous wastes classified by the EPA in Subpart C of 40 CFR 261 have been discussed in Chapter 1 of this book, and consist of solid, liquid, and gaseous streams. They can be further classified as:

- Inorganic or organic
- Metal or non-metal
- Aqueous or non-aqueous
- Halogenated or non-halogenated
- Ignitable or non-ignitable
- Corrosive or non-corrosive
- Reactive or non-reactive
- Toxic or non-toxic

Clearly, many hazardous materials can display more than one of the above properties, and their waste streams become more complex through mixing and contamination as they pass down the waste heirarchy and get farther from the place at which they were generated. It is much simpler to recover materials for reuse when they have not been cross-contaminated with other hazardous wastes, or for that matter any other wastes. Therefore, it is usually desirable to set up recovery and recycling equipment as near the waste generation site as possible, and in any case to segregate waste streams to prevent cross-contamination.

Some waste streams are clearly not compatible and should not be mixed unless special safety precautions have been taken to protect against the hazard generated by the mixing. An example of such incompatible waste streams is provided by an aqueous acid stream and a second stream containing either soluble sulfides or cyanides. Mixing would generate extremely toxic HCN or hydrogen sulfide gases. Such incompatible waste streams usually involve corrosive or reactive wastes as one of the waste streams.

In Chapter 2 the feedstocks to separation processes were classified as homogeneous or heterogeneous. Heterogeneous feeds are usually relatively easy to separate into their respective phases. When phase separation alone provides the necessary recovery process, as is often the case with oily wastewaters, recovery of valuable materials can be quite simple by the use of coalescing and settling devices. However, many waste streams are homogeneous and require more complex processes such as distillation, extraction, adsorption, etc., for recovery of valuable materials. The rest of this chapter will discuss the types of recovery processes commonly used in waste minimization. The types will be categorized on the basis of the phase, i.e., gas, liquid, or solid, of the impurity being recovered as well as the phase of the feed stream from which it is being recovered. The remaining chapters of the book will cover individual process applications in more detail. These are identified below for the various recovery processes.

3.3 RECOVERY PROCESSES FROM GASES

The contaminants in waste gas streams can be gases, liquids, or solids and must be removed from the gas stream prior to discharge to the atmosphere if they provide a hazard to human health or the environment or if their loss represents a major financial burden, as with some very valuable materials. Typical examples of hazardous gases are toxic materials such as carbon monoxide, hydrogen sulfide, and hydrogen cyanide, as well as environmentally undesirable gases such as oxides of nitrogen and sulfur and vapors from hazardous liquids such as paint solvents, gasoline, and halogenated cleaning solvents. The liquids in gases are usually in the form of mists or fogs either caused by entrainment from gas-liquid contacting or by condensation of vapors in the gas stream due to cooling. Finally, the solids are usually in the form of dusts entrained in the gas stream.

3.3.1 Recovery of Gases from Gases

The separation of homogeneous gas mixtures requires either some way of generating a second equilibrium phase or a technique which takes advantage of differing rates of reaction or of mass transport of the components of the gas mixture. At the present time the techniques most frequently used in waste minimization depend on generation of a second phase although membrane processes that depend on relative mass transport rates may become of increasing importance in gas recycling operations in the future.

The common ways of generating the second phase are (1) the use of a solid adsorbent such as activated carbon, or (2) the use of a liquid solvent selective for the undesirable contaminant. These materials can work either by physical or chemical interaction with the contaminants. For example, solid adsorbents can attract an adsorbed phase merely by the physical nature of its surface structure or by building into the adsorbent a chemical functionality which reacts with specific materials in the gas phase to sequester them. Similarly, the addition of a solvent can lead to a physical separation process or a chemical one. For example, in absorption and scrubbing with liquid solvents, the solvent can either provide a liquid phase to permit separation of the gas mixture by physical distillation, or can actually react chemically with one or more of the contaminants in the feed gas to bind them chemically into the solvent phase. Both types of adsorbents and solvents are in common use.

The processes used for recovery of gases from gases are listed in Table 3.1 along with the chapter in which they are dicussed in more detail.

3.3.2 Recovery of Liquids from Gases

As mentioned above, the liquids found in gases are usually in the form of mists or fogs and can be removed effectively by mechanical means. The basic problem with mists and fogs is to coalesce the tiny liquid droplets into larger drops that can be more readily removed from the gas phase by settling. This can be done by the addition of ultrasonic energy to the gas phase, by passing the gas phase through a cyclone to centrifuge out the liquid, by impacting the droplets on a

Table 3.1 Recovery of Gases from Gases

Separation process	Chapter	Chapter title
Absorption	13	Absorption, Stripping, and Scrubbing
Adsorption	14	Adsorption, Ion Exchange, and Reversible Reactions with Solids
Chemical reaction	14	Adsorption, Ion Exchange, and Reversible Reactions with Solids
Permeation	9	Membrane Processes
Scrubbing	13	Absorption, Stripping, and Scrubbing

Table 3.2 Recovery of Liquids from Gases

Separation process	Chapter	Chapter title
Absorption	13	Absorption, Stripping, and Scrubbing
Coalescence	5	Settling and Flotation
Condensation	12	Distillation
Cyclones	7	Centrifuging and Cycloning
Electrostatics	10	Electrostatic Precipitation and Electrochemical Processes
Filtration	6	Filtration and Drying
Scrubbing	13	Absorption, Stripping, and Scrubbing

Table 3.3 Recovery of Solids from Gases

Separation process	Chapter	Chapter title
Cyclones	7	Centrifuging and Cycloning
Electrostatics	10	Electrostatic Precipitation and Electrochemical Processes
Filtration	6	Filtration and Drying
Scrubbing	13	Absorption, Stripping, and Scrubbing

solid surface such as metal mesh demisters and coalescers, or by contacting the droplets with a liquid surface such as exists in scrubbers and absorbers. The processes used for recovery of liquids from gases are listed in Table 3.2 along with the chapter in which they are discussed in more detail.

3.3.3 Recovery of Solids from Gases

The entrainment of solids in gases is in the form of fine particles classed as dusts. Since they form a second phase, as with liquids in gases, they are also separated by mechanical techniques. Large particles can be readily settled out of a gas phase, but dust-like particles generally require some form of mechanical help. This mechanical help can take the form of cyclones, filters, scrubbers, or electrostatic precipitators, and all are in common use. The processes used for recovery of solids from gases are listed in Table 3.3 along with the chapter in which they are discussed in more detail.

3.4 RECOVERY PROCESSES FROM LIQUIDS

The contaminants in liquid streams may also be either gases, liquids, or solids. Furthermore, the liquid streams may be either homogeneous or heterogeneous, depending on whether the contaminants are in solution or form a second phase. The gases are easily settled out by gravity unless they are either dissolved, of very small bubble size, or the liquid is extremely viscous. Heterogeneous liquid phases often exist as clouds of very tiny droplets in a continuous phase of a second liquid. These systems can be very difficult to separate by mechanical means, especially if the densities of the droplets and of the continuous phase are very close to each other. Finally, solid contaminants can also be either dissolved in a liquid phase or exist as a slurry of solid particles in the liquid carrier. The optimum separation process for recovery of each of these contaminants depends on the exact nature of the contaminated liquid. Recovery of the three different phases of contaminant are discussed below.

3.4.1 Recovery of Gases from Liquids

When gases exist as a separate phase in liquids, they can usually be recovered by coalescing the small entrained gas bubbles into large bubbles and letting them settle out of the liquid phase. If necessary, the viscosity of the continuous liquid phase can be reduced by heating to improve both coalescing and settling. On the other hand, if the gases are dissolved in the liquid phase, some sort of carrier phase must be generated to strip the dissolved gases out of the liquid. This can be an inert gas, such as steam, air, carbon dioxide, or nitrogen, blown through the liquid phase as a stripping gas in a stripping column. Alternatively, vapors generated from the liquid by heating it to a boil as in flash distillation

Table 3.4 Recovery of Liquids from Liquids

Separation process	Chapter	Chapter title
Adsorption	14	Adsorption, Ion Exchange, and Reversible Reactions with Solids
Centrifuging	7	Centrifuging and Cycloning
Cyclones	7	Centrifuging and Cycloning
Distillation	12	Distillation
Extraction	15	Solvent Extraction, Solvent Precipitation, and Leaching
Flotation	5	Settling and Flotation
Ion exchange	14	Adsorption, Ion Exchange, and Reversible Reactions with Solids
Permeation	9	Membrane Processes
Settling	5	Settling and Flotation
Solvent precipitation	15	Solvent Extraction, Solvent Precipitation, and Leaching
Stripping	13	Adsorption, Stripping, and Scrubbing
Reverse osmosis	9	Membrane Processes
Ultrafiltration	9	Membrane Processes

Table 3.5 Recovery of Solids from Liquids

Separation process	Chapter	Chapter title
Centrifuging	7	Centrifuging and Cycloning
Chemical precipitation	8	Chemical Precipitation and Sedimentation
Crystallization	11	Evaporation and Crystallization
Cyclones	7	Centrifuging and Cycloning
Drying	6	Filtration and Drying
Evaporation	11	Evaporation and Crystallization
Filtration	6	Filtration and Drying
Flocculation	8	Chemical Precipitation and Sedimentation
Flotation	5	Settling and Flotation
Microfiltration	9	Membrane Processes
Settling	5	Settling and Flotation
Ultrafiltration	9	Membrane Processes

can be used to strip out gases. In this latter case the stripping gases are the vapors generated by the boiling liquid. These processes are discussed in more detail in Chapter 13, "Absorption, Stripping, and Scrubbing."

3.4.2 Recovery of Liquids from Liquids

As with gaseous contaminants, liquid contaminants can also exist in liquid systems as either a separate liquid phase or as a homogeneous solution. When a heterogeneous system is encountered, the recovery processes are mechanical in nature such as settling, flotation, coalescing, cycloning, and centrifuging. On the other hand, if the system is homogeneous, either equilibrium- or rate-controlled processes must be used. The equilibrium processes are distillation, stripping, adsorption, extraction and supercritical extraction, ion exchange, and solvent precipitation. Finally, the rate-controlled processes are membrane diffusion and electrolytic processes. These processes are shown in Table 3.4 along with the chapters in which they are discussed in more detail.

3.4.3 Recovery of Solids from Liquids

Solid contaminants can also exist in liquids either as slurries or in the dissolved state. As with other heterogeneous systems, the slurries are separated by mechanical techniques such as settling, filtration, flocculation, flotation, centrifuging, and drying. With the homogeneous systems, evaporation and crystallization are frequently used to generate a heterogeneous system followed by one of the above-mentioned techniques for heterogeneous systems. These processes are shown in Table 3.5 along with the chapters in which they are discussed in more detail.

3.5 RECOVERY PROCESSES FROM SOLIDS

Solids can be contaminated with relatively small amounts of adsorbed gases and liquids as with spent adsorbents, or with larger amounts of liquids as with filter cakes and sludges. Furthermore, solids can absorb either liquids or gases to give swollen solids and gels. Solid polymers can show this latter behavior. Finally, solids can exist as mixtures with other solids as in the case of many mining and industrial solid wastes. All of these systems are discussed below.

3.5.1 Recovery of Gases from Solids

Adsorbed gases can be recovered from solids by stripping the solids with a condensible gas such as steam. This permits the adsorbed gas to be recovered b, condensing the steam out of it. Alternatively, if the adsorbed gas is readily condensible, as with some vapors, fixed gases such as air, nitrogen, or carbon dioxide can be used for stripping. The desorbed vapors are recovered from the stripping gas by condensing them out of the stripping gas stream.

Dissolved gases require that the structure of the solid be changed to discharge the dissolved gases. This can frequently be done by raising the temperature of the solid during stripping to raise the partial pressure of the dissolved gases. Stripping of gases from solids is discussed in Chapter 14, "Adsorption, Ion Exchange, and Reversible Reactions with Solids."

3.5.2 Recovery of Liquids from Solids

Adsorbed liquids on solids can be recovered by stripping with immiscible condensible gases in much the same way as adsorbed gases. However, they can also be washed off the solids with suitable solvents and then recovered from the resulting mixture with the solvent. Dissolved liquids may be recovered by heating the solids in a dryer under a stream of stripping gas. These techniques are shown in Table 3.6 along with the chapters in which they are discussed in more detail.

3.5.3 Recovery of Solids from Solids

Mixtures of solids use much different processes for separation from those used with liquids and gases. The usual separation processes are based on differences in density, particle size, surface properties, magnetic susceptibility, or electrostatic characteristics of the various solids. For example, with elutriation, in which a stream of gas is pased through a fluidized bed of solids of differing particle sizes and particle densities, the gas carries out with it the lightest and smallest particles first and can be used to make either a density separation or a particle size separation. Surface properties can be used to make a separation as with froth flotation in which dispersed air is bubbled through a slurry of a solids

Table 3.6 Recovery of Liquids from Solids

Separation process	Chapter	Chapter title
Centrifuges	7	Centrifuging and Cycloning
Cyclones	7	Centrifuging and Cycloning
Drying	6	Filtration and Drying
Evaporation	11	Evaporation and Crystallization
Extraction	15	Solvent Extraction, Solvent Precipitation, and Leaching
Filtration	6	Filtration and Drying
Stripping	13	Absorption, Stripping, and Scrubbing

Table 3.7 Recovery of Solids from Solids

Separation process	Chapter	Chapter title
Electromagnetic	4	Separation of Solids
Electrostatic	4	Separation of Solids
Elutriation	5	Settling and Flotation
Flotation	5	Settling and Flotation
Screening	4	Separation of Solids
Sorting	4	Separation of Solids

mixture in water and attaches bubbles of air to the solid particles selectively according to the surface properties of the individual solid particles. This makes the particles with the attached bubbles buoyant and carries them to the surface from which they can be recovered. The process is commonly used on a large scale in the mining industry to separate ores. Finally, electrostatic and magnetic separations of solids are frequently done on a moving belt which sorts materials according to their magnetic properties or their ability to accept an electrostatic charge. These processes are shown in Table 3.7 which also gives the chapters in which they are discussed in more detail.

3.6 HOW TO PICK A GOOD PROCESS

In his text on separation processes, King (1981) gave sixteen good general rules of thumb for decreasing energy consumption in separation processes. These rules of thumb were primarily aimed at process design, but they also give some insights into choosing processes and process equipment. A reduced, somewhat paraphrased list of those items that are more under the customer's control is given in Table 3.8.

Mechanical separations such as settling, screening, filtration, centrifugation, flocculation, etc., are much less energy intensive than separation processes for homogeneous mixtures. Therefore, it is usually preferred to perform the separation of heterogeneous phases such as sludges and emulsions first (Rule 1) before attacking the homogeneous mixtures. The removal of heterogeneous phases greatly simplifies the separation of homogeneous mixtures.

Most of the separation processes encountered in hazardous waste minimization are fairly simple, such as batch distillation, adsorption, settling, filtration, and the like. These have good turndown ratios, i.e., they can operate efficiently over a wide range of operating rates. It is important to have this capability (Rule 2), particularly for continuous processes such as wastewater treating and stack gas scrubbing where flow rates change in response to manufacturing operations.

Table 3.8 Rules of Thumb for Process Selection

1. Do the mechanical separations first if more than one phase exists in the feed.
2. Avoid overdesign and use designs giving efficient turndown. Favor simple processes.
3. Favor processes transferring the minor rather than major component between phases.
4. Favor high separation factors.
5. Recognize value differences of energy in different forms and of heat and cold at different temperature levels.
6. Investigate use of heat pump, vapor compression, or multi-effects for separations with small temperature ranges.
7. Use staging or countercurrent flow where appropriate.
8. For similar separation factors, favor energy driven processes over mass separating-agent processes.
9. For energy driven processes, favor those with lower latent heat of phase change.

Rule 3 is very important in making separation processes economical. It applies to many of the separations involved in waste minimization. For very dilute systems, there is a great advantage to using processes like ion exchange and adsorption for the impurity rather than evaporation of the water. Furthermore, for extraction type processes, it is better to extract the minor component rather than the major component to minimize solvent circulation and regeneration.

High separation factors greatly reduce the number of stages and thus the capital costs required for a separation (Rule 4). They can frequently be increased to essentially infinity by adding a specific reagent that only reacts with one component of the feed mixture. Examples of processes having high separation factors are ion exchange for demineralizing water, chemical complexing for mixtures of olefins and paraffins, adsorption of aromatics from mixtures with saturated hydrocarbons, and to a lesser extent adding a selective solvent to a distillation to convert it to an extractive distillation. Many waste streams, such as solvents in motor oils, have a naturally high separation factor for distillation due to the large difference in boiling point between the homogeneous feed components. These solvents can be recovered in satisfactory purity for recycling in single-stage batch stills and many such operations are already in place.

Heat and refrigeration energy are inherently less expensive than electrical energy or mechanical work. However, as the temperature at which the heat or refrigeration is used departs from ambient temperature, the cost of heat and refrigeration increases. Rule 5 recommends that these effects be considered in choosing a separation process for a given feed mixture.

While mechanical work is more expensive than heat or refrigeration, it can sometimes permit latent heat of condensation to be exchanged for latent heat of vaporization (Rule 6). In such a case, which usually requires a small temperature difference between the heat source and heat sink, vapor compression by mechanical work on the distillate can give it an increased pressure and condensing temperature, actually higher than the temperature in the boiler. This pumping of heat from a lower to a higher temperature permits the vapor to be condensed in the boiler generating the vapor in the first place. A balance between compressor cost and heat cost can determine the desirability of such operation.

Staging and countercurrent flow (Rule 7) can decrease the amount of separating agent needed and increase the purity of the products.

Energy-driven processes are inherently more efficient because they are more nearly reversible. They have the additional advantage that the energy can be readily removed by exchange with another stream and recycled to the process. This can be used to cascade the energy through a series of separations (Rule 8) at gradually reducing temperature from initial source to final sink, using it over and over, thus minimizing the required supply of energy. To do the same type of thing with a mass separating agent requires an additional separation process for each exchange or recovery of the mass separating agent, thereby requiring much more energy.

Rule 9 comes from the fact that in energy-driven processes the energy is largely required to overcome the latent heat of phase change. Other things being equal, this favors heats of fusion over heats of vaporization and choice of selective solvents having lower heats of phase change in their recovery process. The use of multi-effect temperature cascades in the separation process can greatly reduce the advantage offered by using low latent heat processes.

While these rules of thumb are very helpful in preliminary choices of separation processes, it must be remembered that the optimum separation process involves a balance between energy costs and the costs for equipment to minimize energy costs. Therefore, sometimes it is desirable to pay for more energy and avoid the additional capital costs and operating complexity necessary to achieve high thermal efficiency.

REFERENCES

Du Pont (1991). The value of waste, *Du Pont Magazine*, September/October 1991, p. 18.

Freeman, H. M. (ed.) (1988). *Standard Handbook of Hazardous Waste Treatment and Disposal*, McGraw-Hill, New York.

King, C. J. (1981). *Separation Processes*, 2nd ed, McGraw-Hill, New York, p. 687.

4

Separation of Solids

4.1 INTRODUCTION

The processes for separating various types of solid materials from each other are much different from the equilibrium processes used to separate liquids and gases. The processes for solids make use of physical differences in either a bulk property of the solids or in a surface property of the solids. Thus, they are more dependent on physics rather than chemistry and are essentially sorting processes. Sorting is defined as the act of putting the components of a mixture into a given place, container, or rank according to kind, class, or nature. That is, to arrange or separate mixtures of solid materials into homogeneous groups according to the characteristics of the individual components. Typical examples in waste management would be (1) the separation of glass, plastic, paper, and metals from each other, (2) separation of glass bottles into brown, green, and clear glass categories, (3) separation of plastic containers according to the type of polymer used in their manufacture, and (4) separation of scrap metals or metal cans into ferrous and non-ferrous categories. These types of sorting are essential to permit effective recycling programs to be achieved. The sorted products must not be cross-contaminated and must be homogeneous enough to be used as raw

Table 4.1 Vendors for Municipal Waste Recycling Plants

Buhler, Inc. 1102 Xenium Lane Minneapolis, MN 55441 (612) 545-1401	CP Manufacturing 1428 McKinley Ave. National City, CA 92050 (619) 477-3175
Lindemann Recycling Equipment, Inc. 500 Fifth Avenue New York, NY 10110 (212) 382-0630	Lummus Development Corp. P.O. Box 2526 Columbus, GA 31902 (800) 344-0780
Lundell Manufacturing Co, Inc. Box 171 Cherokee, IA 51012 (712) 225-5185	New England CRINC. (Exclusive North American agent for Bezner plants) 74 Salem Road North Billerica, MA 01862 (508) 667-0096

material in a manufacturing process. This has led to the statement that any good recycling program begins with an effective sort. A typical sorting flow plan for recovering recyclable materials from municipal waste is shown in Figure 4.1.

In Figure 4.1 the recyclable municipal trash is dumped on a presorting floor where bundled newspapers and non-recyclable materials are removed. The trash is then put on conveyer belts and fed consecutively through separation devices which remove different materials and put them in separate bins for recycling. Magnetic separation removes tramp iron and steel cans, while screening to remove fines eliminates sand and broken glass. Either air blowing or the Bezner chain curtain sorting machine is used to separate light aluminum and plastic containers from glass bottles and jars. The glass is then sorted by color either automatically or by hand. The large plastic bottles are removed by coarse screening and the aluminum cans are removed by an eddy current magnetic separator. Finally, the plastic is separated by polymer type either manually or automatically. Some vendors of integrated plants for separation of municipal wastes for recycling are given in Table 4.1.

The bulk properties typically chosen as a basis for sorting solids are size, density, magnetic permeability, or electrical conductivity. Solids are sorted by size over a wide size range using screens and rotating perforated drums. A combination of size and density controls the separation of solids in jigs and tables for larger sizes and in fluid beds or by elutriation for smaller sizes. The

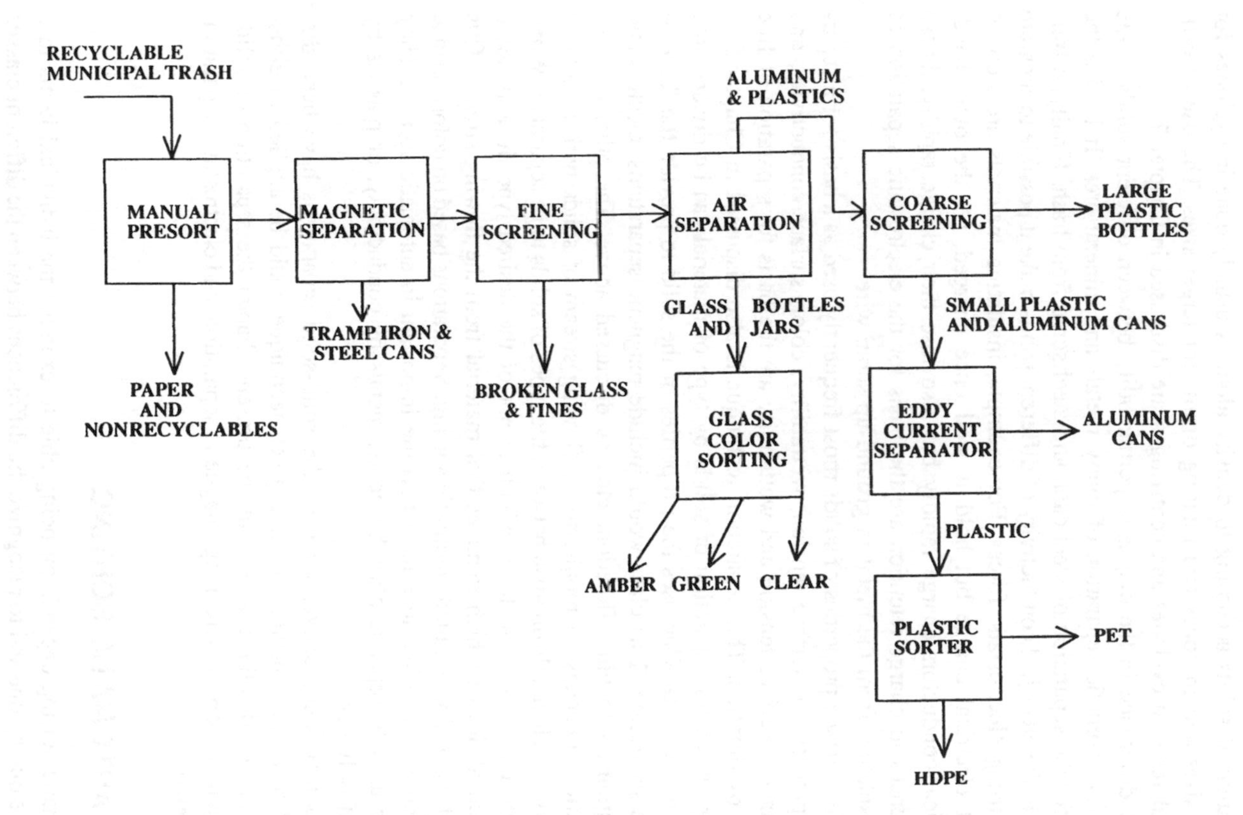

Figure 4.1 Sorting flow plan for recovery of recyclable materials from municipal waste. (Courtesy of New England CRINC.)

separation of solids according to density alone is usually done in cyclones for small sizes and in cones and rotating drums for larger sizes. The use of centrifugal force in cyclones and centrifuges are discussed in Chapter 7.

A difference in the magnetic permeability between different solids is the basis for magnetic separation of many metals and mineral ores. It is also the basis for the separation of steel cans and steel scrap from trash. Finally, differences in the electrical conductivity of different solids make it possible to separate them using electrostatic forces. For example, insulating materials are slow to accept electrical charge but hold it well once charged. On the other hand, electrical conductors charge rapidly but also lose their charge rapidly. These differences in charge retention are the basis for the electrostatic separation of the insulation from the metal in ground-up scrap wire.

The surface properties of solids most frequently used as a basis for a separation process are surface tension, wettability, color, surface conductivity, and reflectance. Surface tension and wettability are the basis for separation of fine solids by flotation. This separation technique is also discussed in Chapter 5.

Processes for separation of solids by type of material can be divided into two major classes. One uses the properties of the solid to provide the force to do the separation. This class would include magnetic separations, electrostatic separations, screening, flotation, etc., as discussed above. The other class of separations depends on monitoring a flowing stream of solids with a suitable detector which can distinguish between the types of solids to be separated. When the detector senses a material which is not of the desired type, it activates a mechanical device which removes that material from the flowing stream. This type of separation is called automatic sorting. Separations based on color, surface conductivity, or reflectance have become important in automatic sorting. They can be used to separate glass by color, metals by conductivity, or plastics by type of polymer.

As mentioned above, some of the solid-solid separations have been discussed in earlier chapters when a given technique could be applied to either solid- or liquid-solid mixtures. In the present chapter the separation of solids by automatic sorting, screening, magnetic separation, and electrostatic separation is covered.

4.2 AUTOMATIC SORTING

Automatic sorting depends on being able to examine the individual items in a mixture one by one and to recognize the differences between the different classes of components. Also needed is a mechanism for placing the components of the same class together in a separate group. This group must be separate from both other separated groups and from the feed stream being sorted. Historically, the most common sorting machine has been the human being. The eyes provide the recognition device and the hands provide the mechanism for picking out the individuals of the same class and putting them together. If the feed mixture being sorted is made up of items of a size readily sortable by hand, it is usually spread out on tables or on moving belts. The sorters can then see, pick out, and segregate the members of the classes being sorted. As electrical, optical, and magnetic techniques for rapidly recognizing different materials have become available, automatic techniques for sorting have begun to replace people at the sorting tables and reduce the drudgery of this type of work. However, in the growing effort to recycle certain materials present in household solid wastes, the homeowner is frequently the sorter. He or she has been either requested or required by law to sort out newspapers, different kinds of glass, plastics, and steel and aluminum cans so that they may be economically recycled. Alternatively, more and more municipal trash collection systems are providing automatic sorting of mixed trash after collection.

4.2.1 Fundamentals of Automatic Sorting

As mentioned above, automatic sorting requires that the elements in the feed mixture be exposed one at a time to a sensing device that can discriminate among the feed components. Also needed is a suitable mechanical method to remove selected particles from the feed stream at the appropriate signal from the sensing device. This is usually done by grinding the mixed stream to a relatively narrow size range and then feeding the sized stream from a feed hopper onto a vibrating feeder and thence through an aligning device onto a troughed conveyer belt. The trough in the conveyer belt keeps the individual particles in line until they are projected in a continuous stream over the end roll of the conveyer belt. An optical scanning device looks at the falling particles one by one and feeds its electrical output to an electronic system which decides, based on preset characteristics for decision-making, whether the particle should be removed from the main stream. If so, a fast-acting air jet system is activated to blow the particle to be separated out of the trajectory of the falling feed particles and thus make it fall into a different collecting container. A sketch of this type of operation is shown in Figure 4.2. For sorting larger pieces, the individual item to be removed from the feed stream may be sensed while still on the belt. It is then removed by a push from a piston activated by the sensing signal.

The lower limit of particle size that can be optically sorted economically is about 1/8 in. and the maximum sorting rate is about 80 particles per second. A more typical range of particle weight for effective optical sorting is from 1 to 3 oz. Up to 6 tons per hour can be sorted for this size range material.

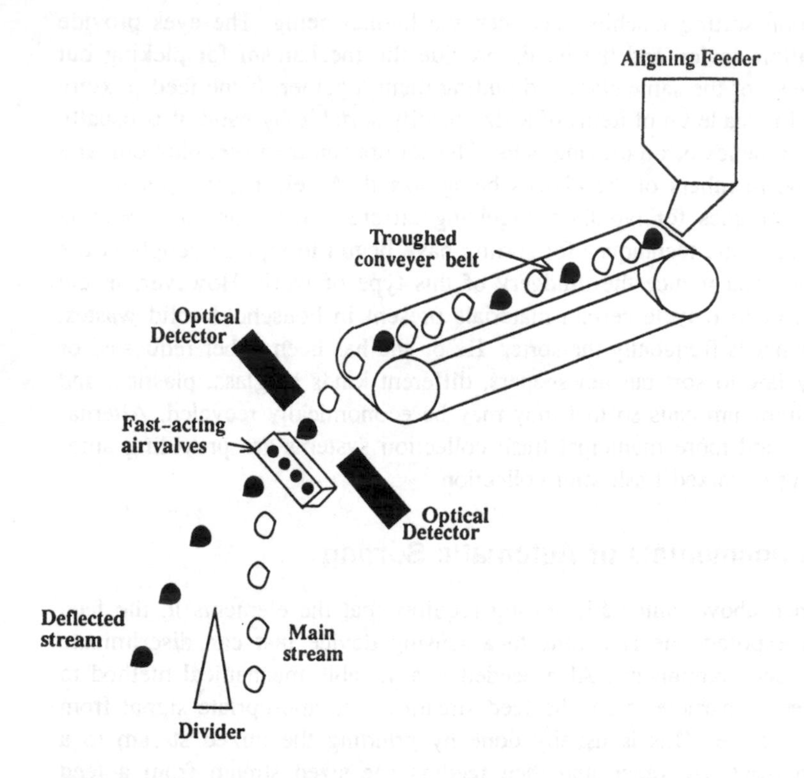

Figure 4.2 Sketch of automatic optical sorting equipment.

Optical properties that can be used for automatic sorting include color, light reflectance, transparence, and fluorescence induced by x-rays or ultraviolet light. In addition to optical properties, differences in electrical conductivity can also be used as a sensing basis for discrimination between different particles. For conductivity sorting, the electrical conductivity of each particle is measured while on the conveyer belt using a brush type contacting electrode. This generates the discriminating signal which is compared to the standard by which acceptance or rejection is determined. A typical conductivity sorter can process up to about 27 tons per hour of particles of 2 to 6 in. size.

4.2.2 Specific Applications of Automatic Sorting

The specific applications of sorting to waste minimization are usually the result of failure to segregate waste streams at the point at which they are generated. For example, sorting is frequently needed for the separations of solid municipal wastes to obtain individual segregated products such as glass, plastic, or aluminum that can be recycled. However, this is usually done manually in this country even by commercial recyclers. Mandatory recycling laws passed by states or local communities usually require the individual homeowner to perform the sorting and segregation of these materials. This is done before setting out the trash for pickup and can reduce the need for extensive sorting by the waste recycler.

Some typical sorting operations required by segregation of municipal wastes for recycling are:

1. Transparent glass from opaque materials
2. Colored glass sorted by color
3. Plastics by polymer type
4. Conducting metals from insulators
5. Aluminum from other metals
6. Different metals from each other

These sorting operations are frequently done primarily by hand, but they are good candidates for automatic sorting. In some cases the sorting equipment already exists. The first three applications can be done by optical sorting using reflectivity, color, or fluorescence excited by x-rays or ultraviolet light, respectively. Application number 4 can be achieved using electrical conductivity in an automatic sorter or by electrostatic separation, discussed later in this chapter. Applications 5 and 6 are achievable using the magnetic separations discussed later in this chapter. These magnetic sorting techniques are also effective for separating aluminum and ferrous metals from other metals and from each other.

Automatic sorting is making some inroads into the mining industry for beneficiating, i.e., enriching, ores. It is expected that innovative developments in that application will translate into corresponding increases in the use of such developments in the waste management industry.

4.2.3 Equipment for Automatic Sorting

Automatic optical sorting equipment was developed primarily for the mining and food industries—in the mining industry for ore beneficiation and in the food industry to assure top quality of fruits, nuts, vegetables, and coffees by color detection. It is just beginning to make inroads into the sorting of glass by color and of plastics by molecular type.

The principal vendors of automatic sorting equipment are given in Table 4.2.

4.2.4 Common Problems and Troubleshooting

The most common problem in automatic sorting is overlap of the categories being sorted. For example, when sorting a mixture of brown and green glass

Table 4.2 Vendors of Automatic Sorting Equipment

Dunkley International, Inc. 1910-T Lake, P.O. Box 3037 Kalamazoo, MI 49003 (616) 343-5583	ESM International, Inc. 10621-T Harwin Drive, Suite 300, Houston, TX 77036 (713) 981-9185
Key Technology, Inc. 150 Avery Street Walla Walla, WA 99362 (509) 529-2161	S&S Electronics of America 1530 East Race Street Allentown, PA 18103 (215) 266-0330
Simco/Ramic Corporation P.O. Box 1666 Medford, OR 97501 (503) 776-9800	Sortex-Scancore, Inc. 33210-T Transit Ave. Union City, CA 94587 (510) 487-8100

by color, the brown glass product may contain as much as 10% green glass, and the green product may contain as much as 10% brown. This overlap is usually made worse by pushing the feed rate too fast, by using a sensing device with too slow a recognition capability, or by too slow a mechanical system for performing the actual particle separation. The quality of the separation can also be improved by using more than one sorting stage. In this way the primary products from sorting can be re-sorted for further purification.

4.2.5 Economics

The economic value of sorting processes depends on the relative values of the separated products as compared to the mixed feed stream. For example, aluminum cans recovered from municipal waste have a recycling value of about $900 per ton. On the other hand, they would cost about $100 to 110 per ton to dispose of in a solid waste landfill in the northeastern United States. The values for other components that can be recovered from municipal wastes are given in Table 4.3. These values vary greatly with geographic area and with distance from the consumer, i.e., freight costs. Similarly, landfilling costs vary widely with geographic area. The value of recycling a material is the sum of the selling price and the savings from not having to dispose of it by another means, such as landfilling or incineration, minus the cost of separating it. Thus for newspapers, which have essentially zero selling price, a saving of about $100 per ton can be achieved by not having to landfill it provided the public does the

Table 4.3 Approximate Value of Components in Municipal Waste

Component in municipal waste	Approximate selling price, dollars per ton
Aluminum	800–900
Corrugated paper	20–50
Newspaper	0–25
Clear glass	40–65
Amber glass	20–55
Green glass	0–20
Steel cans	40–70
HDPE plastic	60–140
PET plastic	40–200

From *Recycling Times* (1992).

sorting. If the public does not provide the sorting operation, this saving is reduced by the cost of the sorting.

The cost of separating solid municipal wastes into recyclable streams is about $26 to 28 per ton when using mechanized sorting. Therefore, even materials having a low selling price as recycled materials produce large savings by recycling. The disposal cost is often the tail that wags the dog. Furthermore, costs of disposal are expected to rise in the future due to further restriction on landfills. This should further increase the value of recycling.

4.3 SCREENING

Screening is the separation of a mixture of various grain sizes into two narrower size range fractions by means of a surface containing uniform size holes. The holes are usually square, but other shapes of holes, particularly rectangular, are frequently used. The screen surface can consist of woven wire, silk or plastic cloth, perforated or punched plate, grizzly bars, or wedge wire sections. Part of the feed consisting of small enough particles can pass through the holes. The rest of the feed consisting of larger particles is retained on the screen surface. The material passing through the screen is called the *undersize* or *minus material* for the particular screen size. The material retained on the screen is called the *oversize* or *plus material*. The size distribution can be determined for the original feed by using a series of screens of successively smaller sized holes. The mixture is initially fed onto the screen having the largest holes, the undersize goes to the screen with the next largest hole size, and so on until the feed has passed through the entire series of screens. The oversize on each screen and the undersize from the smallest and final screen are weighed and reported as percentage of original feed versus screen hole size or mesh.

A standard test called *sieve analysis* has been developed to determine size distributions of solid materials. This is used to follow the effectiveness of operation of screening devices as well as the effectiveness of crushing and grinding equipment. This test uses a series of standard metal mesh sieves arranged vertically in order of increasing mesh or number of wires per inch (decreasing hole size) from top to bottom. The stack of sieves is clamped into a shaker, the feed is introduced into the top sieve, and the stack is shaken for a standardized time and intensity. The amounts left on each sieve and the final undersize are weighed and reported as percent of feed.

For the sieve test to be reproducible and reliable, the test sieves must be standardized with standard size openings in the wire mesh. In response to this requirement, a number of different series of standard sieves have been developed which differ mainly in how the size is identified. The U.S. Sieve Series is based on hole sizes for successively larger screens being in the ratio of the fourth root of 2 or about 19% larger than the previous one. The individual sieves in the series are identified by their hole sizes in millimeters for sizes above 1 mm and in microns for hole sizes below 1 mm. The Tyler Series of Standard Sieves actually preceded the U.S. Series and has the same fourth-root-of-2 size ratio as the U.S. Series, but the individual sieves are identified by mesh. This is the number of wires per inch rather than size of hole opening. Furthermore, the Tyler series only goes up to 1.05-in. hole openings while the U.S. Series extends to 4.24 in. These sieve series are shown in Table 4.4.

Full-size screening operations can be used in much the same way as the above sieve analysis to fractionate a mixed feed into a group of relatively uniform sized fractions. This can be done by passing the feed across a series of screens having decreasing size holes. The oversize and undersize can be collected as desired. The full-size screens are shaken in a reciprocating, rotary, or vibrating motion as the feed passes over the screen surface. This is done to provide a good separation efficiency.

4.3.1 Fundamentals

While screening or sieving separates a mixture of solid particles according to size, it does not do this very precisely. This is because particle shape affects the size of a particle which will pass through a given size hole. Most screening operations are preceded by some sort of size reduction operation such as crushing, grinding, or milling. This produces a mixture of particulate materials which vary in size, shape, density, and abrasive properties. The first thing to do with such a mixture is to get a sieve analysis to determine how much of this mixture falls into the various size fractions of interest to the screening operator. This

Table 4.4 U.S. Sieve Series and Tyler Equivalents

Sieve designation		Sieve opening		Nominal wire diameter		
Standard	Alternate	mm.	in. (approx. equivalents)	mm.	in. (approx. equivalents)	Tyler equivalent designation
107.6 mm.	4.24 in.	107.6	4.24	6.40	.2520	
101.6 mm.	4.00 in.[b]	101.6	4.00	6.30	.2480	
90.5 mm.	3½ in.	90.5	3.50	6.08	.2394	
76.1 mm.	3.00 in.	76.1	2.50	3.00	.2283	
64.0 mm.	2½ in.	64.0	2.50	5.50	.2165	
53.8 mm.	2.12 in.[b]	53.8	2.12	5.15	.2028	
50.8 mm.	2.00 in.	50.8	2.00	5.05	.1988	
45.3 mm.	1¾ in.	45.3	1.75	4.85	.1909	
38.1 mm.	1½ in.	38.1	1.50	4.59	.1807	
32.0 mm.	1¼ in.	32.0	1.25	4.23	.1665	
26.9 mm.	1.06 in.	26.9	1.06	3.90	.1535	1.050 in.
25.4 mm.	1.00 in.[b]	25.4	1.00	3.80	.1496	
22.6 mm.[a]	⅞ in.	22.6	.875	3.50	.1378	.883 in.
19.0 mm.	¾ in.	19.0	.750	3.30	.1299	.742 in.
16.0 mm.[a]	⅝ in.	16.0	.625	3.00	.1181	.624 in.
13.5 mm.	0.530 in.	13.5	.530	2.75	.1083	.525 in.
12.7 mm.	½ in.[b]	12.7	.500	2.67	.1051	
11.2 mm.[a]	7/16 in.	11.2	.438	2.45	.0965	.441 in.
9.51 mm.	⅜ in.	9.51	.375	2.27	.0894	.371 in.
8.00 mm.[a]	5/16 in.	8.00	.312	2.07	.0815	2½ mesh
6.73 mm.	0.265 in.	6.73	.265	1.87	.0736	3 mesh
6.35 mm.	¼ in.[b]	6.35	.250	1.82	.0717	
5.66 mm.[a]	No. 3½	5.66	.223	1.68	.0661	3½ mesh
4.76 mm.	No. 4	4.76	.187	1.54	.0606	4 mesh
4.00 mm.[a]	No. 5	4.00	.157	1.37	.0539	5 mesh
3.37 mm.	No. 6	3.37	.132	1.23	.0484	6 mesh
2.83 mm.[a]	No. 7	2.83	.111	1.10	.0430	7 mesh
2.38 mm.	No. 8	2.38	.0937	1.00	.0394	8 mesh
2.00 mm.[a]	No. 10	2.00	.0787	.900	.0354	9 mesh
1.68 mm.	No. 12	1.68	.0661	.810	.0319	10 mesh

(continued)

Table 4.4 (Continued)

Sieve designation		Sieve opening		Nominal wire diameter		
Standard	Alternate	mm.	in. (approx. equivalents)	mm.	in. (approx. equivalents)	Tyler equivalent designation
1.41 mm.[a]	No. 14	1.41	.0555	.725	.0285	12 mesh
1.19 mm.	No. 16	1.19	.0469	.650	.0256	14 mesh
1.00 mm.[a]	No. 18	1.00	.0394	.580	.0228	16 mesh
841 micron	No. 20	.841	.0331	.510	.0201	20 mesh
707 micron[a]	No. 25	.707	.0278	.450	.0177	24 mesh
595 micron	No. 30	.595	.0234	.390	.0154	28 mesh
500 micron[a]	No. 35	.500	.0197	.340	.0134	32 mesh
420 micron	No. 40	.420	.0165	.290	.0114	35 mesh
354 micron[a]	No. 45	.354	.0139	.247	.0097	42 mesh
297 micron	No. 50	.297	.0117	.215	.0085	48 mesh
250 micron[a]	No. 60	.250	.0098	.180	.0071	60 mesh
210 micron	No. 70	.210	.0083	.152	.0060	65 mesh
177 micron[a]	No. 80	.177	.0070	.131	.0052	80 mesh
149 micron	No. 100	.149	.0059	.110	.0043	100 mesh
125 micron[a]	No. 120	.125	.0049	.091	.0036	115 mesh
105 micron	No. 140	.105	.0041	.076	.0030	150 mesh
88 micron[a]	No. 170	.088	.0035	.064	.0025	170 mesh
74 micron	No. 200	.074	.0029	.053	.0021	200 mesh
63 micron[a]	No. 230	.063	.0025	.044	.0017	250 mesh
53 micron	No. 270	.053	.0021	.037	.0015	270 mesh
44 micron[a]	No. 325	.044	.0017	.030	.0012	325 mesh
37 micron	No. 400	.037	.0015	.025	.0010	400 mesh

[a] These sieves correspond to those proposed as an international (I.S.O.) standard. It is recommended that wherever possible these sieves be included in all sieve analysis data or reports intended for international publication.

[b] These sieves are not in the fourth-root-of-2 series, but they have been included because they are in common usage.

Source: ASTM E-11-61.

allows the operator to determine the efficiency of his screening operation and to know when modifications are needed.

The efficiency of screening can be based either on what goes through the screen or what stays on the screen, i.e., on either undersize recovery or undersize removal. These efficiencies are shown in Eqs. 1 and 2.

$$\text{Undersize efficiency} = \frac{\text{Weight feed through screen}}{\text{Weight undersize of feed}} \tag{1}$$

$$\text{Oversize efficiency} = \frac{\text{Actual weight oversize}}{\text{Weight on screen}} \tag{2}$$

An economic compromize between screening time and screening efficiency dictates efficiencies of 85 to 95%.

When a feed mixture is spread onto a screen surface, the fine particles in the feed tend to move downward as the bed moves across the screen. This segregation or stratification has an important influence in determining the efficiency of screening. Thus, for any combination of screening device and feed mixture, there is an optimum bed depth for effective screening. This is because too thin a bed prevents stratification and causes competition for the screen surface between coarse and fine particles. This causes fines to be entrained with the coarse stream. Stratification helps to keep the screen surface covered with fines and reserves it for particles which pass through the screen. This maximizes flow through the screen. Too thick a bed depth makes the bed so heavy that the screen motion is dampened. This again hinders stratification and leads to entrainment of fines in the oversize stream. The optimum bed thickness is that which permits good stratification of the bed.

Screen capacity calculations can be based on any of overflow, flow-through, or total feed. However these calculations are not precise enough to give the same answer for the different calculation methods. The flow-through method seems to have become the preferred one. The equation giving required screen area in terms of flow-through rate and unit capacity of the screen is taken from Matthews (1972) and given as Eq. 3.

$$A = \frac{0.5\, C_{\mathrm{t}}}{C_{\mathrm{u}} F_{\mathrm{oa}} F_{s}} \tag{3}$$

where:

A = screen area in square feet
C_{t} = flow-through rate
C_{u} = unit capacity
F_{oa} = open area factor
F_{s} = slotted opening factor

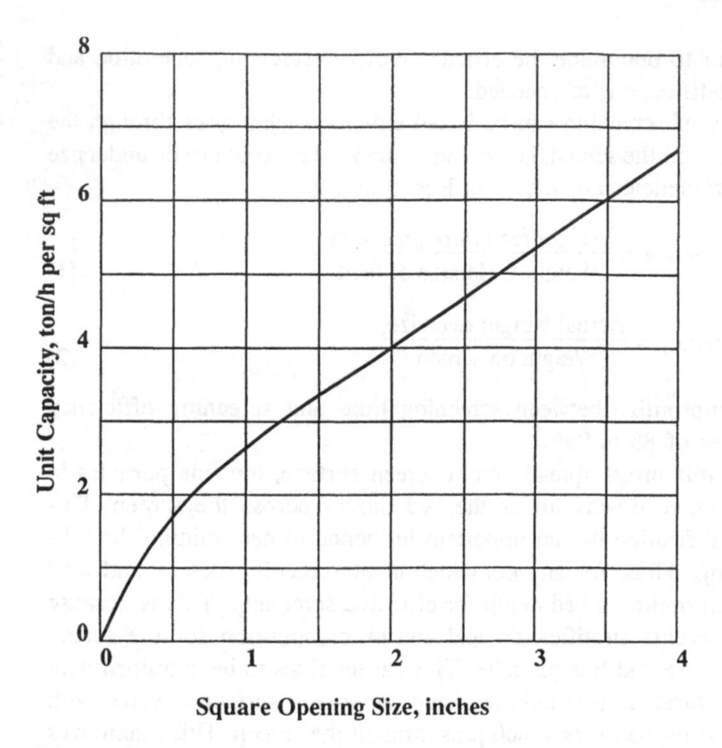

Figure 4.3 Unit capacity for square opening screens feeding crushed stone. (From Matthews, 1972.)

The flow-through rate is determined from the total feed rate to the screen and the fraction of the feed which should pass through the screen as determined from sieve analysis. The unit capacity is determined from Figure 4.3 where crushed stone is the standard. The capacity for other materials is made by using a correction factor based on the bulk density of the feed. The value for crushed stone is multiplied by the ratio of the bulk density of the actual feed to that of crushed stone (100 lb/ft^3) to get the capacity for the actual feed.

The open area factor corrects for the fraction of total screen area which is open to the flow of solids, i.e., the total hole area per square foot of screen. The hole area can be calculated by subtracting the total cross-sectional area of the wires or rods forming the screen from the area of the screen. The fraction of open area is obtained by dividing the open area by the total screen area.

Finally, the slotted opening factor is a correction term for screens that do not have square openings and have the effect of reducing the screening area required for a given flow through the screen. These correction factors are given in Table 4.5 and indicate a 10 to 40% reduction in screen area is achievable

Table 4.5 Slotted Opening Correction Factors for Sieve Capacity Calculations

Screen	Length/width ratio of slot	Slotted opening factor, F_s
Square and slightly rectangular openings	< 2	1.0
Rectangular openings	2–4	1.1
Slotted openings	4–25	1.2
Parallel rod decks	≥ 25	1.4

by using slotted openings in the screen. However, some sacrifice in sharpness of separation is the price of this increase in capacity.

4.3.2 Specific Applications

The applications of screening are many and varied and should be considered whenever separation by size produces a desired segregation. For example, in trash sorting, fine screening can be used to remove fines and broken glass from the trash stream. Furthermore, coarse screening can be used to recover large plastic bottles from a mixture of aluminum and plastic containers in recyclable municipal waste.

There are also many applications throughout industry related to size reduction of solid materials by crushing and grinding. Screening can be used in a scalping operation to protect size reduction mills from oversize material, which might damage the mill. It can also be used to remove fines from the feed to the mill to avoid wasting mill capacity. In addition, screening can be used to separate the mill product into narrow size ranges for specific uses as well as for recovery of oversize for recycle to the mill feed.

4.3.3 Equipment

There are three basic types of screening equipment: grizzlies, revolving screens or trommels, and shaking or vibrating screens. There is some overlap in the size of raw material that can be separated by these different units and their normal ranges of application are shown in Figure 4.4.

Grizzlies consist of a set of parallel rods, bars, or rails with spaces between them running the entire length of the rods, bars, or rails. They may be level or inclined and are used primarily for "scalping," where a small amount of oversize is removed from a predominantly fine feed. A typical feed to a scalping operation might contain about 5% oversize.

The trommel was named from the German word for drum and is a cylindrical screen, which is inclined away from horizontal and rotated at low speed.

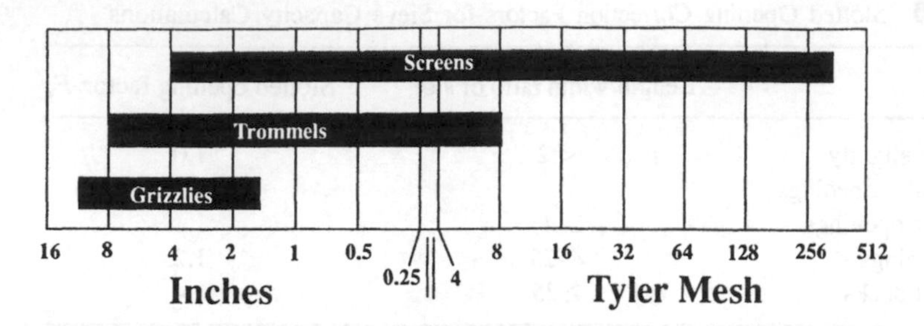

Figure 4.4 Effective range of particle sizes for different types of screening units.

It is fed internally at one end and the feed proceeds through the trommel under the influence of gravity. The rotation of the drum presents a fresh screening surface to the feed and provides flow-through area often enough for the undersize particles to pass out through the screen. A simple sketch of a trommel is shown in Figure 4.5.

Screens come in many different types depending on how mechanical motion is applied to the screens and how the solid mixture is moved through the apparatus. Some of these types are vertically vibrating, horizontally vibrating,

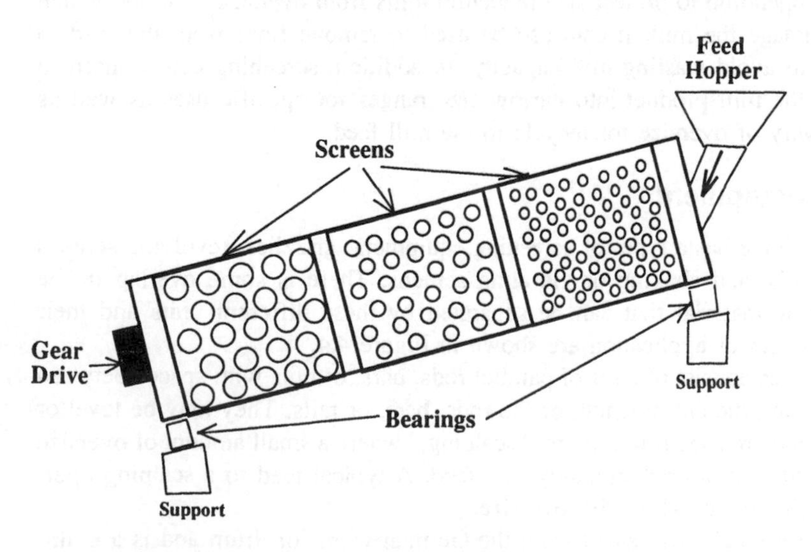

Figure 4.5 Sketch of trommel screen apparatus.

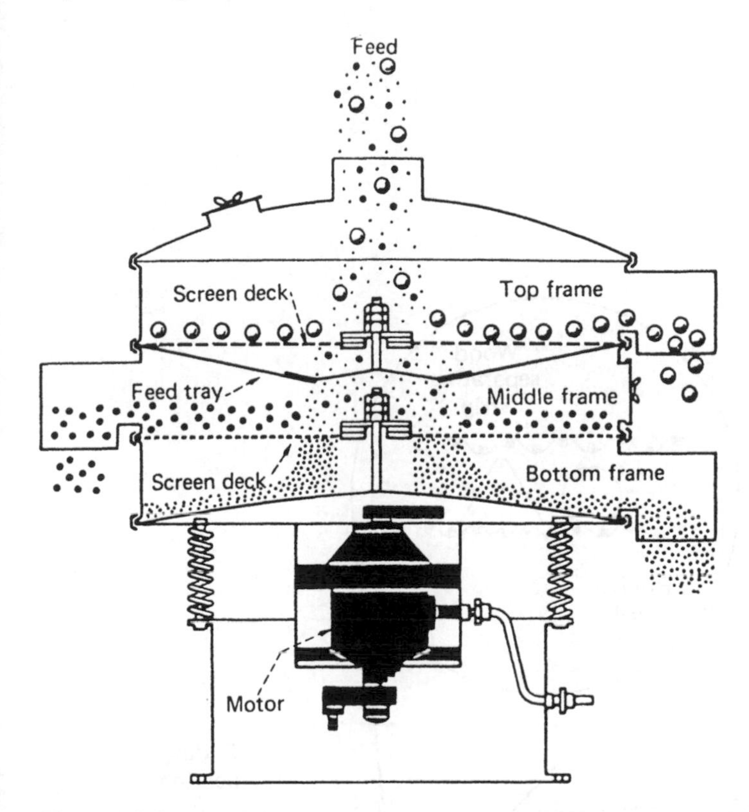

Figure 4.6 Sketch of a two-deck sifter screening apparatus which produces three product size ranges. (From Stone, 1979.)

reciprocating or shaking, centrifugal, and rotating screens. The benefits of mechanical motion are primarily increasing capacity and improving the sharpness of separation by improving the flow of undersize through the screen. Inclining the screens helps in the same way.

Inclined vibrating screens are used to separate a wide range of particle sizes in high-capacity applications as well as for scalping and trash removal. Reciprocating or shaking screens are used both for accurate sizing of large lumps as well as for general size separations. Sifter screens are used for the finest size separations and come with circular, gyratory, or circular-vibratory motion. An example of a two-deck sifter screening apparatus taken from Stone (1979) is shown in Figure 4.6. Three products of differing size ranges are produced by this unit.

A typical gyratory screener is sketched in Figure 4.7. Flow across the screen is driven by gravity. The bouncing rubber balls used in the gyratory screener

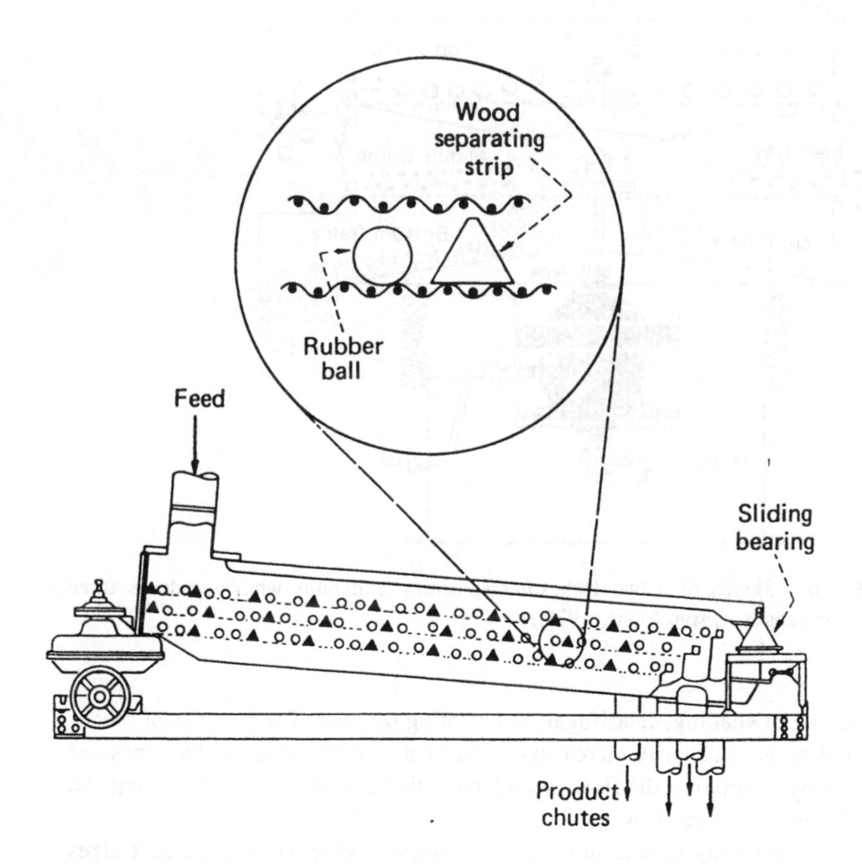

Figure 4.7 Sketch of a typical gyratory screener. (Courtesy of Rotex, Inc.)

Table 4.6 Vendors of Screening Equipment

Andritz Sprout-Bauer, Inc. Process Equipment Division Muncy, PA 17756 (717) 546-8211	Derrick Manufacturing Corp. 590 Duke Road Buffalo, NY 14225 (716) 683-9010
FMC Corporation Material Handling Equip. Div. FMC Building Homer City, PA 15748 (412) 479-8011	Kason Corporation, Inc. 1301 East Linden Ave. Linden, NJ 07036 (908) 486-8140
Midwestern Industries, Inc. 915 Oberlin Rd., S.W. P.O. Box 810 Massillon, OH 44648-0810 (216) 837-4203	Rotex, Inc. 1384 Knowlton Street Cincinnati, OH 45223 800-243-8160
SWECO, Inc. 7120 New Buffington Road Florence, KY 41042 (606) 727-5147	Witte Company, Inc. P.O. Box 47 Washington, NJ 07882 (908) 689-6500
W.S. Tyler, Inc. P.O. Box 1889-T Salisbury, NC 28145-1889 (800) 342-2682	

are to minimize blinding of the screen surface. A list of vendors for screening equipment is given in Table 4.6.

4.3.4 Common Problems and Troubleshooting

Blinding

One of the most common problems in screening is blocking or blinding of the screen. This occurs when particles get stuck in the screen openings and can be either temporary or permanent. The temporary blinding is minimized by vibrating the screens and using bouncing rubber balls to strike the active screens (see Figure 4.7) to dislodge the particles stuck in the holes. When the particles cannot be readily removed, they eventually completely block the openings in the screen and prevent any separation by the screen. This hard blinding requires replacement of the screen and its rate of formation is a function of hole size and particle shape (Beddow 1980).

Safety and Noise

Screening operations generate dust and, depending on the feed materials, the dust may be potentially either explosive or toxic. These possibilities must be carefully evaluated and suitable preventive measures taken to insure safety. Such measures might include pressure-proof enclosures and inert gas blanketing to prevent dust explosions. Adequate isolation, ventilation, and filtration are needed for recovery of toxic dusts.

Noise is generated in a screening plant by the mechanical devices used to provide motion to the screening surfaces and the impact of the feed material on the moving screen surfaces. The mechanical devices can be quieted by (1) enclosing them in insulated boxes and (2) dampening with insulation the vibration of flat surfaces which project the noise. The impact noise from the screens may be quieted by (1) using all rubber or rubber-clad screen decks and (2) suitably isolating with soft rubber gaskets the large solid flat noise-emitting surfaces from sources of vibration.

In addition to noise suppression, the use of rubber-coated screens has the added benefits of (1) reducing wear on the screens and thus extending operational life of the screens, (2) reducing dust formation and blinding of the screens, and (3) reducing sticking of damp materials to the screen surface. Furthermore, rubber-coated screens give improved stratification in the feed bed and thus improve screening efficiency.

Screening Damp Materials

Damp feed materials present a formidable challenge to screening operations because of their tendency to stick together. In addition to the obvious solution of drying the feed before presenting it to the screening operation, the use of heated screens can also remedy the difficulties. The best way to heat the screen is to pass low-voltage electric current directly through the screen wires using the resistance of the wires to generate the heat. The use of low voltage provides safety for the operators and the use of screen wire resistance to generate heat makes low voltage possible at high currents, thus providing adequate heat. The low voltage is provided by a step-down transformer, which generates a high current for connection across the screens.

An unexpected benefit from direct heating of the screens is that the screen tends to be selectively overheated in regions of the screen blinded by wet material. This is because the cooling of the screen by thermal conductivity to the feed bed and undersize is reduced in the blinded regions of the screen. This overheating of the blinded regions evaporates the moisture in the blinding material and allows the screen motion to shake loose the dried residue, unblinding the holes in the screen and providing a self-cleaning characteristic. Some suppliers of screening equipment are given in Table 4.6.

4.3.5 Economics

The economics of screening are highly variable and depend on the nature of the material being screened, the required degree of separation and the complexity of the equipment needed to make the separation. Scalping for removal of a small amount of highly oversized material from a uniform stream of fine material is inexpensive. However, when fine powders must be separated according to size as in the powdered food, chemicals, and powdered metals industries, costs may be much higher. For this reason it is best to contact a manufacturer of screening equipment for cost estimates on the given application before purchasing such equipment.

4.4 MAGNETIC SEPARATION

Magnetic separation techniques have been used commercially for about a century with particular emphasis on either (1) the removal or recovery of strongly magnetic materials such as tramp iron in solids processing, (2) the separation of magnetic from non-magnetic ores in the mineral and mining industries, (3) separation of ferrous from non-ferrous metals in the metal scrap industry, and (4) recovery of steel cans from municipal trash. With the development of techniques and apparatus for handling weakly magnetic materials the range of applications has become much broader and more sophisticated. For example, white sizing clays can have brown iron-stained impurity particles removed based on the magnetic properties of the iron on the brown particles. This improves the brightness and whiteness of the clay product. Other examples are the removal of sulfur contained as iron pyrite in powdered coal and the recovery of iron from coal fly ash.

A relatively new addition to the arsenal of magnetic separations is the eddy current separator. This unit can separate non-ferrous metals from non-metallic materials after the ferrous metals have been removed. It is useful for recovering aluminum cans from household or industrial waste and for concentrating metals obtained from car-shredding operations.

The major applications of magnetic separation to waste minimization are in (1) the recovery of magnetic materials from waste streams to permit their recycling as raw materials and (2) removal of non-ferrous metals from non-metals. A typical example is the recovery of steel cans from municipal wastes followed by recovery of aluminum cans from the non-magnetic tailings. There are many other applications in both the scrap metal and process industries.

4.4.1 Fundamentals

In automatic sorting, the properties of the individual components of a feed mixture permit recognition of the members of a subclass. However, they do not provide the actual separation mechanism. With magnetic and electrostatic separations the properties of the individual solid particles actually cause the separation to occur. This is due to the forces generated on the individual particles due to the presence of the electrostatic or electromagnetic fields. For magnetic separations these forces may be attractive, repulsive, or zero depending on the direction of the field and the magnetic properties of the materials in the feed mixture.

When a particle is introduced into a magnetic field it becomes a magnet. The extent of its magnetism varies widely depending on the magnetic properties of the material from which it is made. This effect is called induced magnetism. The material is called paramagnetic if it is attracted by a magnetic field and diamagnetic if repelled. Substances that can be permanently magnetized to form magnets that do not require the presence of an external magnetic field to be magnetic are called ferromagnetic. Ferromagnetic materials include only iron, nickel, cobalt, and some rare earth elements. Obviously, these are the materials from which permanent magnets are and should be made.

The strength, M, of the magnetism induced in a particle depends on the intensity, H, of the magnetic field and the magnetic susceptibility, X, of the material comprising the particle. This relationship is shown in Eq. 4.

$$M = XH \tag{4}$$

Paramagnetic materials have positive susceptibilities and the induced magnetism increases the flux intensity within the particle. In contrast, diamagnetic substances have negative susceptibilities, and the induced field in the particle reduces the magnetic flux intensity in the particle. This is illustrated in Figure 4.8.

The behavior of ferromagnetic materials is also shown in Figure 4.8, which illustrates that the susceptibility of such materials is both positive and large. However, unlike paramagnetic materials, where magnetization does not become asymptotic as field strength is increased, ferromagnetic materials show definite asymptotic behavior. This behavior is called magnetic saturation and can occur at even moderate field strengths. It can give the condition shown in Figure 4.8 for dilute ferromagnetic materials, where the magnetization is larger at high field strengths for paramagnetic materials than for dilute ferromagnetic materials. This unusual behavior is due to crossing of the magnetization curves for these two materials.

The force exerted on a particle in a magnetic field is exerted on the magnetic dipole, i.e., the magnetic poles induced in the particle. This causes the particle to turn as necessary to align itself magnetically parallel to the lines of force generated by the field. If the field is uniform, the forces on the magnetic poles of the particle will be equal and in opposite directions, giving a net force of

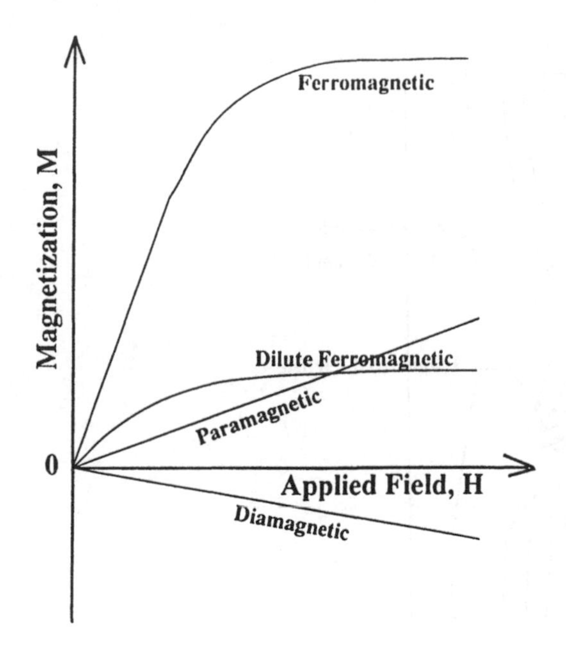

Figure 4.8 Magnetic behavior of materials. (From Beddows, 1981.)

zero on the particle. However, if the field is not uniform and has a gradient in strength, the net force on the particle will be in the direction of the higher field strength. The magnitude of this net force is proportional to the magnetic dipole moment of the particle, and thus its size, as well as the strength of the magnetic field gradient. The effect of magnetic field gradients has been used for the collection of solid particles from flowing liquid or gaseous streams. This is done by inserting metal mesh collection masses into the applied magnetic field. These collection masses are woven or knit from fine ferromagnetic metal wire with diameters on the same order of magnitude as the particles to be collected. The use of fine wire maximizes the field gradient that can be obtained and thus the magnetic force on the particle. Figure 4.9 from Wechsler (1984) models the force balance on a single particle in the presence of the field gradient generated in the vicinity of a single filament of a collection mass.

For a particle to be collected on the filament wire, the magnetic attractive force in the direction of the wire must be greater than the sum of all forces tending to pull the particle away from the wire. The forces to be overcome may include gravity, fluid drag, buoyancy, momentum, etc.

While the above discussion is based on the bulk magnetic properties of solids being constant for a given material, it should be remembered that mag-

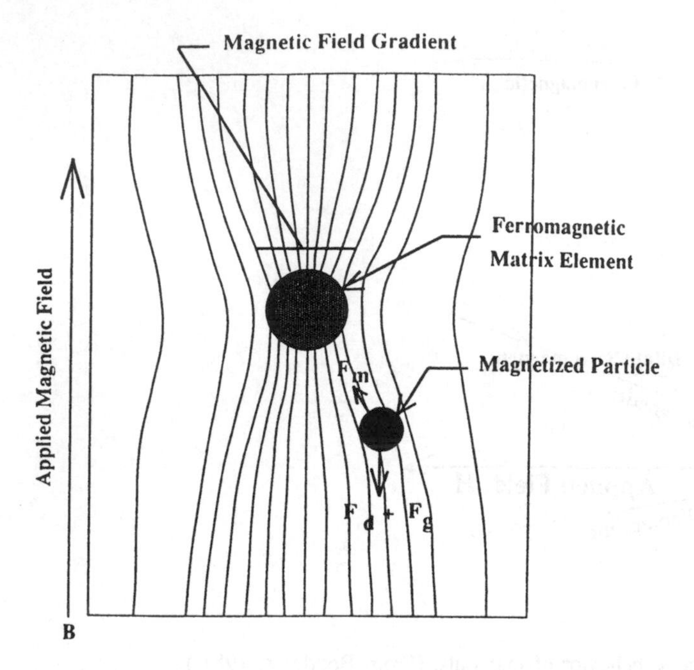

Figure 4.9 Model of particle capture forces. F_m, Magnetic force; F_d, drag force; F_g, gravitational force. (From Wechsler, 1984.)

netic properties of materials can change. This change can be due to exposure to mechanical work, physical and chemical treatment, high temperature, pressure, and induced electrical eddy currents. Therefore, magnetic separations can sometimes be greatly improved by pretreating feeds to alter the magnetic properties of the feed particles prior to the magnetic separation.

The eddy current separator is illustrated in Figure 4.10 and makes use of a temporary magnetic push to separate non-magnetic metals from non-metals. It consists of a non-magnetic, non-metallic hollow drum at the end of a belt conveyer. The drum has inside it a rapidly revolving permanent magnet rotor with alternating polarity permanent magnets. The permanent magnet rotor induces eddy currents in the metallic material passing over the drum on the belt. These currents set up their own magnetic fields, which push against the magnetic fields of the permanent magnets of the rotor. The resulting force pushes the metallic materials away from the drum while the non-metallic materials stay on the belt until they fall off due to gravity. The eddy current separator has greatly expanded the range of applications of magnetic separations.

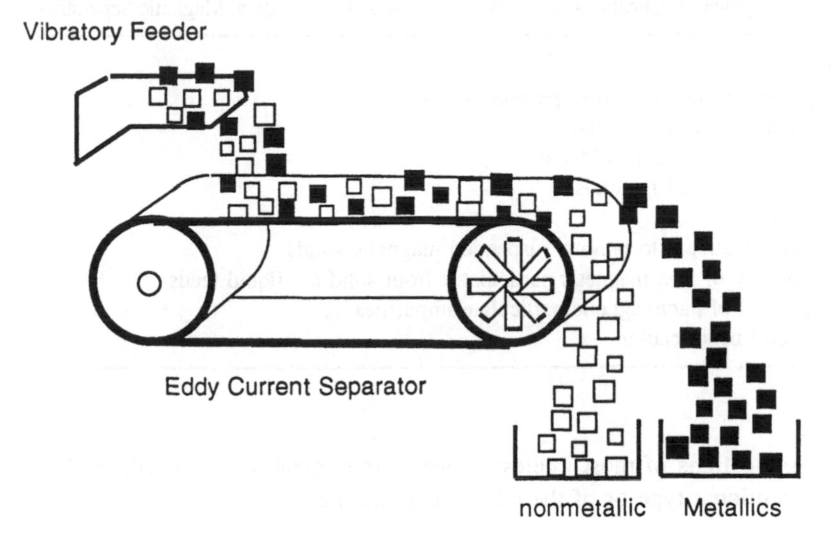

Figure 4.10 Eddy current separator. (Courtesy of Eriez Magnetics.)

4.4.2 Specific Applications

The applications of magnetic separation are divided into three classes: conventional separation processes, high gradient separation processes, and eddy current separation processes. The conventional processes separate magnetic from nonmagnetic materials and the desired product may be either one. Typical examples of conventional and high-gradient magnetic separations are given in Table 4.7.

Conventional magnetic separations include (1) removal of tramp iron from all sorts of process streams, (2) recovery of iron and steel from scrap metals, (3) recovery of tramp iron and steel cans from solid waste, and (4) beneficiation of minerals. Process streams from which tramp iron is removed occur in the food, pharmaceutical, chemical, metalworking, and mineral industries.

The high-gradient applications require the higher forces that can be generated with high-gradient fields. This is to overcome the drag on the magnetic particulates by flowing liquid or gaseous streams. The high-gradient fields are also needed for separation of more weakly paramagnetic materials found in wastes and ores.

As mentioned above, the eddy current processes are used in waste reduction to recover aluminum cans for recycling and in recovering non-ferrous metals from car shredding operations. However, they are also effective for separating metallics from electronic scrap, removing aluminum contaminants from glass cullet, and separating non-ferrous metallic contaminants from foodstuffs.

Table 4.7 Typical Applications for Conventional and High-Gradient Magnetic Separators

Conventional
- Tramp iron removal from process streams
- Separation of scrap metal
- Iron recovery from solid waste
- Beneficiation of minerals

High gradient
- Water treatment to remove suspended magnetic solids
- Recovery of fine magnetic particulates from solid or liquid feeds
- Removal of paramagnetic particulate impurities
- Mineral beneficiation

The applications of most interest in waste minimization are usually either of the conventional type or of the eddy currrent type.

4.4.3 Equipment

Conventional Separators

Just as the applications of magnetic separation fall into the categories of conventional, high-gradient, and eddy current, so do the equipment designs. The conventional separators include grate types, drum types, and belt types. The grate and drum types usually get their magnetic field from permanent magnets imbedded in the grate bars or in the magnetic drum as shown in Figure 4.11.

Grate Separators: These units use stacks of disk-shaped ceramic permanent magnets in tubes to form a grate through which the mixed feed passes. Magnetic materials are attracted to the tubes and held there. Grates are used primarily as in-line magnetic traps for removal of coarse or fine tramp iron from either wet or dry process streams. Such use occurs in the food, chemical, recycling, and paper industries. Periodically, flow must be stopped to remove the grate tray and clean the magnet tubes. Alternatively, the separator design can include self-cleaning features. Grate type units are most effective on feeds that are dilute in magnetic materials. This is because of the resultant reduction in the required frequency of cleaning.

Drum Separators: These make use of a magnetic roll containing imbedded fixed magnets with the adjacent magnets alternating in polarity at the roll surface. The mixed feed falls onto the top of the roll and is carried by the rotation of the roll to where the non-magnetic material falls off into a collection bin. The magnetic materials stick to the roll until they are mechanically removed and fall into a second receiving bin beyond the point at which the tailings fell

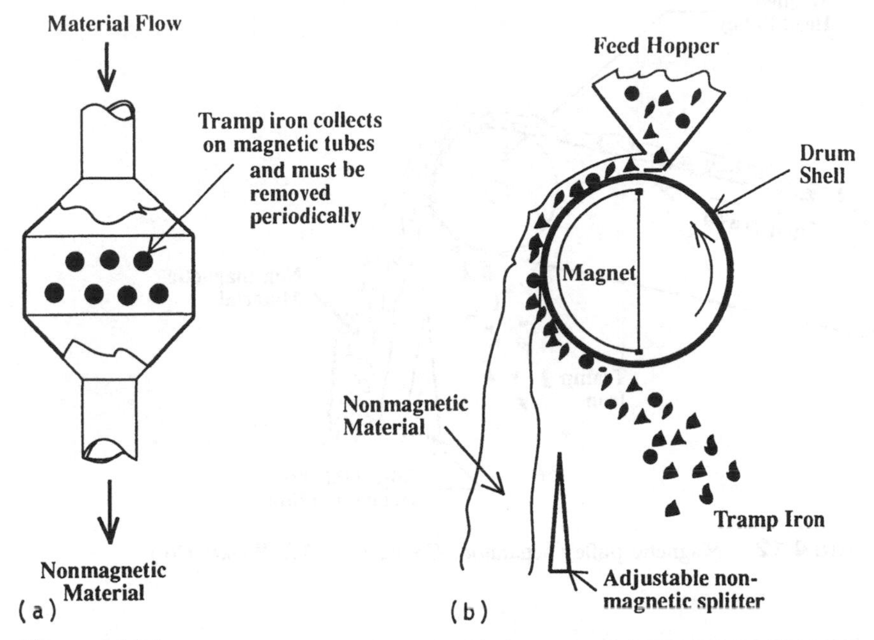

Figure 4.11 (a) Conventional grate and (b) drum magnetic separators. (Courtesy of O.S. Walker Co.)

off. A variation of the drum separator is called the magnetic pulley type separator and is shown in Figure 4.12.

This separator combines the effect of the magnetic drum with the convenience of belt feeding the raw mixture to the drum. It has become the most common magnetic separator in commercial use.

Belt Separators: These make use of the lifting power of a magnetic field to pick up magnetic materials from a mixture moving on a continuous conveyer belt below it. The magnetic material is deposited on the underside of a second traveling belt. It adheres to the second belt until it moves out of the magnetic field which lifted it. At this point the magnetic material drops off the removal belt and is collected as the magnetic fraction. The tailings stay on the feed conveyer belt until dumped off at the end roll of the conveyer. Figure 4.13 shows that the magnetic lifting belt can operate either in line with or at right angles to the feed belt. The cross belt configuration is used when more than one stage of magnetic separation is desired. Different lifting forces can be obtained by changing the gap between magnetic poles or by changing the current to the electromagnets used with the lifting belts.

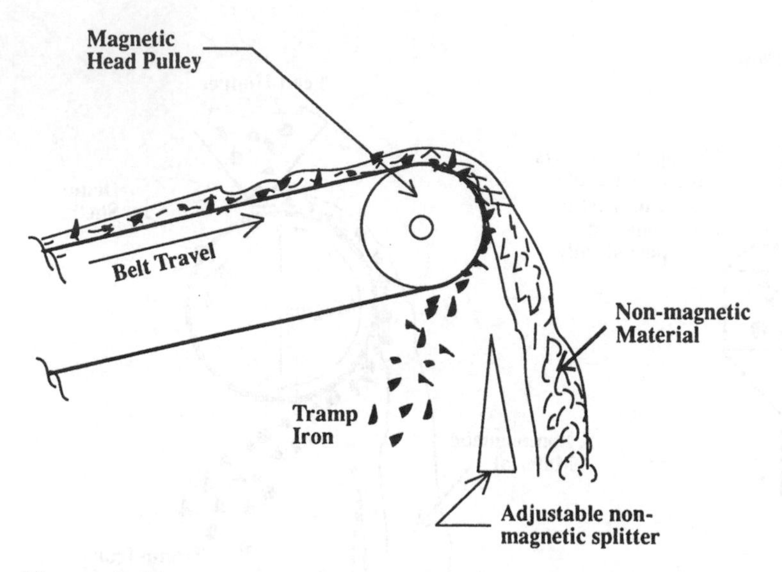

Figure 4.12 Magnetic pulley separator. (Courtesy of O.S. Walker Co.)

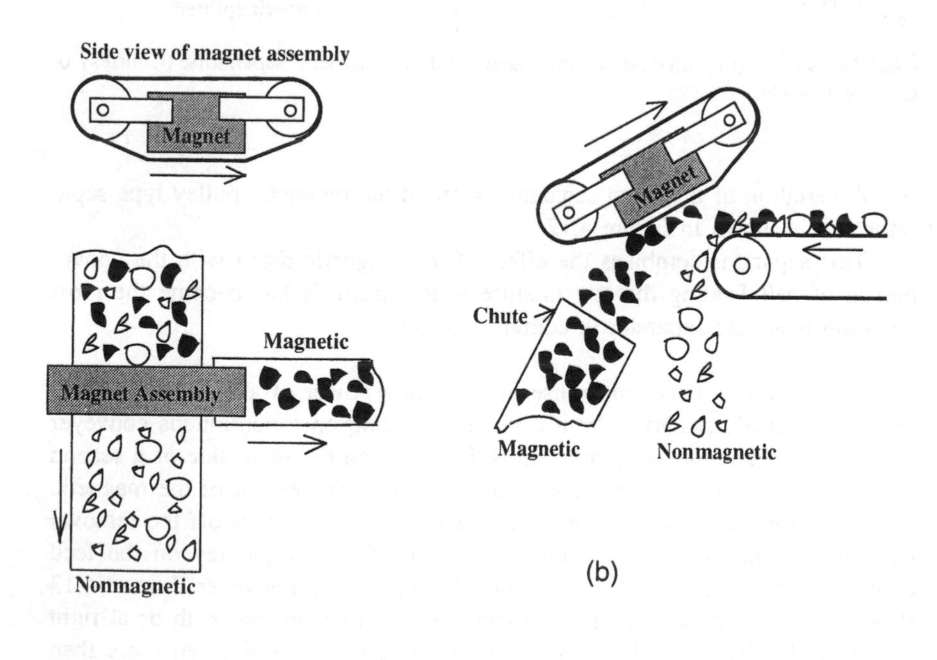

Figure 4.13 Lifting-belt magnetic separators. (a) Cross-belt design, (b) in-line design. (Courtesy of O.S. Walker Co.)

High-Gradient Separators

High-gradient magnetic separation is regarded by Gaudin (1974) as an alternative to flotation and leaching for the treatment of finely divided solids in water. It has been used extensively in the kaolin industry for cleaning slurries of clay. However, a design developed by the Franz company has proved effective for separation of dry solids entrained in air or fed by gravity. It can separate paramagnetic contaminants from dry minerals down to 325 mesh in size. This design is illustrated in Figure 4.14.

The Franz Ferrofilter uses a solenoid to generate the magnetic field. The filter is periodically cleaned by shutting off the solenoid current and flushing the magnetic materials out of the collection matrix. Thus, it is a batch operation. However, continuous high-gradient magnetic separators have been designed. They make use of a carousel operation with a wheel containing flow cells with steel wool matrices in them. The cells are rotated into and out of the flow path of the feed mixture. This flow path is in a high intensity magnetic field. This generates a high-gradient field around the steel wool filaments in the cell in the flow path and thus collects magnetized particles. When the cell rotates out of the magnetic field, the particles trapped by the steel wool are no longer held

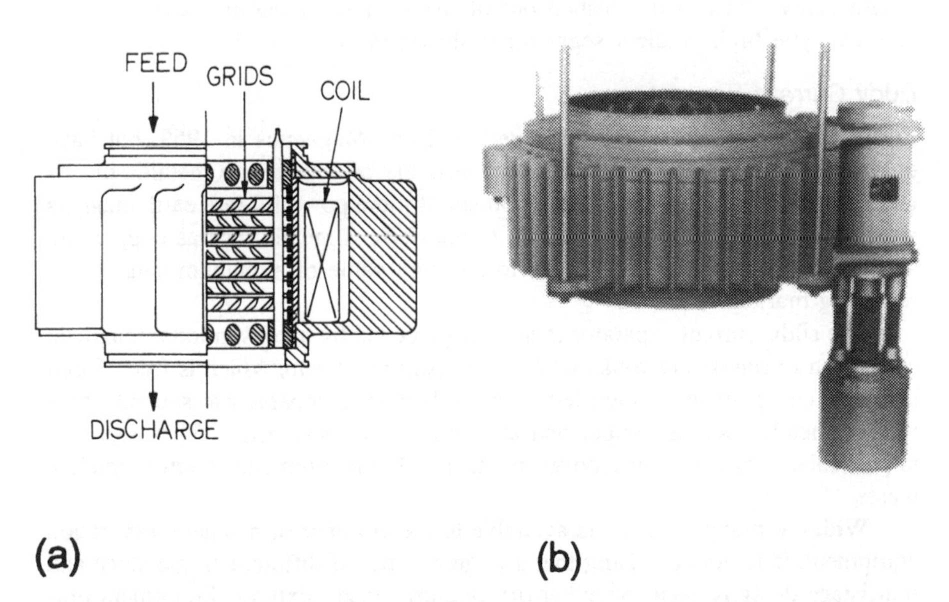

Figure 4.14 Magnetic section of Frantz Ferrofilter. (a) Cutaway view of high-gradient matrix; (b) external view of suspended unit. (Courtesy of S.G. Frantz Company.)

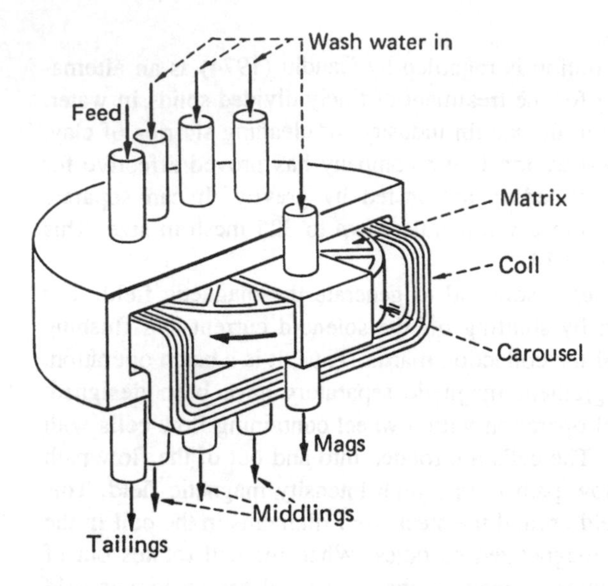

Figure 4.15 Carousel high-gradient magnetic separator. (From Beddow, 1981.)

magnetically and can be flushed out of the steel wool matrix. A sketch of a carousel-type high-gradient separator is shown in Figure 4.15.

Eddy Current Separators

Eddy current separators were patented by Eriez Magnetics in 1969, but have only recently come into general use for recovery of non-ferrous metallic materials. This is largely due to the development of more powerful rare earth magnets which give a stronger magnetic push to non-ferrous metals such as copper and aluminum. However, it is also due to the high value of aluminum cans in the recycling market.

The eddy current separator is usually preceded by a conventional magnetic separation to remove ferrous metals. The resulting stream, which is free of iron and steel components, is then fed to the eddy current separator to separate non-ferrous metals such as copper and aluminum from non-metals. This approach is particularly effective in recovering steel and aluminum cans from municipal waste.

With the many parameters available to the designer of magnetic separation equipment, it is not surprising that a large number of different single-stage and multistage designs exist for either dry or slurry feed mixtures. Equipment providers can and do customize their basic designs to fit specific applications. A listing of some of the major suppliers of magnetic separation equipment is given

Table 4.8 Vendors of Magnetic Separation Equipment

Bunting Magnetics Co. Box 468 Newton, KS 67114 (800) 835-2526	Carpco, Inc. 4120 Haines Street Jacksonville, FL 32206 (904) 353-3681
Dings Magnetic Group 4740 West Electric Ave. Milwaukee, WI 53219 (414) 672-7830	Engineered Magnetics 55 Stephens St. Belleville, NJ 07109 (201) 759-0818
Eriez Magnetics P.O. Box 10608 Erie, PA 16514 (814) 833-9881	Magnetic Separation Systems, Inc. 624 Grassmere Park Dr., No. 8 Nashville, TN 37211 (615) 781-2669
Magni-Power Company P.O. Drawer 122 Wooster, OH 44691 (216) 264-3637	O.S. Walker Company 10 Rockdale Street Worcester, MA 01606 (508) 852-3674
S.G. Franz Co. Inc. P.O. Box 1138 Trenton, NJ 08606 (609) 882-7100	

in Table 4.8. A vendor should be contacted for advice when a specific magnetic separation is being considered.

4.4.4 Common Problems and Troubleshooting

As with all sorting procedures, the most common problems in magnetic sorting are (1) incomplete removal of a component from the feed stream or (2) entrainment of undesirable components along with the material being separated. For example, if too deep a bed of material is maintained on a conveyer belt, an overhead lifting magnet may have difficulty in removing all the magnetic materials without carrying some non-magnetic material along with it. This can be at least partially overcome by decreasing the feed rate to the conveyer, thereby presenting a thinner bed to the magnets. Furthermore, both recovery and product purity can be increased by adding more magnetic separation stages to the product and the tailings streams obtained from the primary separation.

4.4.5 Economics

The costs of magnetic separation clearly depend on the complexity of the feed mixture to be separated and the required magnetic purities of the product streams. For example, a small amount of tramp iron can be effectively removed from a flowing stream of non-magnetic material, such as powdered foods, using a single-stage magnetic grate. This is inexpensive. However, at the other end of the cost scale, a mixture containing large amounts of strongly magnetic, weakly magnetic, and non-magnetic materials along with non-ferrous metals is frequently encountered in waste streams. This type of feed mixture may require many stages of many different types of magnetic separation to produce good magnetic homogeneity in the separated products. This is relatively expensive. Between these extremes there are many different applications. Proper selection of equipment for a given magnetic separation requires the advice of a vendor of such equipment. Similarly, cost data can only be obtained after the equipment requirements are specified.

4.5 ELECTROSTATIC SEPARATION

Electrostatic separation is much like magnetic separation except that static charge and electrical field take the place of magnetic induction and magnetic field. The electrostatic forces are generated on a particle by selectively charging it with either positive or negative charge and then placing it in an electrostatic field which either attracts or repels the particle. Charges of the same sign repel and charges of opposite sign attract just as with magnetic poles. Applying the static charge to the particles in a feed mixture is a necessary step prior to introducing the feed mixture into the electrostatic field where the separation is carried out. The effectiveness of the separation depends on the static charges on individual particles being separated so that each particle has either positive, negative, or zero charge. Furthermore, the charge on each particle must last long enough for the separation of the differently charged particles to be made.

The types of materials that can be separated electrostatically are classed as conductors and non-conductors. The non-conductors are also called insulators or dielectrics. Obviously there is a spectrum of materials between these two extremes and it is the degree of difference between the materials in the feed that determines the ease of electrostatic separation of the feed components. The larger the difference, the easier the separation. Good conductors gain charge rapidly when in contact with charged surfaces. They also lose charge rapidly when in contact with oppositely charged or grounded conducting surfaces. In contrast to the conductors, the insulators are slow to gain charge from charged surfaces and also slow to lose charge to grounded or oppositely charged surfaces.

From the above discussion it is clear that a suitable feed for electrostatic separation must have the conductor particles and the insulator particles mechanically separated in the feed stream. This is usually done by a size reduction operation such as milling or grinding. The resulting particles are small enough that a single particle is either a particle of insulator or a particle of conductor but does not contain both. As a result the electrostatic separation can be quite clean with both high yield and high purity of products.

In addition to the nature of the feed materials, the nature of their gaseous environment during charging and separation also influences the effectiveness of techniques for electrostatic separation. For example a high humidity enhances the leakage of charge from the individual particles and thus the stability of charge on feed particles is a function of relative humidity. These factors must all be considered in designing electrostatic separation equipment.

4.5.1 Fundamentals

The three main methods for charging the feed particles are contact electrification, conductive induction, and corona discharge or ion bombardment. No matter what technique is used for charging the particles, the surface area of a particle and the maximum achievable charge per unit surface area limit the charge that can be placed on a particle. The electrostatic separation depends on generating sufficient force on the particle from the interaction of the electrostatic field on the static charge of the particle to overcome gravity, momentum, or drag forces on the particle and remove it from the uncharged particles in the feed mixture. Since static charge is proportional to surface area of a particle while inertial forces are a function of volume of the particle, there is an upper limit in size, depending on particle shape, to separation by electrostatic means. This limit is about 1/16 in. for granular materials and up to 1 in. for thin platelets and long needles. On the other extreme of particle size, very small particles are strongly affected by the drag forces exerted on them by a moving carrier fluid. Thus, separation of fine particles entrained in air is very difficult for particles less than 200 mesh in size and essentially impossible at particle sizes of 325 mesh.

Charging Mechanisms

Contact electrification occurs when different materials rub against each other and electrons are transferred from particles of one material to particles of the other, giving a negative charge on the particles receiving the electrons and a positive charge to the particles losing the electrons. If the particles are poor electrical conductors, the individual particles can hold their charge for a relatively long time. Introducing this mixture of charged particles into an electrostatic field will push the positively and negatively charged particles in opposite directions leading to their separation. Fluidizing a bed of fine particles with dry air is an effective

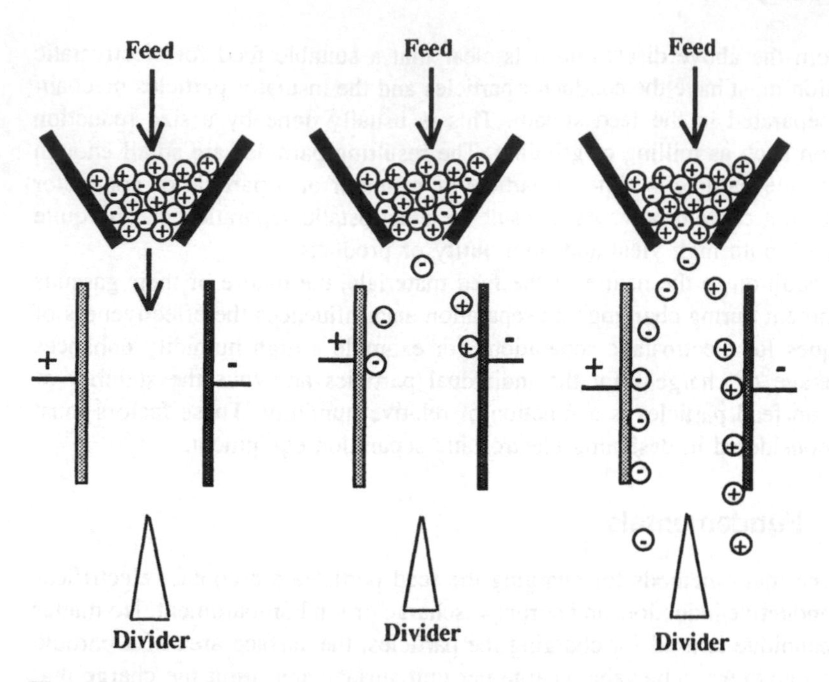

Figure 4.16 Sketch of a free-fall electrostatic separation. (Courtesy of Carpco, Inc.)

way of charging particles by contact electrification. Grinding and size reduction also generate charged particles but usually not as effectively as a fluidized bed.

The first electrostatic separators used the contact electrification method of charging followed by free fall of the charged particles between charged vertical plates, which provided a horizontal electrostatic field. This is illustrated in Figure 4.16. While falling through the field the positively charged particles were deflected from their freefall path toward the negatively charged plate and the negatively charged particles toward the positive plate by the electrostatic forces on them. Thus they could be collected in separate containers. Unfortunately, the contact charging method requires humidity control and a complex feeding mechanism to prevent variation in degree of charging. Furthermore, the separator requires many stages to get effective separation. This complexity and the associated costs led to the demise of freefall separators using contact electrification in the late 1940s. They were replaced by units using ion bombardment or conductive induction as the charging technique.

Conductive induction is the process by which uncharged particles become charged by contact with an electrically charged surface. This is illustrated in Figure 4.17a. Good conductors, shown as the black particle, assume the charge and polarity of the surface rapidly while the white non-conductors (dielectrics

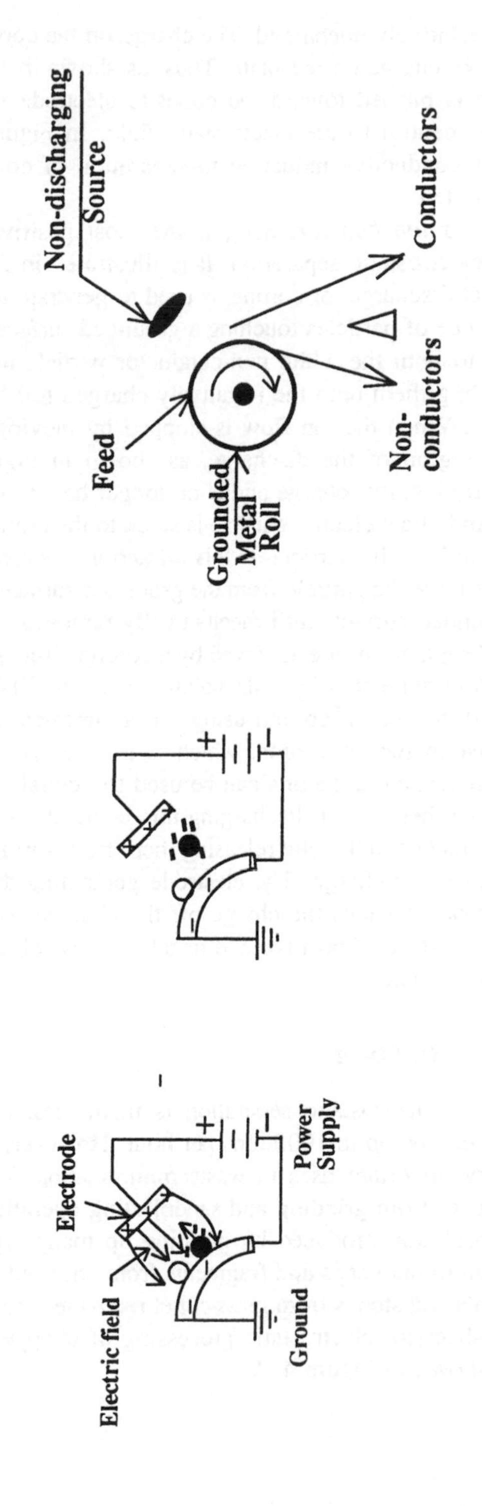

Figure 4.17 Charging by conductive induction. (a) Charging dark conducting particle, (b) pushing conductor off of roll, (c) practical separation of conductors and non-conductors. (Courtesy of Carpco, Inc.)

and insulators) remain relatively uncharged. The charge on the conductor particle is repelled by the like charge on the plate. Thus, as shown in Figure 4.17b, the conducting particle is pushed toward the opposite electrode while the dielectric particle will be neutral to the electrostatic field. In Figure 4.17c, the practical application of conductive induction to separation of conductors and non-conductors is illustrated.

Corona discharge, or *ion bombardment*, is the most positive method of charging particles for electrostatic separation. It is illustrated in Figure 4.18a. A high-voltage electrical discharge, or corona, is used to generate ions from the atmosphere in the presence of particles touching a grounded surface. These ions give a positive charge to both the white non-conductor particle and the black conductor particle, holding them onto the negatively charged roll by the forces indicated by the arrows. When the ion flow is stopped by moving the surface of the roll out of the region of the discharge, as shown in Figure 4.17b, a conductor particle rapidly loses its charge and is no longer held to the grounded surface. On the other hand, the dielectric particle is stuck to the grounded surface by the positive charge still on the particle. If this attraction is strong enough to offset forces trying to remove the particle from the grounded surface, the particle will remain on the grounded surface until mechanically removed. Figure 4.18c illustrates how practical separations are achieved by a corona discharge machine.

The corona is usually generated by a dc voltage of up to 50,000 V either positive or negative with respect to ground using a fine tungsten alloy wire as an electrode. In contrast to the dc corona, which is a very effective way of generating charged particles, an ac corona can be used to neutralize the charge on particles. This makes it helpful for discharging non-conductor particles after the separation has been made and thereby releasing them from surfaces to which they were stuck by their static charge. The electrode generating the ac corona is called a wiper electrode. It wipes the charge off the charged particles in its corona by generating a mixture of positive and negative ions, which neutralize the residual ions on the particles.

4.5.2 Specific Applications

The major application of electrostatic separation is in mineral beneficiation, where recovery of ores can be up to 1000 tons per hour. However, electrostatic separation also has very important uses in waste minimization. These are (1) recovery of metal powders from grinding and sandblasting operations, (2) separation of metal and insulation produced by grinding up many types of scrap wire, (3) separation of aluminum caps and fragments from chopped PET bottles, and (4) removal of metals and stones from glass cullet recovered from municipal wastes. A typical flowsheet for electrostatic processing of chopped PET from plastic soda bottles is shown in Figure 4.19.

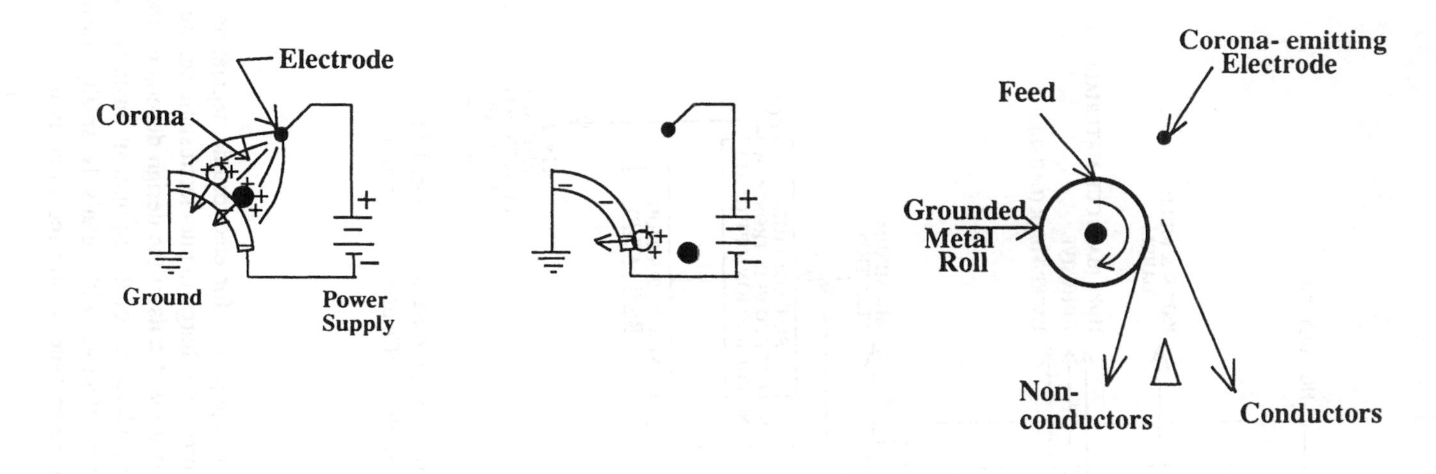

Figure 4.18 Charging by corona discharge. (a) Light non-conductor and dark conductor particles charged by corona. (b) Discharged conductor falls off roll. Non-conductor held on roll. (c) Practical separation of conductors and non-conductors. (Courtesy of Carpco, Inc.)

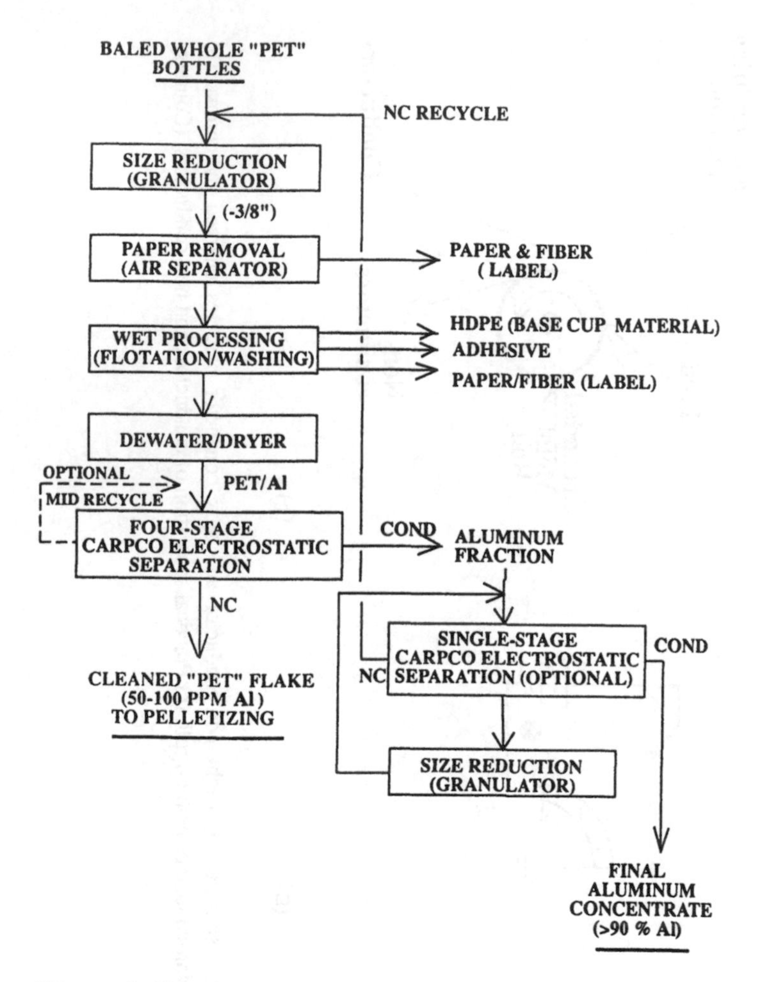

Figure 4.19 Typical flowsheet for reclaiming chopped PET. COND, Conductors; NC, non-conductors; MID, middlings. (Courtesy of Carpco, Inc.)

4.5.3 Equipment

Essentially all the modern equipment for electrostatic separation uses either conductive induction or corona discharge to put electrostatic charge on the feed particles. However, after that the details of the design depend on the particular application as well as the preferences of the equipment manufacturer. Electrostatic separations are only effective with relatively small, uniformly sized particles. This means that some type of size reduction is usually needed as a pretreating step. This is desirable not only to mechanically free conductors from insulators as is done by chopping scrap wire or plastic bottles, but also to generate a feed with a relatively narrow size range. The integration of size reduction equipment with electrostatic separation should be discussed with equipment suppliers to make sure they are compatible.

The major supplier of electrostatic separation equipment for the waste industry is:

Carpco, Inc.
4120 Haines St.
Jacksonville, FL 32206
(904) 353-3681

4.5.4 Common Problems and Troubleshooting

The most common problems encountered in electrostatic separation usually have to do with the inadequacy of feed preparation. For example, if the feed particles are too small, it may be impossible to get adequate electrostatic charge on a particle to overcome the drag of the carrier fluid on the particle. If the particles are too large, the inertial forces on the particle such as momentum and gravity may be too large to be overcome by the electrostatic forces on the particle. Similarly, wet solids can stick together preventing selective charging of individual particles as well as forming heavy clumps of particles. By properly drying, shredding, and sizing the feed prior to the electrostatic separation, these problems can be avoided.

Another factor which can make the separation unsatisfactory is inadequate mechanical separation of the insulators from the conductors in the feed preparation. For example, when grinding insulated wire, if some particles contain both metal and insulation, they will reduce the effectiveness of the separation. Fixing the grinding operation will fix the separation.

4.5.5 Economics

Like all other separation processes, the cost of electrostatic separation depends on both the complexity of the feed and the product purities required. High product purities at high recovery usually require multistage operation with the accompanying higher equipment cost.

4.6 SUMMARY

The separation processes discussed in this chapter can be used individually for specific applications such as (1) optical sorting for separation of colored glass, (2) screening for separation by particle size, (3) separation of metals and ores by magnetic properties, and (4) separation of conductors and insulators by electrostatic properties. However, the various individual separation techniques are frequently combined into a separation scheme to process complex mixtures. The flow scheme shown in Figure 4.1 uses all of these separation techniques to recover valuable materials from a municipal trash feed stream. This not only reduces the amount of trash that must be disposed of by expensive landfilling or incineration. It also provides income from the sale of the recovered materials which can be recycled.

REFERENCES

Beddow, J.K. (1980). *Particulate Science and Technology*, Chemical Publishing Company, New York.

Beddow, J. K. (1981). Dry separation techniques, *Chemical Engineering*, *88*:70 (Aug. 10, 1981).

Gaudin, A. M. (1974). Progress in magnetic separation using high intensity, high gradient separators, *Mining Congress International*, Jan. 1974; 18-21.

Matthews, C. W. (1972), Screening, *Chemical Engineering*, July 10, 1972; 76-83.

Recycling Times (1992). "Waste Age's Recycling Times," May 5, 1992.

Salimando, J. (1989). Rhode Island's state of the art plant, *Waste Age*, September, 1989.

Stone, L. H., (1979) Upgrading circular vibrating screen separators, *Chemical Engineering*, Jan. 15, 1979; 125-130.

Wechsler, I. (1984). *Perry's Chemical Engineer's Handbook*, 6th ed., (R. H. Perry and D. W. Green, eds.), McGraw-Hill, New York, p. 21-35.

5

Settling and Flotation

5.1 INTRODUCTION

Gravity settling is a commonly used process for separating insoluble particles of solids or liquids from a liquid or gaseous carrier fluid. It uses as a driving force for separation the effect of gravity on the difference in density between the dispersed particles and the carrier fluid in which the particles are dispersed. If the particles are denser than the carrier fluid, they sink to the bottom of the settling chamber. If they are less dense, they rise to the surface of a liquid carrier. Gravity settling is a subcategory of the more general field of sedimentation, which includes the slower processes of hindered settling, flocculation, coagulation, clarification, and thickening. These latter processes are discussed in Chapter 8 along with chemical precipitation. Furthermore, the enhancement of gravity separation by the use of centrifuges and cyclones is discussed in Chapter 7. The present chapter discusses only the relatively simple separations by gravity used for removal of suspended particles from wastewaters and separation of dusts and mists from gas streams. These are free-settling cases for dilute suspensions where the particles do not interfere with each other during settling.

Flotation is also a gravity separation to remove insoluble particles from a liquid. However, it uses bubbles of gas, usually air, that attach to the insoluble particles and lift them to the surface of the liquid by buoyancy. The particles can then be removed from the surface by various skimming devices. By selectively attaching the bubbles to one particular kind of particle in a mixture, separation can be made between particles lifted to the surface and particles settled to the bottom. This technique is used extensively in the mining industry for ore beneficiation, but less extensively in waste treatment.

In addition to its uses in water and wastewater treatment, as well as dust and mist collection, gravity settling has many uses in the process industries. Typical of such processes are those which mix two phases and then separate them to get a change in composition. Such processes are solvent extraction, distillation, absorption, scrubbing, stripping, and the like. The role of settling in these processes is discussed in the chapters dealing with them in detail. Similarly, a wide variety of foam and bubble-induced separations also exist for removing surface active agents from aqueous streams. These areas will also not be discussed in this chapter.

5.2 SETTLING

5.2.1 Fundamentals of Settling

The dynamics of a single particle moving through a fluid under the influence of gravity is well treated by Sakiadis (1984). As with all rate processes, the settling rate is proportional to a driving force divided by a resistance. Figure 5.1 illustrates the dynamics of particle motion for a spherical droplet. Figure 5.1a shows a droplet settling in a flowing fluid while, in Figure 5.1b the droplet is settling in a quiescent fluid. In both cases the driving force for vertical settling is the density difference between the particle and the surrounding fluid multiplied by the particle volume. If the particle is less dense than the continuous phase, the particle will rise. If it is denser than the continuous phase, the particle will fall. The resistance to particle motion is the viscous drag on the moving particle. These forces in a vertical direction remain the same whether or not other forces are causing the droplet to move horizontally, as long as the flow is not turbulent. Turbulent flow causes the occurrence of high-velocity vortices, which can drag the particle in many directions, disrupting its smooth settling in a vertical direction. After a brief equilibration period, the particle moves at a constant vertical speed called the terminal velocity. This terminal velocity is reached relatively rapidly as compared to total settling time and is frequently used in sizing settling chambers. The terminal velocity is described by Eq. 1 for hard spherical particles.

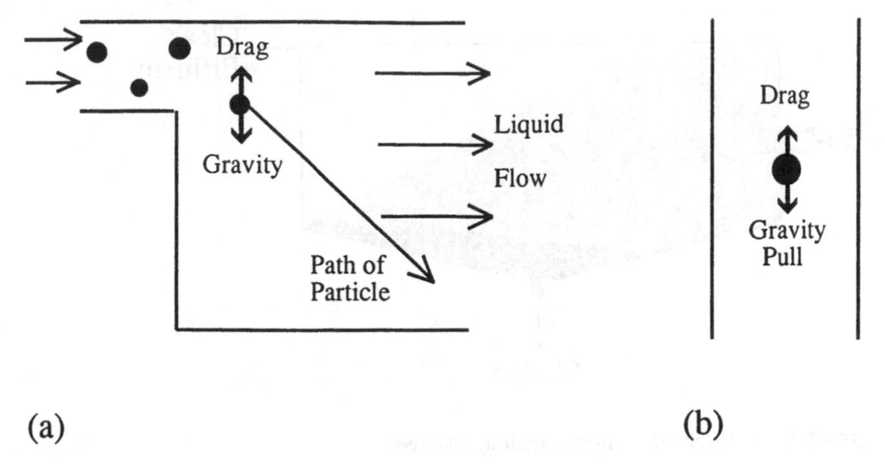

Figure 5.1 Model of forces on a settling particle. (a) Horizontal liquid flow, (b) quiescent pool.

$$\text{Terminal velocity} = K\frac{D^2(\rho_p - \rho)}{\mu} \quad (1)$$

where:

D = particle diameter
ρ_p = particle density
ρ = carrier fluid density
μ = carrier fluid viscosity
K = proportionality constant

Equation 1 shows that terminal velocity is directly proportional to the density difference between the phases, directly proportional to the cross-sectional area of the particle, and inversely proportional to the viscosity of the carrier phase. The constant, K, has included in it the effects of particle shape, the flow regime of the carrier phase, i.e., the degree of turbulence, and the effect of flow regime on the viscous drag relationships.

The terminal settling velocity can be used to size a gravity settler because it determines the time required for a particle to settle a given distance. If the slowest moving particles in the feed settle at a terminal velocity of 0.1 ft/sec, 10 sec of holding time is required for each foot of depth in the settling tank. Thus, if the tank is 6 ft deep, the residence time in the tank must be 60 sec or 1 min for a particle starting at the surface to reach the bottom. If the flow rate of the feed stream is 500 gal/min, the tank must then have a capacity of at least

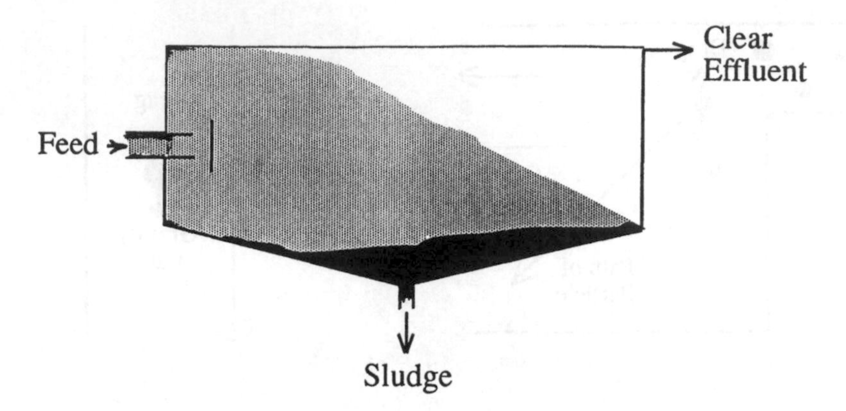

Figure 5.2 Sketch of a simple settling chamber.

500 gal. These relationships are described by Eq. 2 for the simple settler shown in Figure 5.2.

$$\frac{\text{Volume of feed}}{\text{Area of settler}} < \text{Terminal velocity} \tag{2}$$

where:

volume is in cubic feet per minute
area is in square feet
terminal velocity is in feet per minute

From the terminal velocity equation it is clear that large particles of high density settle rapidly in a low viscosity carrier fluid. Conversely, small particles with a low density difference settle very slowly in a viscous medium. The sign of the density difference determines whether the particle will rise or fall. Finally, if there is no density difference available, no settling will take place. The magnitude of the differences in settling velocity that can be encountered is illustrated in Table 5.1 taken from terminal velocity graphs in Sakiadis (1984).

Table 5.1 shows that the settling velocities in water can vary from 0.00001 ft/sec up to 4 ft/sec as particle size and particle density increase. At the first settling rate a particle requires a retention time of 100,000 sec or almost 1 day to settle 1 ft. Conversely, at 4 ft/sec settling is very rapid.

The effect of particle size is much greater than the effect of particle density. It is also linear with the square of the particle diameter at the smaller particle sizes as would be expected from Eq. 1. Comparison of settling in water with settling in air shows much higher terminal velocities in air. This is largely

Table 5.1 Terminal Settling Velocities of Spherical Particles in Air or Water

Water at 70°F				
	Terminal velocity, ft/sec, at particle density, g/ml:			
Particle diameter, μm	1.05	1.25	2.0	5.0
10 (0.01 mm)	10^{-5}	5×10^{-5}	2×10^{-4}	7×10^{-4}
100 (0.1 mm)	10^{-3}	4×10^{-3}	0.02	0.06
1000 (1 mm)	0.05	0.16	0.4	1.0
10,000 (1 cm)	0.4	0.9	2	4

Air at 70°F				
	Terminal velocity, ft/sec, at particle density, g/ml:			
Particle diameter, μm	0.5	1.0	2.0	5.0
10 (0.01 mm)	5×10^{-3}	10^{-2}	2×10^{-2}	5×10^{-2}
100 (0.1 mm)	0.4	0.8	1.6	3.0
1000 (1 mm)	10	14	21	40
10,000 (1 cm)	40	55	80	110

Source: Sakiadis (1984).

because of the low viscosity of air as compared to that of water. The viscosity of air at 70°F is 0.018 c.p. while that of water is 0.98 c.p.

Because of the high terminal velocities of large particles of dense materials, the settling chambers for these materials can be deep and still be effective. However, with small particles the settling rates are much lower. Thus, the small particles only settle at rates of 0.1 to 0.001 ft/sec in water and even slower in a more viscous carrier. This limits the particle size that can be removed by gravity settling chambers on a practical basis to about 10 μm or larger and preferably 100 μm or larger. To get good removal of small particles, either long residence times are needed or shallow settling zones must be provided. Settling ponds with very long holding times are often used with slow settling materials. To get short settling distances, settling chambers are frequently designed with slanted parallel plates or tubes to shorten the settling distance and the time required for particle removal. Such units provide good clarification of the effluent stream in a small volume. A sketch of how such units work is given in Figure 5.3.

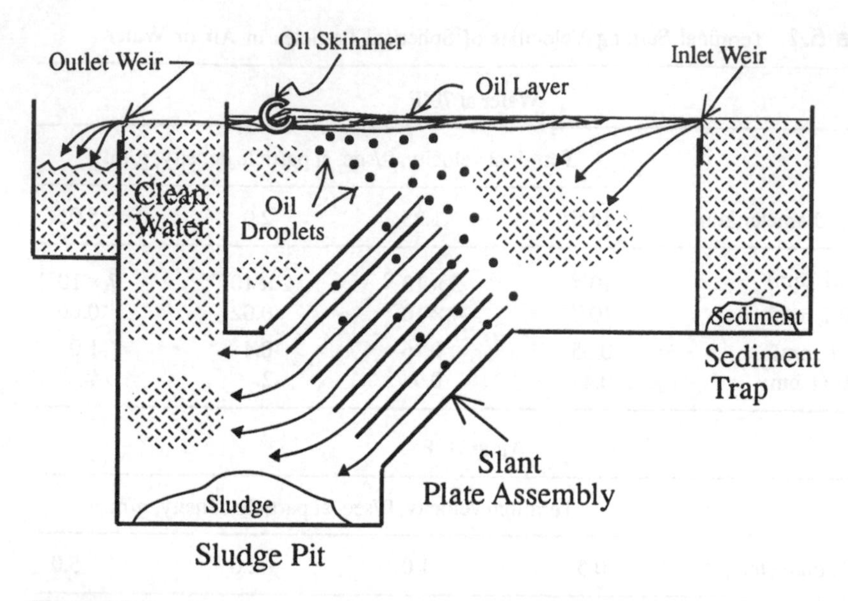

Figure 5.3 Representation of a lamella plate separator.

In Figure 5.3 the particles lighter than water, such as oil droplets, float upward until they reach the underside of the inclined plates. There they tend to coagulate and flow up the underside of the plate to the surface of the liquid in the settler where they collect as a layer. The particles heavier than water settle to the top side of the plates and tumble down the plates until they reach the bottom of the settler. Here they collect as a sludge layer. The oil and sludge layers are either continuously or intermittently removed depending on settler design.

5.2.2 Specific Applications

In general, gravity settling is most effective for particles in the 100 to 1000 μm particle size range. However, with very large settling chambers, i.e., long retention times, or with the addition of coalescing devices, particles as small as 1 μm can be removed. Some typical applications of gravity settling are described below.

Grit Removal in Water Treating

One of the simplest and most effective separations achieved by gravity settling is the removal of high-density grit from water in either potable water or wastewater treating. In these cases, the solid particles are composed of sand, gravel, rust, and the like. They are usually of sufficient density and particle size to give rapid settling rates. Thus, the settling chamber for grit removal at the front end of a water treating plant is usually only a small fraction of the total settling volume.

Metal Chips from Cutting Fluids

Because of their high density, metal chips carried away by coolants and lubricants from metalworking tools are easily removed by gravity settling. Such tools are lathes, drills, milling machines, saws, etc. Usually the settling chamber is built right into the machine tool coolant sump by the manufacturer. The clarified coolant is then continuously recycled to the cutting tool. The chips are periodically removed as necessary.

Oil and Grease Removal

Oil and grease are frequently encountered in wastewater streams from aqueous washing operations. They are also present in wastewater streams from petroleum production, ship's bilge and ballast water, food processing, and metalworking. In simple gravity settling operations these materials rise to the surface of the water and can be either skimmed off mechanically or allowed to overflow with the surface effluent from the settler. This is illustrated in Figure 5.4. Figure 5.4a shows a settling chamber where both heavy solids and light oil are being removed from a wastewater stream. The lighter oil and grease rise to the surface of the settler and the heavier sludge settles to the bottom. From there, the sludge is removed by mechanical rakes. The oil and grease can be removed by various types of belt or disk skimmers and recovered.

For greases and cooking fat the settler is usually operated above the melting point of the light material so that the droplets are liquid and spherical. In addition, a coalescing device such as a bed of metal mesh or a stack of chevron trays, as illustrated in Figure 5.4b may also be used to increase the size of the droplets and make them settle faster.

In some cases the floating of the oil and grease to the surface can be enhanced by blowing air bubbles into the settler. This will be discussed under flotation later in this chapter.

Dusts

Dusts are dispersions of dry solid particles in a gaseous carrier. They are frequently generated by grinding or size-reduction operations, or by attrition of solid particles in moving or fluid bed operations. They are collected from the gas streams in a variety of ways such as filtration, scrubbing, electrostatic precipitation, cycloning, and gravity settling. The oldest of these is gravity settling which uses a large-volume quiescent chamber to let the dust particles settle out of the gas. Dusts normally settle quite slowly because of their small particle size. If they settled rapidly, they wouldn't be carried by the flowing

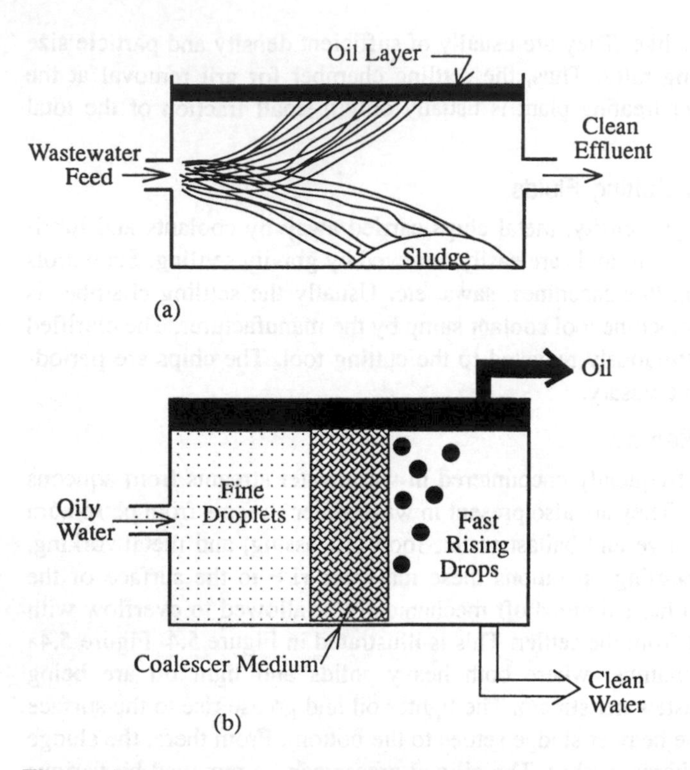

Figure 5.4 Diagram of a gravity settler for wastewater treatment. (a) Simultaneous oil and solids removal, (b) oil-water separation.

gas in the first place. Dust particles are generally only a few μm in size. From Table 5.1 it is estimated that a 1 μm particle would have a terminal settling velocity of about 0.0002 ft/sec. This would require a residence time of about 5000 sec for a particle to settle 1 ft. Such slow settling would require extremely large settling chambers for the high gas flows encountered in modern processing operations. Thus, simple gravity settling has lost popularity in favor of bag filters, cyclones, scrubbers, and electrostatic precipitators in modern dust collection.

Mists

Mists are very much like dusts, except that the particles are liquid and therefore spherical. They can be generated by condensing vapors, by droplet formation when bubbles collapse in boiling, by splashing, or by entrainment from mixing liquids with gases. While the droplets are of the same order of magnitude of particle size, although slightly smaller than with dusts, they can frequently be coalesced as a pretreat to the final settling. This is done in the same way as for oil removal from water, using mesh or stacked trays as the coalescing medium.

5.2.3 Equipment

The equipment used for settling varies with the ease with which the particles settle and the time required to get a clarified effluent. For systems having large particles, low viscosity, and high density differences, very simple settling chambers of moderate size can do a good job of clarification. These settlers are usually simple rectangular pools or cylindrical tanks. However, small particles dispersed in a viscous medium require long residence times, short settling distances, or even gravity enhancement by the use of centrifugal force as in centrifuges or cyclones. A variety of settlers are available in either standard or customized designs in a wide range of sizes. They may or may not be equipped with mechanical devices for assisting with coalescing liquid droplets or for providing continuous removal of bottom sludge or surface oils and greases. For the larger sizes, the units probably require custom design. Some of the vendors of gravity settling equipment are given in Table 5.2.

5.2.4 Common Problems and Troubleshooting

The effectiveness of gravity settling is usually measured by the clarity of the effluent stream from the settler. The major difficulties causing poor clarification of the effluent are:

- Short residence time
- Excessive turbulence in the settler
- Low density difference
- Formation of emulsions

These will be discussed in the order in which they are listed.

Short Residence Time

Too short a residence time occurs when the settling depth requires more time than the flow rate through the settler will permit. This means that the area of the settler is too small for the flow rate of the feed. The options available to improve clarification in this case are (1) increase the surface area of the settler, (2) reduce the feed rate, (3) install baffles to decrease the settling distance, or (4) install a coalescing bed for liquid particles to increase particle size and make them settle faster.

Table 5.2 Equipment Suppliers for Gravity Settling

ACS Environmental
303 Silver Spring Road
Conroe, TX 77303
(800)-359-3923

AFL Industries, Inc.
3661-C West Blue Heron Blvd.
Riviera Beach, FL 33404
(407) 844-5200

Alloy Hardfacing & Eng. Co.
1201 Clover Drive,
South Minneapolis, MN 55420
(612) 881-7515 X-43

Dorr-Oliver, Inc.
P.O. Box 3819
Milford, CT 06460-8719

Graver Water
2720 US #22
Union, NJ 07083
(908) 964-2400

Great Lakes Environmental
463 Vista
Addison, IL 60101
(708) 543-9444

Highland Tank & Mfg. Co.
Box 338
Stoystown, PA 15563
(814) 893-5701

Humboldt Decanter, Inc.
3200 Points Pkwy (Norcross)
Atlanta, GA 30092
(404) 448-4748

Inlay, Inc.
P.O. Box 461
Hope, NJ 07844
(201) 459-5677

Kisco Water Treatment Co.
305 West Marquette Ave.
Oak Creek, WI 53164
(414) 764-5700

Lancy International, Inc.
181 Thom Hill Road
Warrendale, PA 15086-7527
(412) 772-0044

National Fluid Separator, Inc.
827 Hanley Industrial Court
St. Louis, MO 63144
(314) 968-2838

Parkson Corporation
2727 NW 62nd Street
P.O. Box 408399
Fort Lauderdale, FL 33340
(305) 974-6610

Permutit
30-T Technology Drive
Warren, NJ 07059-0920
(908) 668-1700

Pha Sep
1110 Jenkins Road
Gastonia, NC 28052

Pollution Control Engineering
3233 Halladay
Santa Ana, CA 92705
(714) 641-1401

Western Filter Co.
P.O. Box 16323 M/S 14
Denver, CO 80216
(303) 288-2617 X-31

Excessive Turbulence in the Settler

Excessive turbulence occurs when the feed is introduced directly into the settling zone as high-velocity jets from feed pipes. For this reason, most settlers are eqipped with a specially designed inlet zone to remove all eddies and turbulence before the feed enters the quiescent settling zone. The design of these inlet zones usually takes the form of devices to reduce the velocity of the incoming feed by introducing it over a large inlet area. Such things as diffusers, distributor plates, gravel or wire mesh beds, or stacks of horizontal baffles are regularly used.

Low Density Difference

When the density difference between the particles and the carrier phase is too low to get adequate particle removal in a given settler, one or more of the following remedies may be helpful.

1. Residence time can be increased.
2. Settling distance can be decreased.
3. Density difference can be increased.
4. Particle size can be increased.
5. Gravity can be enhanced.

The first of these is done by decreasing flow rate, the second by installing baffles, the fourth by agglomerating the particles, and the fifth by installing cyclones or hydroclones to provide centrifugal force. The third one, increasing the density difference, can only be done in certain cases. For example, if an oil has a density too close to that of water to settle effectively, it may be possible to reduce its density. This can be done by using a lower density oil in the process from which the dispersion was generated. However, a more common generic solution is to provide a coalescer to make the drops larger so that the low density difference is adequate.

Formation of Emulsions

Occasionally emulsions form in oil-water mixtures if surfactants are present and mixing is severe. This can occur when an oil-water mixture is passed through a high-pressure pump. If a surfactant is present to stabilize the emulsion, it may not settle under the influence of normal gravity. In such cases it may be necessary to resort to the sedimentation techniques discussed in Chapter 8 or to the centrifugal separations discussed in Chapter 7.

5.2.5 Economics

Gravity settling is a relatively inexpensive and low capital cost separation process. However, very large units may require a lot of ground area. The basic unit is a pool, pond, or tank for which the size and complexity are set by the terminal settling velocity of the particles in the dispersion. For rapid settling rates, the tank can be small and plain. However, as the settling rate decreases, the requirement for long residence times, as well as for coalescing devices and slanted settling plates or tubes, adds considerably to the cost. Furthermore, the degree of difficulty in removing the settled sludge and surface layers affects the design and thus the cost of continuous removal equipment. Ponds are usually used for slow settling silts and sludges. They are merely allowed to fill up with settled solids as they accumulate over long periods of time. On the other hand, rapid settling usually requires continuous removal of the settled materials to keep the settling volume from filling with settled materials and thus reducing residence time. Vendors should be contacted for cost information on specific applications.

5.3 FLOTATION

In contrast to gravity settling, where the driving force for particle motion is the density difference between the phases, flotation uses the buoyancy provided by the attachment of small bubbles of a gas (usually air) to the dispersed particles. This lifts them to the surface of the liquid in the separation chamber. When air cannot be used because of its oxidizing properties, natural gas or nitrogen can be substituted for it. The bubbles usually attach to hydrophobic particles such as oils and grease and can thus be used to remove them from mixtures with water. However, some solid particles are also hydrophobic or can be made so by surface treatment with additives. This permits them to be removed from aqueous systems by flotation along with oils and grease. It also permits their removal from hydrophilic particles in a mixed suspension in water. Thus, a flotation unit can lift hydrophobic particles and allow hydrophilic particles to sink at the same time. This provides a separation in a single unit of hydrophilic and hydrophobic solids from each other as well as from the carrier liquid. This technique is widely used in the mining industry for the enrichment of mineral ores. However, the flotation discussed in this chapter will be largely confined to clarification of aqueous effluents.

There are two major types of flotation, named for the manner in which the gas bubbles are produced. One is called dissolved air flotation (DAF) while the other is induced air flotation (IAF). With DAF the air is dissolved in the feed stream under pressure and then released in the form of tiny bubbles as the pressure is reduced in the flotation chamber. In IAF the air is dispersed in the water using motor-driven mixers, which draw air from the atmosphere and disperse it under the surface of the feed to be separated. IAF is the older process and has been widely applied to separation of oil and water in the petroleum and marine industries as well as in the beneficiation of minerals.

5.3.1 Fundamentals

In order for flotation to occur, gas bubbles must attach to the particles to be floated and stay attached long enough to get the particles to the surface and keep them there until they are collected. This means the bubbles must stick pretty tenaciously to the particles. This is not a problem with oils and greases because gas bubbles naturally stick to strongly hydrophobic surfaces. However, with less hydrophobic surfaces, such as occur with many solids, additives may be needed to modify the surface and make it more hydrophobic to get satisfactory bubble adhesion. The common additives used are collectors (or promoters) and frothers. The collectors are surface modifiers to make the surface more hydrophobic and thus attach more readily to the bubbles. The frothers are usually detergents, which modify the bubble size generated by the dispersers.

Flotation processes are generally categorized by the two major techniques used to generate the gas bubbles needed. These techniques are the dissolved air flotation and induced, or dispersed, air flotation mentioned above. Flotation will be discussed below under these two headings. Dissolved air flotation is frequently used for clarification of residual hazes in wastewaters, while induced air flotation is often used for recovery of bulk materials as in the mining industry.

Dissolved Air Flotation

Dissolved air flotation, as its name implies, generates tiny bubbles by dissolving air under pressure into the feed stream and then releasing it as bubbles by reducing the pressure. This takes advantage of the effect of pressure on the solubility of gases in liquids. At high pressure more gas is soluble in the liquid than at lower pressures. This means that the extra gas dissolved at high pressure must come out of solution as gas bubbles when the pressure is reduced. Bubbles generated in this way are μm sized and are very efficient at attaching to dispersed particles.

The amount of gas that can be released from the DAF unit is the difference between the amounts of gas required to saturate the feed at the high and low pressures. This is directly proportional to the difference in absolute pressures, i.e., gage pressures plus 14.7 lb/in.^{2}. This comes from the fact that Henry's Law predicts that the amount of gas dissolved in a liquid at saturation is proportional to the absolute pressure of the gas. The proportionality constant varies with temperature and decreases with increasing temperature. When the low pressure is one atmosphere, this relationship reduces to Eq. 3.

$$Q = Q_S \frac{fP}{14.7} - Q_S \tag{3}$$

where:

Q = gas released at atmospheric pressure
Q_S = saturation gas content at atmospheric pressure
f = fraction of saturation reached in pressurizer
P = absolute pressure in pressurizer

Equation 3 states that the theoretical amount of the dissolved gas that is released on reducing the pressure is the amount put in at the pressurizer minus the amount retained in solution in the flotation unit. These values of Q can be in any consistent units of gas weight or gas volume per unit of liquid volume.

The liquid loading generally encountered in DAF is between 2 and 3.5 gal/ft^2 while the air-to-solids weight ratio is normally 0.02 to 0.04. The feed can be pressurized with gas in a variety of flow plans. First, the entire feed stream can pass through the pressurizer. Alternatively, only part of the feed stream, or even a recycle stream, can be fed to the pressurizer and then mixed with the main feed as the pressure is released through a back pressure regulator. The method used depends on how much air is needed to do the flotation, as well as the pressure ratings available in the pressurizing equipment. When the particle loading is heavy, the entire feed is passed through the pressurizer; for lighter loadings only part of the feed needs to be directly pressurized. The recycle technique is used when the feed components can't stand the intense mixing conditions in the pressurizer. The recycle stream used is taken from the clarified product.

As discussed by Kominek and Lash (1979) and illustrated in Figure 5.5, air can attach to the particles by either (1) using the particle as a nucleus for bubble formation as the air comes out of solution or (2) by sticky collisions between already formed bubbles and the particles. Figure 5.5a shows the first type of attachment and 5.5b shows the second. Both types of attachment occur with DAF, while the second type is more predominant with IAF.

Induced Air Flotation

Induced, or dispersed, air flotation is brought about by pulling atmospheric air into motor-driven dispersers and breaking it up into appropriately sized bubbles under the surface of the liquid feed. Both the air and the liquid feed pass through the disperser to get good contact between the air bubbles and the particulate matter. These bubbles then attach to the particulate material and float it to the surface where it collects as a layer of froth. The froth is then skimmed from the surface of the water and the clarified water is withdrawn from well below the surface.

From Kominek and Lash (1979), the air is applied at rates from 1 to 10 ft^3 per gallon of liquid. They also quote liquid retention times of about 4 min, which permits minimal space requirements.

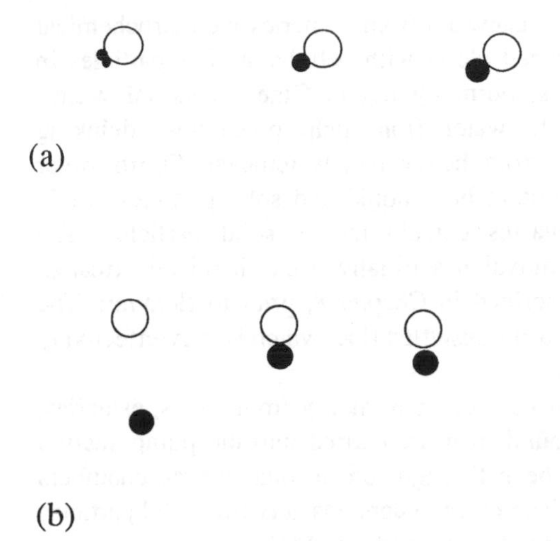

Figure 5.5 Mechanisms of gas attachment to particles. (a) Particle nuclei for bubble formation. Air bubbles shown in black. (b) Collision with existing air. Air bubbles shown in black.

5.3.2 Specific Applications

The specific applications listed in this section will exclude those used for minerals beneficiation even though that is probably the largest single use. In mineral beneficiation two kinds of solid particles are separated from each other by causing one to float and the other to sink using induced air flotation. This is usually done by controlling the surface properties, i.e., wettability, of the solids by the use of additives.

The kinds of applications listed here are generally those used for final clarification after a coarse separation by settling has already been achieved. This can involve air flotation of dispersed liquid droplets, while at the same time settling solid particles from aqueous streams. Flotation is particularly helpful when particulates are less than 5 μm in size or there is no density difference between the particles and the carrier fluid. The major areas of application other than mineral beneficiation are:

- Industrial wastewater
- Intake water for industrial use
- Transportation wastewater
- Food processing wastewater
- Municipal sludge thickening

The industrial wastewaters include wastewater from refineries and petrochemical plants as well as oily brines produced along with petroleum. The particles in these streams are oils and greases, normally liquid. Other industrial wastes include storm water runoff, white water from pulp processing, deinking wastewaters, and aqueous wastes from heavy metals removal. Storm water runoff from industrial sites can contain both liquid and solid particles, while the white water and deinking wastes contain mostly solid particles. The wastewater from heavy metals removal has usually been chemically treated, flocculated, and coagulated, as described in Chapter 8, prior to flotation. The solid particles still present are primarily unsettled floc, which is very effectively removed by dissolved air flotation.

The intake water for industrial use comes primarily from rivers, estuaries, and the sea. The solid particles found in it are carried into the pump suction by the inlet water. Usually it has been through screens and settling chambers before it gets to the flotation units. Therefore, it contains very fine solid particles of silt or sand, which can be effectively removed by DAF.

Transportation wastewater is generally oily water coming from ship's ballast, ship's bilges, railroad and aircraft maintenance, and automotive manufacturing machine shops. It has usually been through a settling chamber, most probably equipped with a coalescing bed of some kind. Therefore, the residual oil droplets in the water are very fine and very responsive to flotation.

The food processing wastewaters usually contain particles of oils, greases, or fatty solids. They may come from any of the following activities.

- Vegetable oil processing
- Stockyard and feedlot runoff
- Fat rendering
- Poultry processing
- Meat packing
- Canning
- Seafood processing
- Prepared foods
- Pet foods

Again, flotation is very effective for cleaning up wastewaters from these operations.

Municipal sludge thickening is used to increase the solids content of the sludge from municipal water treating plants. The fine particles of floc generated by the water treatment are very dilute but can be effectively concentrated by dissolved air flotation up to about 4% solids. These DAF units are usually designed on a case-by-case basis.

There are probably many other smaller applications of flotation for cleaning up wastewaters which can take advantage of the ability of flotation to reduce oil content in water to the order of 10 parts per million.

5.3.3 Equipment

The equipment used in flotation varies in the way the bubbles are generated, how the froth is skimmed and separated, and what, if any, pretreatment is required. The various vendors of flotation equipment supply both standard designs in a variety of sizes, as well as custom designed units for specialized applications. Table 5.3 lists some of the major vendors of flotation equipment. It is best to contact a vendor with a specific application in mind and let the vendor recommend the appropriate equipment. Frequently lab or pilot plant tests may be needed to insure that the recommended equipment will work satisfactorily.

A typical dissolved air flotation unit for oily wastewater is shown in Figure 5.6. This unit consists of a cylindrical tank equipped with a rotating rake at the bottom and a skimmer blade at the surface of the liquid. The rake scrapes the settled solids to the sludge outlet and the skimmer blade pushes the floated oil into the oil discharge sump. The clarified water product is recovered from a vertical location near the middle of the tank and passed up over an overflow launder to the discharge. The illustration shows operation with the dissolved air being supplied to a clarified recycle stream in a pressurized aeration tank. The pressure is released through a backpressure regulator just as the recycle stream is mixed with the feed entering the flotation tank. The reduction in pressure at this point generates the flotation bubbles. These air bubbles stick to the oil droplets, carry-

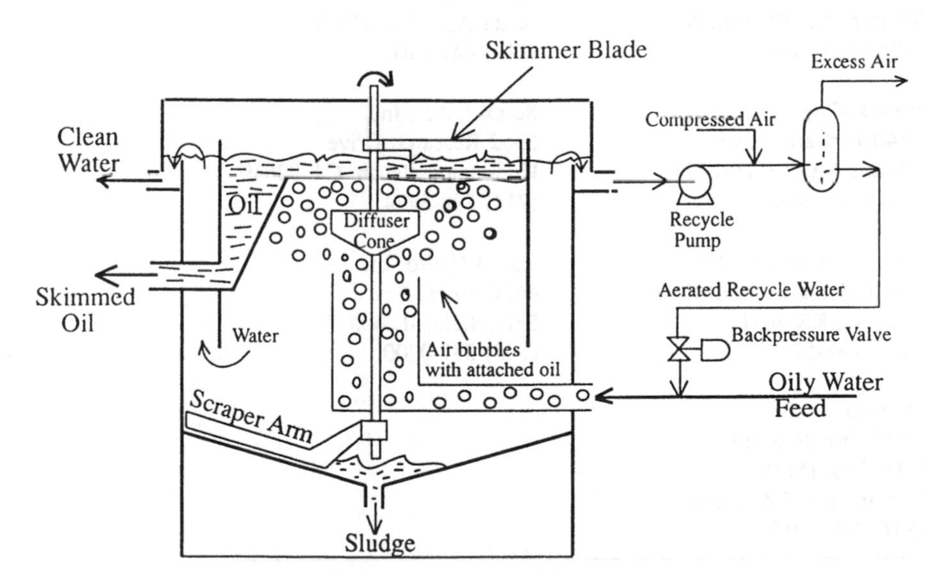

Figure 5.6 Example of a dissolved air flotation system.

Table 5.3 Equipment Suppliers for Flotation

ACS Environmental 303 Silver Spring Road Conroe, TX 77303 (800) 359-3923	Aeromix Systems, Inc. 2611 North Second St. Minneapolis, MN 55411-1633 (612) 521-1455
AFL Industries, Inc. 3661-C West Blue Heron Blvd. Riviera Beach, FL 33404 (407) 844-5200	Graver Water 2720 US #22 Union, NJ 07083 (908) 964-2400
Inlay, Inc. P.O. Box 461 Hope, NJ 07844 (201) 459-5677	Krofta Engineering Corp. P.O. Box 972 Lenox MA 01240 (413) 637-0740
Lancy International, Inc. 181 Thom Hill Road Warrendale, PA 15086-7527 (412) 772-0044	Microlift Systems 107 East Walnut P.O. Box 678 Sturgeon Bay, WI 54235-0678 (414) 743-7022
Permutit 30-T Technology Drive Warren, NJ 07059-0920 (908) 668-1700	Pollution Control Engineering 3233 Halladay Santa Ana, CA 92705 (714) 641-1401
Redux Corp. 1840 Fenpark Drive Fenton, MO 63026-2922 (314) 343-0030	Serck Baker, Inc. 5352 Research Drive Huntington Beach, CA 92649 (714) 898-3474
Smith & Loveless, Inc. 14040 Santa Fe Trail Drive Lemexa, KS 66215 (913) 888-5201	Tenco Hydro, Inc. 4620 Forest Ave. Brookfield, IL 60513 (708) 387-0700
Wemco 1796 Tribute Road P.O. Box 15619 Sacramento, CA 95852 (916) 929-9363	

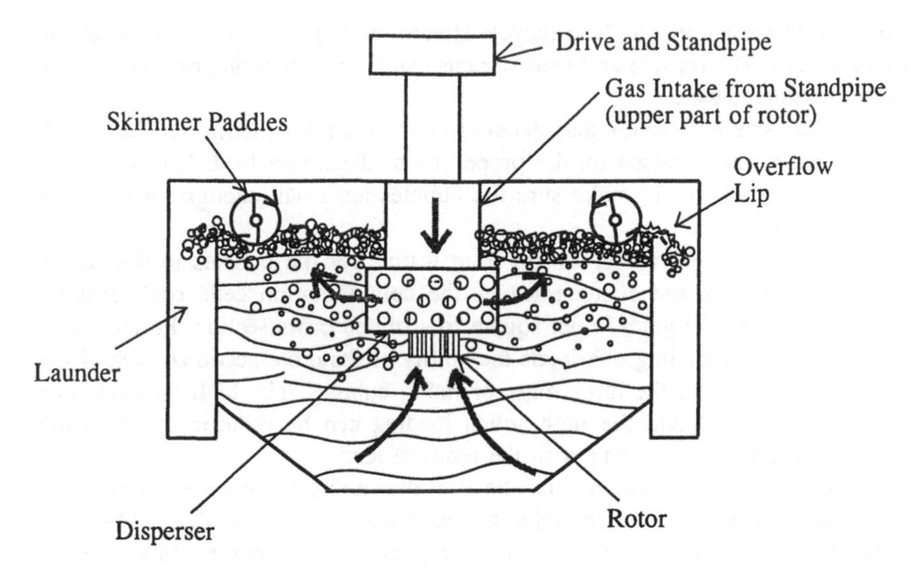

Figure 5.7 Example of an induced air flotation unit.

ing them to the surface, and permit the unattached solids to settle to the bottom and be scraped into the outlet. Depending on the surface properties and size of the solid particles, they may either settle to the bottom or be carried with the oil.

Figure 5.7 shows a typical induced air flotation unit. In this unit, air at atmospheric pressure is drawn down a standpipe by the slight vacuum created by a mechanically driven aeration rotor. The rotor acts like a pump, forcing the air and liquid feed through a disperser, which breaks the air into tiny droplets. These droplets attach to the particles in the feed and carry them as a froth to the surface of the liquid in the flotation tank. From there the froth is skimmed into collecting launders and removed from the cell.

5.3.4 Common Problems and Troubleshooting

Flotation units generally are automatically controlled and give trouble-free operation with a minimum of attention. However, occasionally minor problems can occur. These generally show up as inadequate clarity of the effluent stream. The causes are usually related to bubble generation and attachment, particle loading of the feed, or air and liquid flow rates through the unit.

For dissolved air flotation, the adequate amount of air must be dissolved during pressurization to give an air bubbles-to-particles weight ratio of 0.02 to 0.1 in the flotation cell. Partial plugging or clogging of the air injector in the pressurizer can cause this to be too low. Regular cleaning of the injector, or injection of the air into a clean recycle stream instead of the feed, can alleviate this problem. The recycle and feed streams can be mixed during or immediately after pressure release.

Bubble size in flotation also depends on the interfacial tension at the liquid-gas interface and depends on the properties of the liquid feed. It can be controlled with additives to make sure the bubbles are small enough for effective particle attachment.

Too high a feed rate gives too short a time for the bubbles to rise to the top surface before the effluent leaves the cell. Flotation cells can normally accommodate 1 to 3 gal/min per square foot of cell cross-section. Furthermore, too high a solids loading overloads the ability of the cross-sectional area of the cell to accommodate the large mass of rising bubbles. The high feed rate can be merely reduced and the high solids loading can be reduced by recycling some clarified product along with the reduced feed.

With induced air flotation units, the air is drawn directly into a mixer-disperser where it is broken up into bubbles by mechanical agitation. The bubble size depends on mixer design, intensity of mixing, and feed properties. Mixer design and intensity of mixing are determined by the equipment supplier's design. However, the properties of the feed mixture may vary from time to time and require the use of additives to keep the unit operating satisfactorily. A good source of information regarding types and quantities of additives is the equipment supplier.

5.3.5 Economics

Because flotation requires only short residence times, the units are relatively small. This means that the cost is not so much in size of tanks but rather in equipment for making bubbles and controlling levels and flow rates. This makes cost relatively independent of size of units, but very dependent on the number of stages of flotation needed. Most non-mineral applications require only a single stage of separation. Thus, capital cost is low.

Since no phase change is required in flotation, no latent heat is required for a phase change. This means that thermal energy costs are also low. However, mechanical energy is needed for bubble generation, either as mixing or as pressure to dissolve gas.

Manpower requirements are also low because the units are automatically controlled and require only a small amount of maintenance time. Thus overall, flotation is an inexpensive process.

5.4 SUMMARY

Gravity settling can be used to remove insoluble solid or liquid particles dispersed in a carrier fluid. The carrier fluid is then recovered particle-free. The particles can be denser than the carrier, in which case they settle to the bottom, or they can be less dense, in which case they float to the surface. The settled layers are then removed by either skimming the surface or scraping the bottom. Coalescing media or parallel trays can be used in the settler to make it more effective.

If the particles are too small to settle effectively or if there is no density difference to cause settling, flotation can be used to carry them to the surface attached to small bubbles of gas. The buoyancy provided by the bubbles of gas (usually air) can lift particles even denser than the carrier fluid to its surface where they are collected as a froth. The bubbles stick best to hydrophobic surfaces and are thus especially effective at removing oils and grease from wastewaters. The bubbles are generated either by dissolving air in the feed under pressure and then releasing the pressure or by dispersing atmospheric air into the liquid by mechanical mixing. Both methods are in wide use with many different applications.

REFERENCES

Kominek, E. G., and Lash, L. D. (1979), Sedimentation, *Handbook of Separation Techniques for Chemical Engineers*, (P. A. Schweitzer, ed.), McGraw-Hill, New York, pp. 4-113 to 4-133.

Sakiadis, B. C. (1984). Fluid and Particle Mechanics, *Perry's Chemical Engineers' Handbook*, 6th ed., (R. H. Perry and D. W. Green, ed.), McGraw-Hill, New York, pp. 5-63 to 5-68.

6

Filtration and Drying

6.1 INTRODUCTION

Filtration and drying are widely used in waste minimization either to recover valuable materials for recycling, to remove toxic materials from otherwise benign wastes, or simply to minimize disposal volume and its attendant cost. Some examples are (1) the pretreatment of plating wastes for the removal of metals and cyanides prior to disposal to a wastewater treatment plant, (2) sludge dewatering prior to landfilling, (3) floc and sediment removal in water treating, (4) removing particles from recyclable oils, and (5) removing dusts and mists from exhaust gases, to name just a few.

Filtration and drying are similar processes at first glance because they both involve separation of liquids from solids. Drying always involves liquids, while filtration can also include recovering dusts and mists from gases. Filtration depends on mechanical methods for separating the solids from the liquid, while drying depends on evaporation to separate the liquid from the solids. Filtration is usually cheaper since no heat of vaporization is involved. However, it cannot produce a solid product totally free from liquid. Very dry filter cakes contain 10 to 20% liquid, while liquid contents of 60 to 80% are much more common.

Therefore, to produce dry solids, filtration is often used as a pretreating step to concentrate solids from a slurry, followed by drying of the filter cake to remove the last of the liquid from the solids. A few manufacturers offer filter-dryer combination units, which can be very advantageous when applicable. However, they do require very specific properties of the solids in the feed slurry.

There are so many different techniques for filtering and drying that complete coverage is well beyond the scope of this book. For example, filter media alone can vary from deep beds of granular materials through sintered metals and ceramics, various cloths, felts, fibers, and wire screens to semi-permeable plastic membranes. Therefore, if the reader wishes to study this subject in more depth, Sections 19 and 20 of *Perry's Chemical Engineers' Handbook* (Perry and Green, 1984) and Section 4 of Schweitzer's (1979) *Handbook of Separation Techniques for Chemical Engineers* are recommended as starting points. This chapter is largely based on these two reference books.

6.2 FILTRATION

Filtration is defined as the separation of a fluid-solids mixture by passing the mixture through a porous barrier, which retains most of the solid particles and passes most of the carrier fluid through the barrier. The carrier fluid can be either a liquid or a gas or vapor. However, filtration processes can be further subdivided into three broad categories: (1) cake filtration, (2) depth filtration, and (3) surface filtration. This chapter will not discuss ultrafiltration or microfiltration, which are discussed in Chapter 9, Membrane Processes.

In cake filtration, the porous barrier holds back the solid particles, which build up into a steadily thickening cake as the filtration progresses. The cake must be porous enough to allow the fluid to flow through it as the cake builds up.

In depth filtration, the porous barrier is a relatively thick layer of coarse material. This allows clear fluid to pass through while trapping the solid particles in the relatively coarse spaces between the particles forming the filter bed.

In surface filtration, the porous barrier is a membrane of controlled pore size. The solid particles are stopped if they are too large to pass through the pores in the barrier. Thus, the filtration in this case is essentially a straining process, much like screening. However, because the filtration barrier tends to blind (that is, plug) the pores with solid particles, the filtration rate decreases as the filtration progresses and the filter is completely plugged before a cake can be formed.

6.2.1 Fundamentals of Filtration

The existing theories of filtration are inadequate as a sole basis for the design of filtration equipment. However, they are useful for predicting the effects of changes in operating conditions and thus in seeking optimum conditions for a given feedstock. Therefore, in practice, experimental filter rates are determined on the slurry in question and the theory is then used to develop a design from those experimental points.

Filtration theory consists of various ways to solve the general rate equation used in chemical engineering. That is, the rate is equal to the driving force divided by the resistance. The filtration rate can be expressed as the volume of clear filtrate collected per unit time and the driving force is the total pressure drop across the filter medium and the cake deposited on it. Both these quantities can be easily measured. However, the difficulty with theoretical prediction of filter performance lies in the prediction of the resistance to flow offered by the filter. This filter resistance is the sum of the resistances of the filter cake and the filter medium, the porous barrier. In cake filtration the resistance of the porous barrier is normally small compared to the resistance of the filter cake and can usually be neglected. This leaves the resistance of the filter cake as the major resistance to flow.

Cake Filtration

Unfortunately, the resistance of a filter cake is usually not a simple term to estimate. The cake usually consists of a mass of solid particles of various sizes with small channels for the clear liquid to flow through. These small channels mean that the liquid flow is always in the viscous region and flow resistance is proportional to liquid viscosity. It is also proportional to the thickness of the cake and the specific resistance of the cake, which is a constant for a particular cake in its instantaneous condition. However, the thickness of the cake builds up with time and the specific resistance varies with pressure due to the possible compressibility of the filter cake. The differential relationship for flow rate per unit area as a function of driving force over resistance is given by Carman (1938) in Eq. 1.

$$\frac{dV}{Ad\theta} = \frac{\Delta P}{\mu\,[\,\alpha\,(W/A) + r]} \tag{1}$$

where:

$dV/d\theta$	=	volumetric flow rate
A	=	area of filter
ΔP	=	pressure drop across filter
μ	=	viscosity of fluid
α	=	average specific cake resistance
W	=	mass of accumulated dry solids from V
r	=	resistance of filter medium

The specific average cake resistance, α, is not a constant and depends on the pressure as shown in Eq. 2.

$$\alpha = \alpha' P^s \tag{2}$$

In Eq. 2, the exponent s depends on the cake compressibility, varying from 0 for rigid incompressible cake to 1.0 for very highly compressible cakes. Examples of incompressible cakes are fine sand and diatomite, while gelatinous precipitates such as metal hydroxides form highly compressible cakes. Most industrial slurries give values of s between 0.1 and 0.8.

Tiller and Crump (1977) proposed that filtration flow characteristics can be divided into three categories based on the pumping method used to generate flow. These three categories are:

1. Constant pressure filtration
2. Constant rate filtration
3. Variable pressure, variable rate filtration

The first type of filtration occurs when the driving force for flow is a compressed gas at constant pressure, as when using a slurry blowcase. The second occurs when positive displacement pumps, such as piston or moving cavity pumps, are used. The third type occurs with centrifugal pumps where flow decreases as back pressure builds up due to increasing cake thickness or plugging of the filter medium. Looking at Eq. 1 in the light of these restrictions tells us something about the effects of operating variables for the different filtration modes.

Pressure: For incompressible granular or crystalline solids, where s equals zero, the filtration rate is almost linear with pressure drop as would be predicted from Eq. 1. Furthermore, the filtration rate for slimy or flocculent precipitates, where $s = 1$, are almost unaffected by an increase in pressure drop, again as would be predicted by the equation.

Cake Thickness: The average flow rate during a cycle of filtration is inversely proportional to the cake thickness and directly proportional to the square of the filtration area.

Viscosity: Filtration rate is inversely proportional to the viscosity of the filtrate. This is most simply altered by changing the temperature of the slurry, as indicated below. However, it can also be reduced by diluting the slurry with a lower viscosity diluent. This is only desirable if the diluent can be readily removed from the product filtrate or the filtrate is not the desired product.

Temperature: The viscosity of most liquids is reduced dramatically as temperature is increased. Thus, filtration rate is greatly increased as temperature is increased. However, the effects of increased temperature are limited by cost of heating, vapor pressure of the filtrate, and the effect of temperature on solubility of the solids in the filtrate.

Particle Size: The effect of particle size on filtration rate is shown through its effect on the specific cake resistance as well as on the compressibility *s*. Therefore, close control of particle size in the slurry feed to the filter is necessary. Small particles result in lower filter rates and higher liquid content of the filter cake. Therefore, it is desirable to avoid any size reduction due to violent pumping or mixing action. Furthermore, any techniques that increase particle size can be used to increase filter rate. This can often be done by adding flocculants or coagulants to the feed slurry to collect small particles into larger aggregates.

Filter Medium: The optimum filter medium requires as open a weave as possible to reduce plugging while, at the same time, as tight as necessary to prevent bleeding of fine particles. Bleeding usually stops after a thin cake has formed because the small particles are then caught in the cake. Thin, pliable filter cloths are less prone to plugging than thick, stiff cloths, although they are more susceptible to bleeding.

Solids Concentration: Apart from the obvious effect that thicker slurries build up filter cake faster, slurry concentration can also affect cake resistance and plugging of the filter medium. These factors favor higher slurry concentrations. They can be achieved, if economically justified, by pretreating the feed with various types of thickeners.

Filter Aids: When problems such as slow filtration rate, poor filtrate clarity, or rapid medium blinding occur, filter aids can often relieve the problem (Cain, 1979a). These materials are granular or fibrous solids which can form a highly permeable cake which traps very fine particles or slimy, deformable flocs. The two most popular are diatomaceous silica and perlite. They can be either precoated on the filter at startup, added to the filter feed throughout the filtration, or both. Adding filter aid during the filtration is called body feed and increases the porosity of the new cake laid down throughout the filtration.

Depth Filtration

Theory for depth filtration has been developed only for deep bed sand filters, although it is assumed that it applies equally well to cartridge filters. However, for cartridge filtration, the most reliable scale-up method is to run a test on a single cartridge and then use as many cartridges in parallel as are needed to get the desired filter rate (Nickolaus, 1979).

The basic equation developed for removal of turbidity from a liquid stream in a sand bed is given by Cain (1979b) as Eq. 3.

$$\frac{-dC}{dL} = \lambda C \tag{3}$$

where:

C = concentration of particles in suspension as volume fraction
L = distance into the filter from surface
λ = filter coefficient in dimensions of reciprocal length

However, λ varies both with time of filtration and depth in the sand bed. Thus, a solution to Eq. 3 is difficult to obtain. For design purposes it is best to contact a vendor and have test runs made on the slurry in question. The vendor can then recommend the proper filter dimensions for the desired application.

Surface Filtration

According to Cain (1979b), very little theory has been developed for surface filtration. Therefore, filter surface area is determined by running filtration experiments with the slurry in question on a test filter matrix with small enough pores to give the required filtrate clarity. The volume of liquid that passes through the matrix before it completely plugs can be scaled up in direct ratio to the surface area of the actual filter to the area of the test filter.

6.2.2 Specific Applications

There is a potential application for filtration any time a solid needs to be removed from a liquid or gas stream. Thus, the applications are numerous and (1) vary from very small to very large, (2) include purification or concentration of input, recycle, and waste streams, and (3) are used either alone or in combination with other separation techniques. The few examples discussed here are for water and sewage treatment, clarifying recycle streams, and for dust collection.

Water Treating

The water-treating applications are probably the largest liquid-solid separations encountered. They include preparation of high-purity potable or industrial water supplies, treatment of wastewaters for process water recycle, treatment of sewage prior to disposal, and dewatering of sludges before disposal. Examples from each of these areas will be discussed briefly.

Potable Water: In preparing potable water from natural supplies, the water is frequently treated with various chemicals and flocculating agents to adjust pH and precipitate trace impurities. The precipitated floc is settled, and the settler effluent is clarified by filtration to remove the last traces of silt and fine floc. The filters used to do this final filtering are usually deep graded-bed sand or anthracite filters, which trap the fine particles in the feedwater between the grains in the filter bed. When the filter is nearing its capacity for holding filtered solids, the bed is backwashed to remove the filtered solids from the bed as a concentrated stream. This concentrated slurry must be treated, usually by further filtration, before disposal. Potable water deep-bed filters may vary in size and shape from 10-ft-diameter horizontal cylindrical pressure tanks for small municipal water supplies to large concrete open tanks covering acres for water supplies of large cities.

Wastewater Treating: Industrial wastewaters must frequently be pretreated before they can be released to conventional sewage treating plants. For example, wastewaters from electroplating processes are too toxic for biological treatment due to their content of metals and cyanides. These materials are removed by chemically converting the soluble metal salts to insoluble hydroxides or sulfides and the cyanides are destroyed by chemical oxidation. These reactions are discussed more thoroughly in Chapter 8. After the insoluble metal salts have been precipitated, they are removed by settling and/or filtration. The filtration is usually done by means of a plate and frame filter press. The filtrate is then sent to a biological wastewater plant, and if desirable the metals can be recovered from the filtered solids.

Sludge Dewatering: Sludges generated from water and wastewater treating contain large amounts of water, which makes further treatment or disposal expensive. A typical sludge may contain only a few percent solids. This excess water can be removed from the sludge by pressure filtering in plate and frame presses or by squeezing it out between two belts in a compressive belt press. The plate and frame press can make a sludge cake containing 20 to 50% solids, while the belt press can produce a cake of up to 20% solids. In the compressive belt press, the sludge is fed to one end of a continuous moving filter belt and water is removed initially by gravity. The open belt covered with sludge then moves into contact with a second overhead moving belt, which squeezes the water out of the sludge by passing the cake through rollers of ever-increasing pressure. The sludge cake is removed at the end of the filter press and dropped by gravity to a conveyer belt to carry it away from the press for disposal.

Clarifying Recycle Streams

With the increased emphasis on waste minimization, many industries are finding that with relatively little treatment, it is often possible to recover and recycle valuable materials from waste streams. In many cases, all that is needed is removal of solid particles by filtration. However, some composition adjustment of the filtrate by addition of chemicals may sometimes be needed. Examples of this type are:

- Recycling of used oils such as hydraulic oils, heat treating oils, transformer oils, cutting oils, etc.
- Recycling of plating bath solutions
- Recycling of process water as, for example, cooling tower water

The filters used in these operations are usually cartridge filters and the spent filter elements are normally discarded. However, when they contain an unusually valuable component, as in the case of precious metal recovery from plating baths, they are reprocessed to recover the metals.

Dust and Mist Collection

With increasingly stringent regulation of pollution of the atmosphere under the Clean Air Act, emissions of dusts and mists in industrial stack gases are being greatly reduced. The most common devices for collecting these particles are cyclones, scrubbers, filters, and electrostatic precipitators located between the dust and mist generators and the stack. Frequently cyclones are used as a preliminary step for removing the larger particles and then one of the other three techniques is used for final cleanup. When filtration is used, the filters are usually either bag or cartridge filters. Small applications may use a single bag or cartridge, while large volume applications may have literally hundreds of filter bags hanging in parallel from a common header near the roof inside a baghouse. Capacities up to 100,000 ft^3/min of air flow are available.

These applications can vary from improvement of workplace conditions by removal of nuisance dusts from recycled ventilation air to major recovery operations for powdered materials. A few typical applications are:

- Sawdust collection in woodworking operations
- Mist collection in shops doing wet machining
- Collection of spray-dried foods
- Collection of minerals from dryers and smelters
- Collection of powdered chemicals
- Recovery of catalysts from fluid bed operations
- Removal of fly ash from stack gases

Fabrics are available to permit operation of bag filters at temperatures up to about 500°F.

6.2.3 Equipment

There are many different types of filters available for both batch and continuous solids recovery, limited only by the ingenuity of their designers. These units vary in practically every possible way, such as (1) whether batch or continuous, (2) how the slurry and filter medium are contacted, (3) how the pressure gradient is obtained, (4) how liquid is removed from the filter cake (wash or gas blowing), (5) how washing is achieved, if required, and (6) how solids are removed from the filter medium. Under each of these categories there are a number of options, giving a variety much too numerous to cover here. A brief summary of equipment types will be given here, but for more comprehensive coverage either Perry's handbook, Schweitzer's handbook, or filter vendors should be consulted. Table 6.1 lists a few of the very many filter vendors.

Table 6.1 Vendors of Filtration Equipment

Andritz Ruthner, Inc. 1010 Commercial Blvd. S. Arlington, TX 76017-7130 (817) 465-5611	Andritz Sprout Bauer Inc. Sherman St. Muncy, PA 17756 (800) 765-4919
Barneby & Sutcliffe Corp. Box 2526 Columbus, OH 43216 (614) 258-9501	C P Environmental Filters, Inc 1336 Enterprise Drive Romeoville, IL 60441 (708) 759-8866
Dorr-Oliver, Inc. 612 Wheeler's Farm Rd. P.O. Box 3819 Milford, CT 06460 (203) 876-5400	Duriron Co., Eng. Systems Grp. 9542 Hardpan Rd. Angola, NY 14006 (716) 549-2500
Dustex Corp. P.O. Box 7368 12034 Goodrich Dr. Charlotte, NC 28241-7368 (704) 588-2030	Eimco Process Equipment Box 300 Salt Lake City, UT 84110 (801) 526-2000
Faudi-Artech 1221 E. Houston Broken Arrow, OK 74012 (918) 251-0880	Filcorp Industries Box 2304 Concord, NH 03302-2304 (603) 225-6638
Filtra Systems Co. 30000 Beck Rd. Wixom, MI 48096 (313) 669-0300	Great Lakes Environmental 463 Vista Addison, IL 60101 (708) 543-9444
Hewitt Machine Co. P.O. Box 688 375 Byrd Ave. Neenah, WI 54957 (414) 722-7713	JWI, Inc. 2155 112th Avenue Holland, MI 49423 (616) 772-9011
Krystil Klear Filtration R-2, Box 300 Winamac, IN 46996 (800) 333-3458	Larox, Inc. 9730 Woods Drive Columbia, MD 21046 (410) 381-3314

Table 6.1 (Continued)

Mefiag Filtration 44 Campanelli Parkway Stoughton, MA 02072-0507 (617) 344-1700	Monroe Environmental 11 Port Ave. Box 806 Monroe, MI 48161 (800) 992-7707
Netzch, Inc. 119 Pickering Way Exton, PA 19341-1393 (215) 363-8010	Pacific Press Co. 2418 Cypress Way Fullerton, CA 92631 (800) 878-8029
Pall Corporation 2200 Northern Blvd. East Hills, NY 11548 (800) 645-6532 x4987	R & B Filtration Systems Co. 2221 Newmarket Pkwy, Suite 112 Marietta, GA 30067 (404) 955-9335
Rosedale Products, Inc. Box 1085 Ann Arbor, MI 48106 (800) 821-5373	Scientific Dust Collectors 4740 West Electric Ave. Milwaukee, WI 53219 (414) 672-0564
Serfilco, Ltd. 1777 Shermer Road Northbrook, IL 60062 (800) 323-5431	Solberg Manufacturing, Inc. 1151-T West Ardmore Ave. Itasca, IL 60143 (800) 451-0642
Tiggs Corporation Box 11661 Pittsburgh, PA 15228 (412) 563-4300	Torit & Day Donaldson Co. 1400 W. 94th Street Minneapolis, MN 55431 (800) 365-1331
Universal Process Equip. Inc. Box 338 Roosevelt, NJ 08555 (800) 243-6228	US Filter / IWT 4669 Sheperd Trail, Box 560 Rockford, IL 61105-0560 (800) 333-3458
US Filter/Lancy Envir. Systems 181 Thorn Hill Rd. Warrendale, PA 15086-7527 (412) 772-0044	Waste Tech, Inc. 1931 Industrial Drive Libertyville, IL 60048-9738 (708) 367-5150

Table 6.1 (Continued)

Wemco, 1796 Tribute Road P. O. Box 15619 Sacramento, CA 95852 (916)-929-9363	Westech Engineering, Inc. Box 65068 Salt Lake City, UT 84115 (801) 265-1000
W. W. Sly P.O. Box 5939 Cleveland, OH 44101 (800) 334-2957	Zimpro Passavant 301 W. Military Rd. Rothschild, WI 54474 (800) 826-1476

Batch Cake Filters

Nutsche Filter: One of the simplest and earliest batch filters is called the nutsche filter illustrated in Figure 6.1. It gets its name from the German word for sucking because many of them are operated by vacuum (Zievers, 1979). It consists of a tank with a false bottom, which is either perforated or porous and can either support a filter medium or act as the filter medium. The slurry is fed into the filter vessel and the separation is driven by gravity flow, gas pressurization, vacuum, or a combination of these forces. This type of filter is favored for small scale operations, such as laboratory and pilot plant filtrations. However, for large scale operations it requires a lot of floor space for a given filter area and cake removal is awkward. It is amenable to good cake washing either

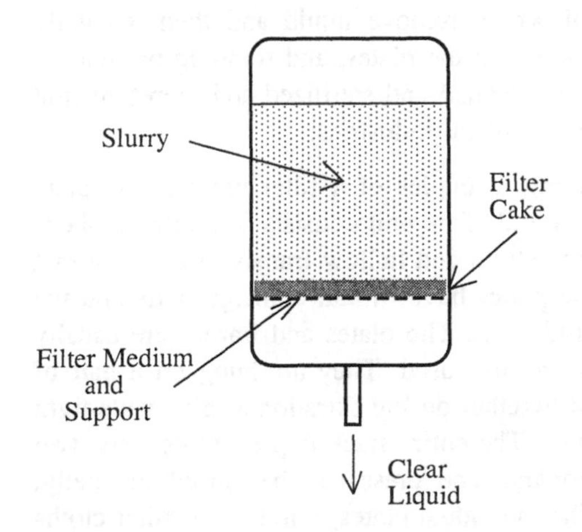

Figure 6.1 Sketch of a simple nutsche filter.

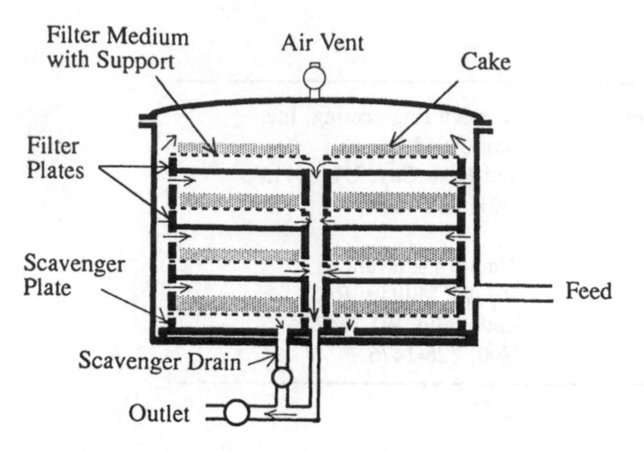

Figure 6.2 Sketch of a horizontal plate filter.

(1) by adding wash solvent before the cake is exposed to air to blow out slurry liquor or (2) by adding a mixer to reslurry the cake in the wash solvent prior to final drying of the cake.

Horizontal Plate Filter: Closely related to the nutsche filter is the horizontal plate filter illustrated in Figure 6.2, which uses a stack of round horizontal filter plates in a cylindrical tank to get a large filter area into a reasonable floor space. Filter paper or cloth is placed on each plate and a precoat of filter aid can by applied if needed. The slurry feed is introduced through a central or annular feed manifold. Filtration is continued until the cake-holding capacity of the filter is reached or until cake resistance limits filter rate to an unsatisfactory level. The cake can be washed or air-blown to remove liquid and then manually recovered by opening the vessel, removing the plates, and scraping or sluicing off the solids. These filters are easily cleaned and sterilized and therefore find extensive use in the food and pharmaceutical industries.

Filter Press: One of the outstanding examples of a filter press is the plate and frame press, illustrated in Figure 6.3. This unit consists of alternate plates covered on both sides by the filter medium and hollow frames in between that provide room for cake buildup. The plates have filtrate drainage ports and the frames have feed and wash manifold ports. The plates and frames are usually rectangular, although other shapes are also used. They are hung on a pair of horizontal support rods and pressed together during filtration to give watertight sealing between all plates and frames. The entire stack is pressed between two end plates, one of which is stationary. The press can be closed manually, hydraulically, or by motor drive. The individual plates, frames, and filter cloths have holes in their four corners to permit manifolding of the press to get

Figure 6.3 Bank of large plate and frame filter presses. (Courtesy R & B Filtration Systems Co.)

continuous channels from the stationary plate to the other end of the press. This permits a variety of feed and filtrate discharge flow options, as well as a variety of cake washing options. Filter presses are made in plate sizes from 4 by 4 in. up to 61 by 71 in. and frame thicknesses from 1/8 in. to 8 in. Operating pressures are commonly up to 100 psig with higher pressures achieved by custom design.

The basic advantages of filter presses are simplicity, low capital cost, and ability to operate at high pressure. They are useful in either cake-filter or clarifying-filter applications. They require little headroom and floorspace per unit of filter area and capacity can be adjusted by adding or deleting plates and frames. They are easily cleaned, the filter medium is easily replaced, and with proper operation they can produce a denser, drier cake than most other filters.

However, filter presses also have several serious disadvantages, including imperfect washing, short filter cloth life, and high labor requirements. While they frequently drip or leak, the major problem is that they must be opened for cake discharge. This permits routine operator exposure to the contents of the filter, which is becoming more of a problem with restricted exposure limits on many materials once considered safe.

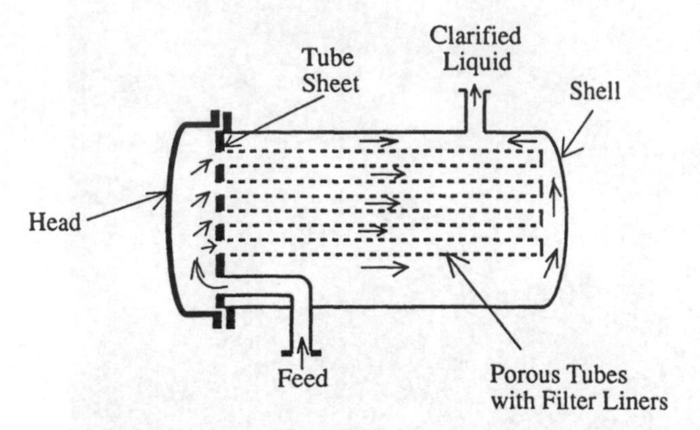

Figure 6.4 Schematic of horizontal tubular filter.

Tubular Filter: The tubular filter (Jacobs, 1984) consists of one or more perforated tubes supported either horizontally or vertically in a tube sheet inside a cylindrical shell having the same axis as the tubes. A sketch of a horizontal unit is shown in Figure 6.4 for operation with the cake inside the tubes. The dead ends of the tubes are closed with a plug and the filter medium is inserted into each tube as a liner. Slurry under pressure enters the tubes through the tube sheet and deposits the solids as a cake on the tube liner. The filtrate flows radially through the liner and perforated tube into the shell. It exits from the shell through the filtrate discharge line. The filtration cycle is over when the tubes are filled with cake or the media have become plugged. The cake can be washed and then air-blown to dry it.

The filter is usually provided with a hinged head to allow easy access to the tube sheet and mouths of the tubes. The cake is then removed in the form of "sausages" by removing the retaining rings and pulling out the liner from each tube. The tubes can also be easily removed for inspection and cleaning since they are sealed into the header with an *O*-ring.

The advantages of the tubular filter are (1) the use of an inexpensive, disposable, easily removable medium, such as filter paper, (2) the filtration cycle can be interrupted and the shell emptied without loss of the cake, (3) the cake is readily recovered in dry form, and (4) the inside of the filter is conveniently accessible. The disadvantages are the necessity for and the labor requirements of emptying and replacing the filter media by hand. Also, there is a tendency for heavy solids to settle out in the header chamber. This accounts for its application as a cleanup filter to remove fines not captured by preliminary filters or for pilot plant and small scale plant operations.

Vertical tubular filters are also available for operation with the cake and filter medium on the outside of the tubes. The cake is normally removed wet with a combination of backwashing and gas blowing and drained from the bottom of the tank.

Pressure Leaf Filter: Pressure leaf filters consist of multiple flat filter leaves supported in a pressure shell. The leaves have filtering surfaces on both faces and can be of a variety of shapes, although usually circular or rectangular. A sketch of a typical filter leaf assembly is shown in Figure 6.5. The filter type is either vertical or horizontal with the axis of the shell determining the type. A variety of different designs of horizontal and vertical leaf filters are available, differing mainly in how the filter cake is recovered. That is, by (1) sluicing with liquid sprays, (2) vibrating the leaves, (3) rotating the leaves against a knife, wire, or brush, or (4) spinning the leaves so that centrifugal force removes the cake.

A filter leaf consists of a heavy screen or grooved plate, which supports on both sides a filter medium of woven fabric or fine wire cloth. Textile fabrics are more common with chemical service, whereas wire screen cloth is more common with filter aid precoat applications. The leaves are supported at their top, bottom, or center, and the filtrate can be discharged from these same locations and manifolded together. The filter medium, whether textile or metal, must be kept as taut as possible during leaf assembly to minimize sagging during cake buildup. Sagging can cause cake cracking or dropping.

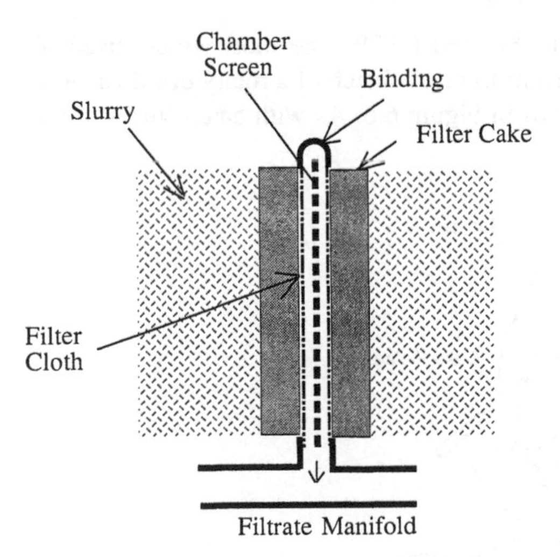

Figure 6.5 Typical filter leaf assembly.

The pressure leaf filter is operated batchwise as far as the solids are concerned. The shell is closed and slurry is introduced in a way that minimizes settling until the shell is full, filtration occurs on the leaf surfaces, and filtrate exits through the filtrate manifold. Filtration is stopped when a cake of the desired thickness has formed, as determined by flow rate or pressure drop. Excessive cake formation causes cake consolidation leading to difficulty in both cake washing and cake discharge. If the cake is to be washed, the unfiltered slurry is drained from the shell at the end of the filtration and the shell is refilled with wash liquid. After washing, the cake is removed by one of a variety of methods depending on filter design.

The advantages of pressure leaf filters are their flexibilty in thickness of cake, their low labor costs, their simplicity, their effectiveness at displacement washing, and the fact that they are totally enclosed, thus limiting exposure to vapors and fumes. Their disadvantages are their requirement for more watchful supervision to avoid cake consolidation or dropping, their inability to form as dry a cake as filter presses, their tendency to form non-uniform cakes unless the leaves rotate, and their pressure limitation to about 75 psig or less.

Continuous Cake Filters

Continuous cake filters are generally applicable when cake formation is fairly rapid. That is, when slurry flow is greater than 1 to 2 gal/min, solids concentration is above 1%, and the solids are greater than 100 μm in diameter. Reasonably low liquid viscosity, below 100 c.p., of the filtrate is also required for maintaining rapid liquid flow through the cake.

Rotary Drum Filter: According to Emmett (1979), the most widely used of the continuous filters is the rotary drum filter. A sketch of a rotary drum vacuum filter with scraper discharge is shown in Figure 6.6. As with batch filters, there

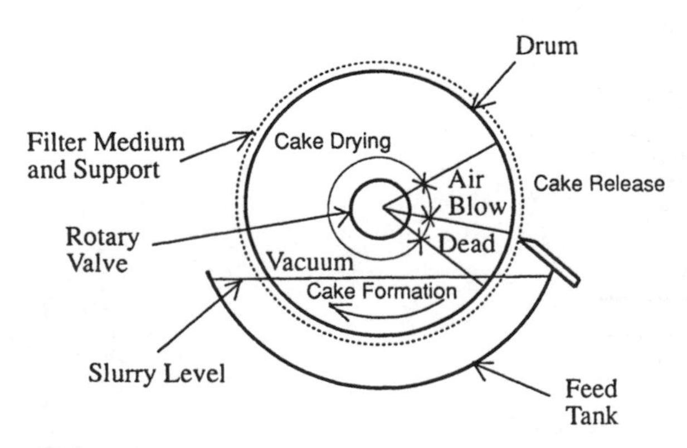

Figure 6.6 Sketch of a rotary drum vacuum filter.

are many different designs available, including both pressure and vacuum operation. However, all of the options use a horizontal axis drum covered on the cylindrical portion by a filter medium. The filter medium is supported on a set of grids which allow drainage of different sections of the drum to different manifolds. Most rotary filters are equipped with a rotary valve in the drum-axis support trunnion to permit removal of filtrate and wash liquid, and to allow introduction of air or gas if needed for cake blowback. The valve controls the relative duration of each part of the filtration cycle as well as any dead time required. Most drum filters are fed by submerging the drum in a slurry trough or filter tank to cover about 35% of its surface. However this can vary from 0 to almost 100% submergence. The trough can be provided with any of a variety of agitators to keep solids in suspension, or the rotation of the drum alone may be adequate to do this.

Most drum filters operate at rotation speeds in the range of 0.1 to 10 rpm. However, they are usually equipped with variable speed drives to allow adjustment for changes in cake formation or drainage rates.

Drum filters are commonly classified by the way cake is discharged from the filter medium, which includes many options.

Rotary Disk Filter: A disk filter is a vacuum filter consisting of a number of vertical disks spaced at intervals on a continuously rotating horizontal hollow central shaft. It is illustrated in Figure 6.7. Each disk is made up of 10 to 30

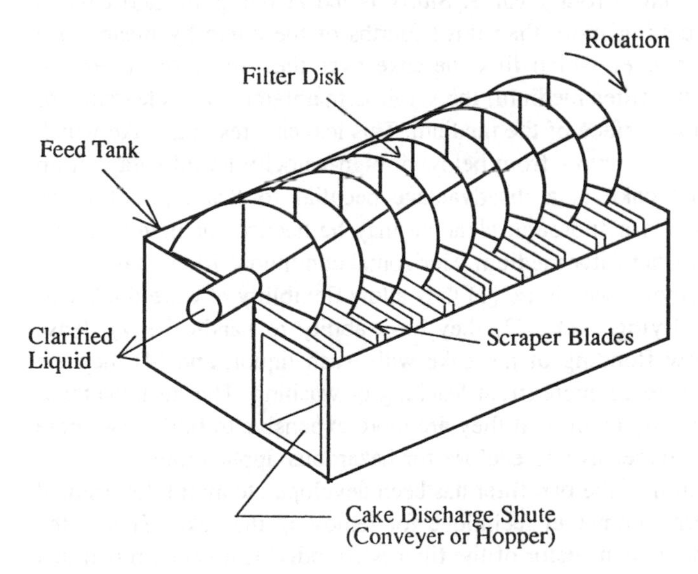

Figure 6.7 Sketch of a rotary disk filter.

sectors ribbed on both sides to support a filter cloth and provide drainage through an outlet nipple into the central shaft. The filter medium is usually a cloth bag slipped over the sectors and sealed to the discharge nipple. Each sector may be replaced individually.

The disks are typically 40 to 45% submerged in a slurry trough which may be agitated to maintain suspension of the solids as with rotary drum filters. Vacuum is applied to the sectors as they rotate into the slurry to allow cake formation and maintained as the sectors emerge from the liquid and are exposed to air. Wash may be applied with sprays, but most applications are for dewatering only. As the sectors rotate to the discharge point, the vacuum is shut off and a slight air blast is used to loosen the cake. This allows scraper blades to direct the cake into discharge shutes located between the disks. Vacuum and air blowback is controlled by an automatic valve as with rotary drum filters.

The vacuum disk filter is the lowest in cost per unit area of all continuous filters, provided no exotic materials of construction are needed. It provides a large filtering area with minimum floor space and is used in large scale dewatering operations with sizes up to 3300 ft^2 of filter area. Its main disadvantages are ineffective wash and difficulty in totally enclosing the filter for hazardous operations.

Table and Pan Filters: These are basically horizontal revolving annular tables with the top surface a filter medium, as illustrated in Figure 6.8. The table is divided into sectors, each of which is a separate compartment, like the sectors of the disk filter. Vacuum is applied through a drainage chamber beneath the table that leads to a large rotary valve. Slurry is fed at one point and cake is removed after completing more than three-fourths of the circle by means of a horizontal scroll conveyer which lifts the cake over the rim of the filter. To prevent damage to the filter medium, the scroll is maintained at a clearance of about 0.4 in. from the surface of the medium. This leaves a residual cake which can be loosened by an air blow from below or high-velocity liquid sprays from above. This residual cake is a disadvantage peculiar to this type of filter. Occasional shutdowns for thorough cleaning may be needed for materials that can cause blinding of the filter medium. Horizontal continuous filters have some major advantages, among which are (1) they allow flexibility of cake thickness, washing time, and drying cycle, (2) they have ability to handle heavy dense solids, (3) they allow flooding of the cake with wash liquor, and (4) they are easily designed for true countercurrent leaching or washing. The disadvantages generic to horizontal filters are that they are more expensive to build, use more floor space, and are expensive to enclose for hazardous applications.

A modified design of the pan filter has been developed to avoid the residual cake left by the scroll conveyer technique for removing the cake. This is the tilting pan filter where each sector of the filter is an individual vacuum pan that

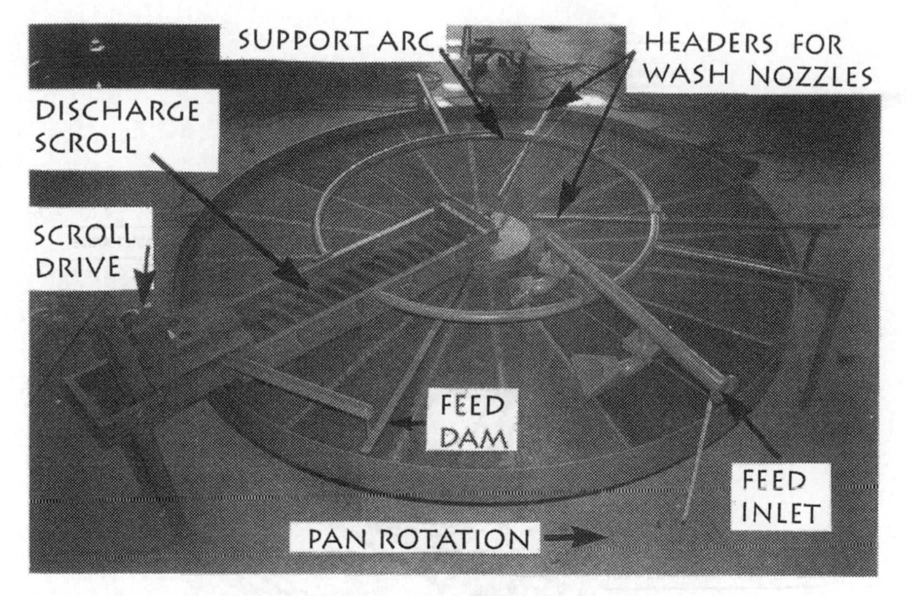

Figure 6.8 Horizontal table filter. (Courtesy of Dorr-Oliver, Inc.)

can be turned over at the cake discharge point and emptied with the help of an air blast. Tilting pan filters have the additional advantages of complete wash containment per sector, good cake discharge, filter medium washing, and availability in large sizes. Their relative disadvantages are higher maintenance cost due to increased mechanical complexity and higher capital costs. Most applications involve dewatering of free-draining inorganic salts.

Horizontal Belt Filter: This filter consists of a filter fabric belt supported by a slotted or perforated elastomer conveyer belt. Both belts are supported by and pass across the surface of a support deck which is compartmented to form separate vacuum chambers to collect filtrate and multiple washes, if desired. An example is illustrated in Figure 6.9.

The belt filter is fed with slurry at one end and the cake is drained, washed, and redrained as the belt moves across the vacuum chambers. The washed and drained cake is then removed at the other end as the belt passes over the end pulley. The filter belt and drainage conveyer belt can be separated over individual pulleys at the end pulley so that the filter belt can be thoroughly washed with sprays before rejoining the conveyer belt at the feed end of the filter. Some horizontal belt filters pass the filter cake between two belts to aid drying by compressing the cake between the belts.

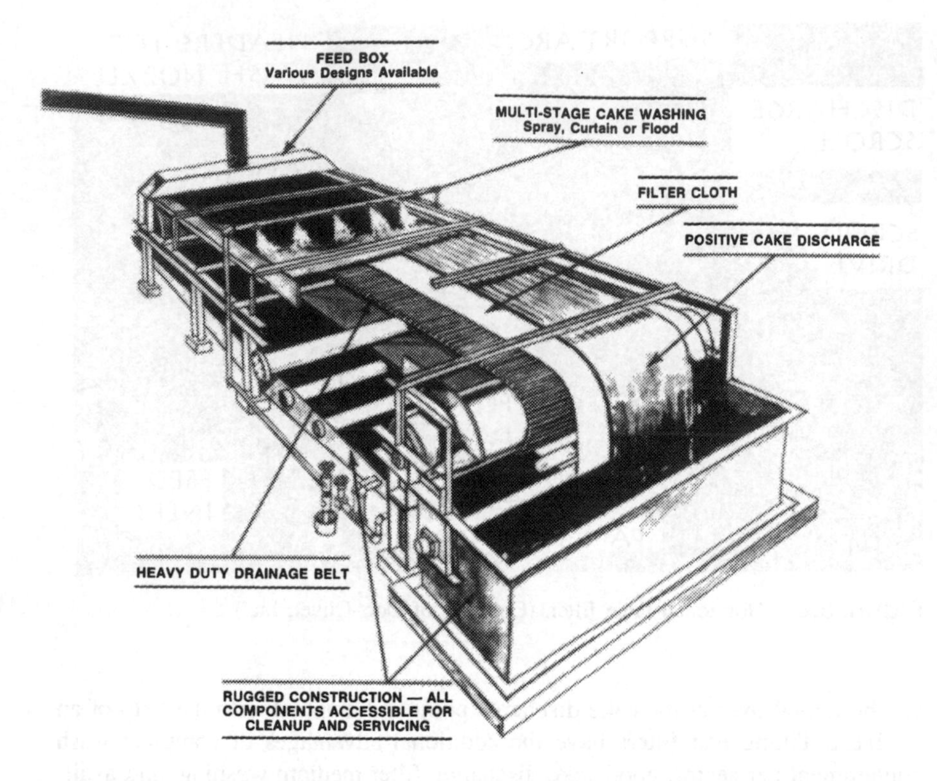

Figure 6.9 Sketch of a horizontal belt filter. (Courtesy of Eimco Process Equipment.)

The advantages of horizontal belt filters are complete cake removal and effective filter medium washing. The major disadvantage is that at least half of the filter medium is idle on the return trip to the feed end of the filter.

Depth Filters

Depth filters can vary from small in-line filters to large open tanks of very large diameter.

Deep Bed Filters: Deep bed filters are strictly for clarification, usually in high-capacity applications such as water treatment. The feed has usually been pretreated by flocculation and gravity settling so that it contains less than 1000 ppm of solids.

The filter bed consists of a filtering medium of granular material such as sand or anthracite coal at least 3 ft deep in a horizontal or vertical tank. A sketch of a horizontal unit is shown in Figure 6.10. The bed is often graded in

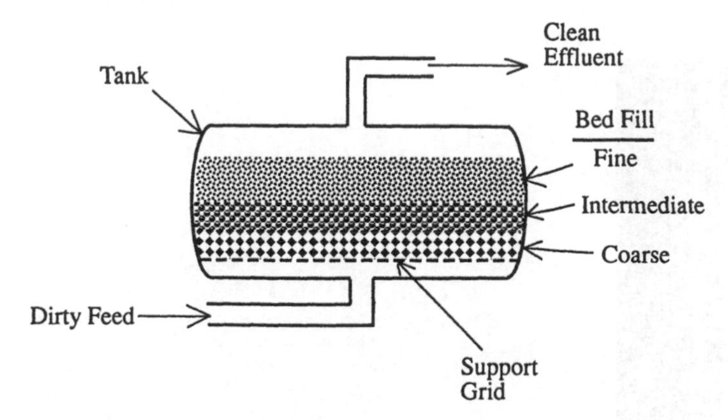

Figure 6.10 Sketch of a deep bed filter.

particle size with the larger particles at the bottom and the smallest ones at the top. The bed is supported on a false perforated vessel bottom or supplied with perforated pipes at the bottom to serve as inlet or filtrate removal devices. Feed can be either upflow or downflow, with upflow designs, giving greater solids handling capacity. In upflow operation, the larger particles in the feed are removed by the coarser particles at the bottom of the bed where there is more interstitial space. This leaves the finer grained solids at the top of the bed only the task of removing the smaller volume of the finest solids in the feed, i.e., polishing the effluent. Downflow operation plugs the finer upper part of the bed more rapidly, requiring more frequent backwashing of the filter bed. Backwash liquid is normally upflow to lift the filtered solids out of the filter, sometimes with an added air sparge for increased agitation. The backwash effluent is much more concentrated than the feed and must usually be further concentrated by more conventional filtration before solids disposal.

Deep bed filters can be large open tanks of concrete, or closed horizontal or vertical vessels that can operate under pressure.

Cartridge Filters: Cartridge filters are units that contain one or more renewable or replaceable cartridges which contain the active filter element. The unit is usually placed in the line carrying the liquid to be clarified. Thus, it is an in-line filter. When the pressure drop across the filter rises to the maximum permitted, the housing must be opened and the cartridge element cleaned or replaced.

Cartridge filters are normally used as polishing filters where the solids content of the feed is less than 0.01% and the solids cake does not have to be handled (Nickolaus, 1979). The particles removed from fluids by cartridge filters vary from submicron sizes to over 40 μm. The filter medium acts as a depth filter in much the same way as a deep bed filter, although on a vastly smaller

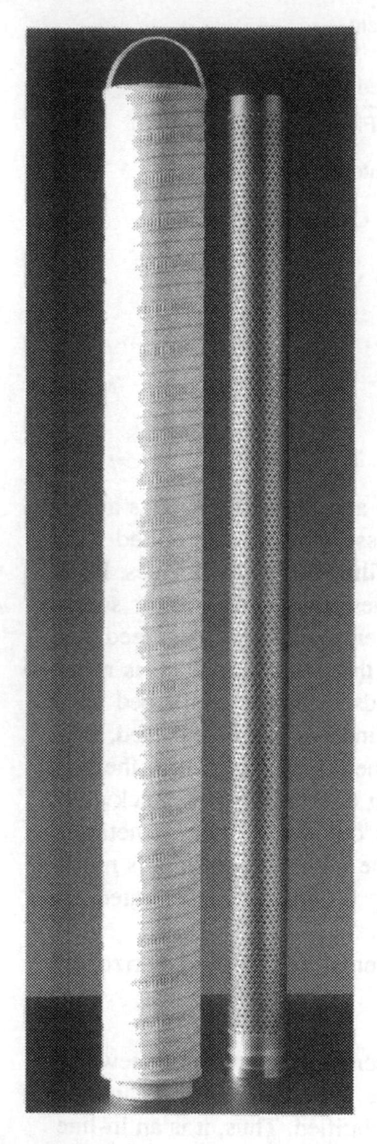

Figure 6.11 Filter cartridge and its metal support. (Courtesy of Pall Corporation.)

scale. That is, the particles are trapped in the interior of the filter medium as the clarified filtrate passes through. It is important to test the chemical compatibility of the cartridges with the slurry medium and the structural compatibility of the cartridge and housing at the temperature and pressure planned for use. Flow rate can be measured experimentally for a single cartridge and scaled up by using the appropriate number of cartridges in parallel. An example of a filter cartridge and its metal support is shown in Figure 6.11.

Cartridge filters can be used to cold sterilize liquid products by removing all the bacteria from the filtrate. However, because this application is so critical, the absolute integrity of the filter and the sterility of both the filter cartridge and housing must be guaranteed. The sterility is achieved by in situ steam sterilization and filter integrity is assured by field testing the filter assembly before and after use.

Bag Filters: Bag filters are usually made out of felt fabrics of various natural or synthetic fibers (Wrotnowski, 1979). The felt serves as a depth medium to trap fine solids from slurries such as paints or used lubricating oils. A variety of particle size ratings from 1 to 200 μm are available for various felt media so that the appropriate bag can be chosen by experimental trial. These ratings are the particle sizes above which all particles are removed by the filter element.

The bag filter can be open with the bag attached to the end of a pipe with a snap ring, as illustrated in Figure 6.12, or enclosed in a housing similar to

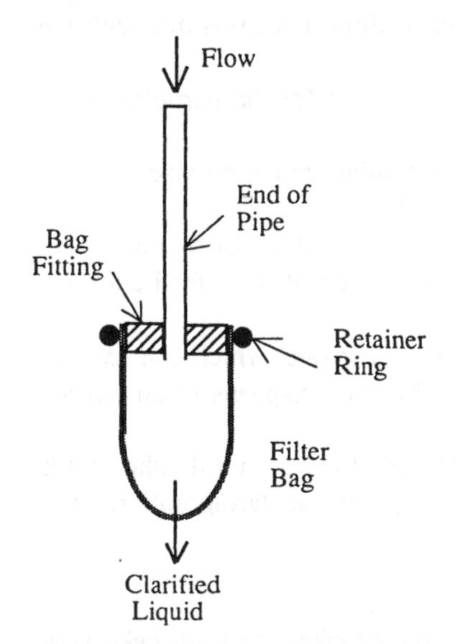

Figure 6.12 Sketch of a simple bag filter.

cartridge filter assemblies. The housing assembly design can be such that the bag is either inside a metal strainer basket which serves as a structural support for the bag or outside the strainer basket, depending on whether feed is inside or outside the bag. In general, bag filters are used for large quantities of liquid containing only small amounts of oversized particles to be removed. This minimizes the frequency of emptying the bags. Where continuous operation is required, several bags in parallel can be used so that one can be emptied while others are on-stream.

Very large-scale bag filters are used for removing fine dusts from gas or vapor streams as an alternative to scrubbing or electrostatic precipitation. A particularly common use is to remove solids from flue gases or ventilation exhausts. In these cases the bag filters are usually final clean-up devices after pretreating with cyclones for bulk removal of solids.

6.2.4 Common Problems and Troubleshooting

The two most basic problems in filtration are unsatisfactory filtrate quality and inadequate filter rate. Some of the causes and remedies for these follow:

Poor filtrate clarity:

- *Cause*: Bypassing of the filter membrane due to membrane failure, leaking seals, etc.
  Remedy: Replace failed membrane(s) and examine all seals for leakage, replacing faulty ones. Check that the filter medium is compatible with the slurry liquid and is not attacked by it.
- *Cause*: Choosing a membrane with too large holes for the particles in the slurry.
  Remedy: Replace membrane with more appropriate retention size.
- *Cause*: Incorrect size or quantity of filter aid.
  Remedy: Filter aid must be matched to the particle size distribution in the feed slurry. It may be necessary to use it as a precoat, as a feed additive, or both.
- *Cause*: Startup with too high a pressure and flow rate, which can extrude particles through the filtration membrane. This often happens when centrifugal pumps are used to feed slurry.
  Remedy: It is better to use a constant reduced flow rate until cake builds up. This is achieved by throttling centrifugal pumps at startup until cake is fully formed.

Low filtration rate:

- *Cause*: Large content of fine particles in slurry makes the filter cake very resistant to liquid flow.

Remedy: Use of filter aid as a body feed can make a more permeable cake. Possibility of increasing particle size by pretreating the slurry should be explored.

- *Cause*: Wrong collection size (too small) of membrane and filter aid, leading to excessive pressure drop and low flow rate.
  Remedy: Sizes must be matched to each other and to the slurry.
- Filter medium is blinding or plugging, which leads to short filtration cycles.
  Remedy: Can be fixed by precoat and/or body feed of filter aid.
- *Cause*: Filter medium incompatible with the carrier liquid of the slurry. Swelling of medium can cause reduction in hole size leading to reduced rates.
  Remedy: Use a compatible filter medium.
- *Cause*: Trapped air in the filter housing can reduce the filter surface in contact with the liquid.
  Remedy: Suitable vents should be provided to allow purge of air, especially on startup.
- *Cause*: Sedimentation of filter aid in filter housing can reduce active filter surface as well as the effectiveness of the filter aid.
  Remedy: A better match of filter aid density with slurry density can reduce sedimentation.

6.2.5 Economics

The estimated cost of a filtering operation must include the installed cost of the filter, its operating life over which costs are amortized, maintenance and operating labor costs, replacement costs for the filter medium, costs of any filter aids used, and any product yield loss occurring during solids recovery.

The cost of the filter alone can vary about 20-fold in cost per square foot of filter area, with the cost increasing as the mechanical complexity of the filter increases. Furthermore, the cost per square foot of filter surface decreases dramatically as size of the filter increases. Hall (1982) has shown that a 100-fold increase in filter size can lead to about a 30-fold reduction in cost per square foot of filter area. To get reliable cost estimates as well as advice on selection and operation of filters, filter vendors should be contacted. Installed costs can be two to three times the purchase price of the filter.

6.3 DRYING

Drying normally involves the removal of a liquid from a solid by evaporation. This is in contrast to filtration where the liquid is removed from the solid by mechanical means. Thus, drying is a thermal process involving heat transfer and vaporization. With direct contact drying, the heat is usually provided by means

Figure 6.13 Static solids bed.

of a hot gas which also acts to sweep the vaporized liquid away from the solid particles. Thus, such drying equipment is basically solids-gas contacting equipment.

Schurr (1984) characterizes drying equipment by the way the solids bed exists in the equipment and how the gas contacts the solids. His definitions are given below and illustrated by the sketches in Figures 6.13 to 6.20. The solids bed can exist as a static bed, a moving bed, a fluidized bed, or a dilute bed.

A *static bed* is a dense bed of particles with the particles resting on each other at the settled bulk density, as illustrated in Figure 6.13. There is no relative motion among the solid particles. This would be typically encountered in a tray dryer where the solids are contained in a tray over which the drying gas flows.

A *moving bed* is a slightly expanded bed where the particles can flow over one another. The flow can be induced either by gravity or by mechanical agitation, as illustrated in Figure 6.14. Mechanical agitation is shown by the rotary kiln in Figure 6.14a and a moving bed using gravity is shown in Figure 6.14b.

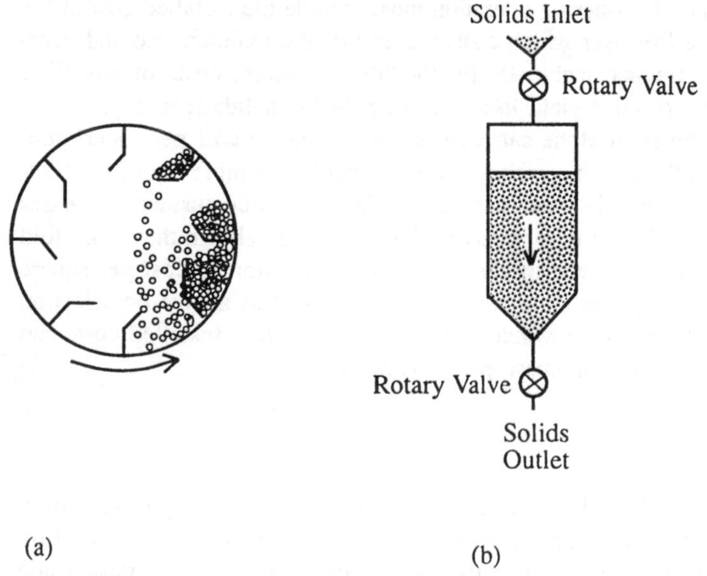

Figure 6.14 Moving solids bed. (a) Rotary kiln, (b) moving bed.

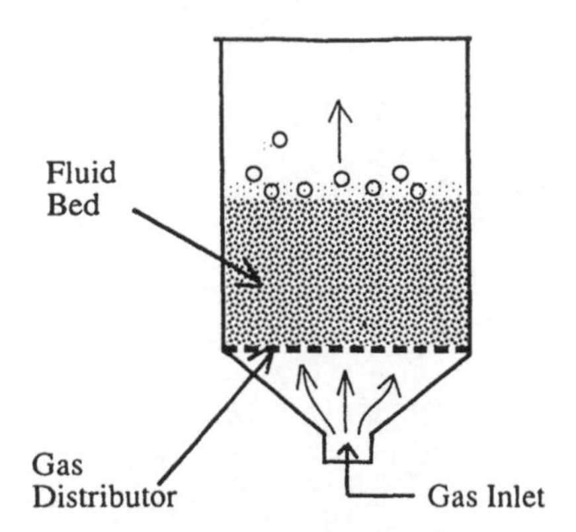

Figure 6.15 Fluidized solids bed.

A *fluidized bed* is in an expanded condition where the upward superficial velocity above the bed is not high enough to carry the particles out of the reactor. However, the higher velocity due to the presence of the solids inside the denser fluidized bed is sufficient to drag the particles upward and greatly expand the bed even though not high enough to carry the particles out of the containment vessel. Thus, the particle motion is one of being lifted by eddies and falling back due to gravity. The critical fluidization velocity must be below the terminal settling velocity of the particles above the bed and higher than the velocity needed to lift the particles in the bed. The result is that the fluidized bed behaves physically like a boiling liquid and has the excellent heat transfer and mixing properties encountered with boiling liquids. A sketch of a fluid bed is shown in Figure 6.15.

A *dilute bed* is in a fully expanded condition where the solid particles are so far apart that they have no influence on each other. Usually, the gas velocity is above the terminal settling velocity of the particles and the particles are carried out of the vessel with the gas. Two notable exceptions to this statement are prilling towers and spray-drying towers, where gravity settling occurs countercurrent to the gas flow. In these cases the dilute bed is achieved by spraying a relatively small amount of slurry or liquid into a large volume of flowing gas. When a dilute bed is used for vapor phase chemical reactions using solid catalysts, the reactor is called a *fast fluid bed* or *transfer line reactor*. A sketch of a dilute bed is shown for a spray dryer in Figure 6.16.

In addition to defining the various ways a bed of solids can exist in a dryer, Schurr (1984) also defined the various ways that a gas may contact a bed of solids. These are parallel flow, perpendicular flow, and through circulation.

Parallel flow occurs when the gas flow is parallel to the surface of the solids bed. The solids bed is usually in a static condition and the contacting is

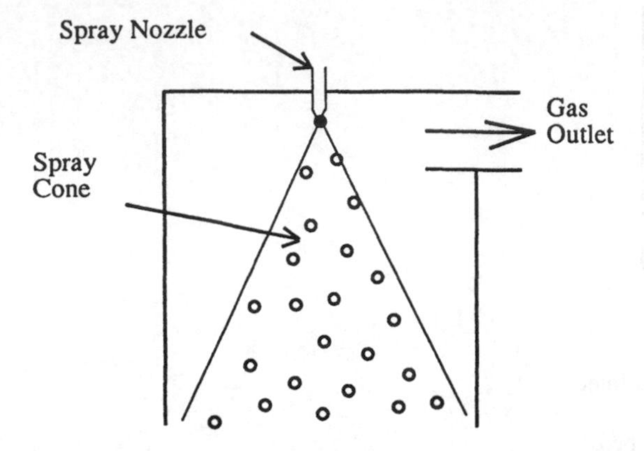

Figure 6.16 Example of a dilute bed in a spray dryer.

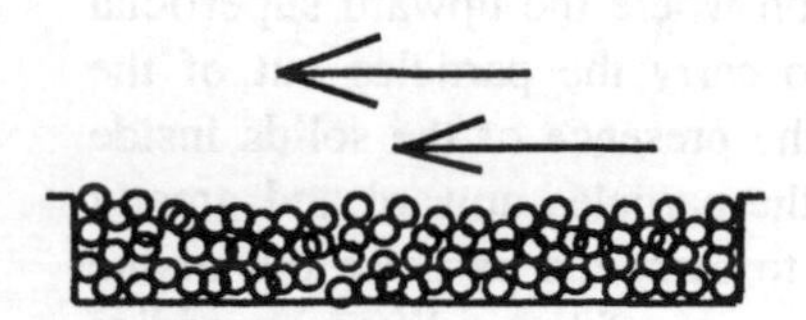

Figure 6.17 Parallel gas flow.

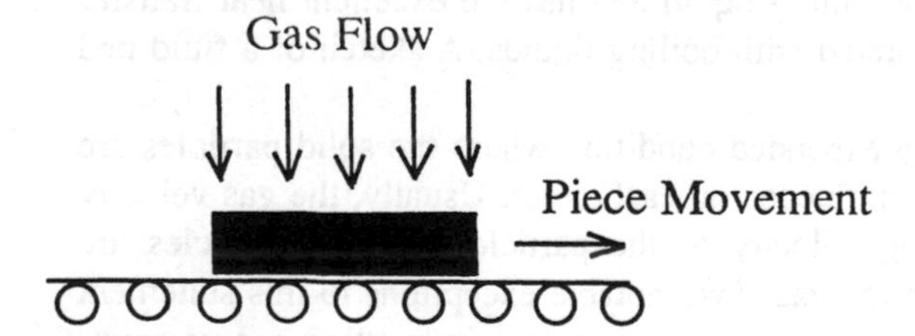

Figure 6.18 Perpendicular impinging gas flow.

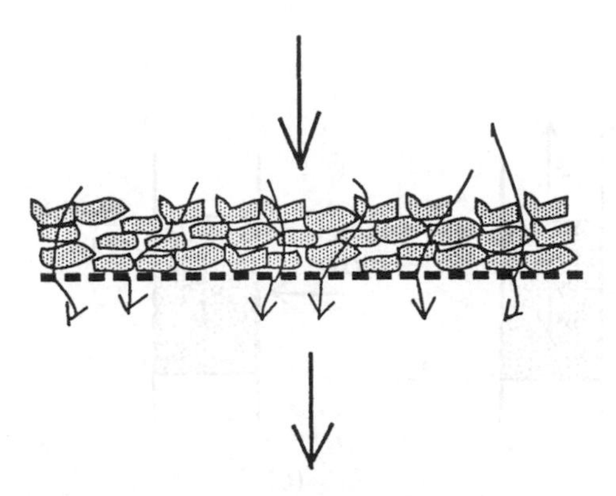

Figure 6.19 Through circulation.

primarily at the interface between the solids and gas. A small amount of gas penetration into the voids among the particles at the surface can occur. Parallel flow is illustrated in Figure 6.17.

Perpendicular flow occurs when the gas flow is perpendicular to the surface of the solids. The solids bed is again usually in the static condition, and the gas impinges on its surface. This is illustrated in Figure 6.18.

Through circulation occurs when the gas penetrates the bed and flows through the interstices among the solids more or less freely around the individual particles. This can take place with solids beds in either the static, moving, fluidized, or dilute condition. This is illustrated in Figure 6.19 for solids on a perforated belt conveyer.

The flow of gas can be either *cocurrent* with the solids, *countercurrent* against the flow of the solids, or *cross-current* perpendicular to the flow of the solids. These three types of gas flow are illustrated in Figure 6.20.

6.3.1 Fundamentals of Drying

The fundamental theory of drying is concerned with getting heat into the wet solids, evaporating the liquid, and getting the resulting vapors out. Thus, it is a problem in both heat transfer and fluid flow. Assuming an adequate supply of heat can be introduced into the wet solids, the temperature and rate at which the liquid vaporizes depends on the vapor concentration in the surrounding atmosphere. If the surrounding atmosphere is pure vapor, the vaporization will take place at the boiling point of the liquid at the ambient pressure. However, if the concentration of the vapor in the surrounding gas is less than 100%, as when an inert purge gas is used, liquid will vaporize at the temperature where the vapor pressure of the liquid equals or is greater than the partial pressure of

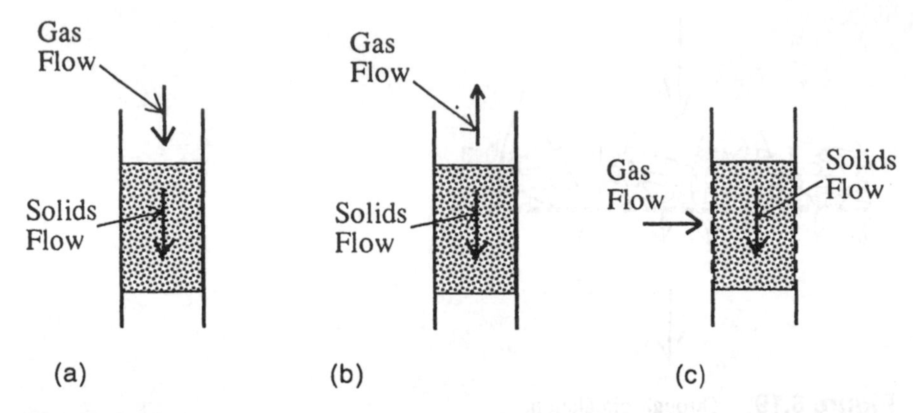

Figure 6.20 Illustration of (a) cocurrent, (b) countercurrent, and (c) cross-flow operation.

the vapor in the purge gas. If the partial pressure of the vapor in the purge gas is greater than the vapor pressure of the liquid, condensation will occur and no drying can be achieved.

In most drying operations water is the liquid being removed and air is the purge gas used. However, the liquid removed can be any of a number of organic solvents, such as benzene, toluene, carbon tetrachloride, etc., and the purge gas can be another inert gas, such as nitrogen, carbon dioxide, etc. In all these cases, the drying temperature can be found using the vapor pressure curves for the liquids being removed. However, for safe operation, care must be taken to match the solvent being removed with a purge gas that is not reactive with it. That is, don't use air to remove combustible or explosive organic liquids. Use an inert gas.

The concept of humidity is widely used in drying calculations and is defined as the weight of vapor that is in a given weight of dry purge gas, that is, the pounds of water per pound of dry air. The use of dry purge gas as a basis gives a constant quantity that can be used as a tracer for the amounts of liquid being removed from the solids under varying conditions. For example, the humidity of the exit gas from the dryer minus the humidity of the feed gas gives the weight of liquid removed per weight of dry purge gas. The saturation humidity is the maximum weight that can be carried at a given temperature and total pressure, i.e., the purge gas is saturated with vapor. The humidity is calculated from the vapor pressure of the liquid, the total pressure, and the molecular weights of the evaporating liquid and purge gas as shown in Eq. 4 for water and air.

$$H = \frac{p_w}{18} \bigg/ \frac{p_a}{28.9} \tag{4}$$

where:

H = humidity (pounds water/pounds dry air)
p_w = partial pressure of water
p_a = partial pressure of air (18 and 28.9 are the molecular weights of water and air

If we let p be the partial pressure of the vapor and P be the total pressure, then the partial pressure of the dry purge gas is $(P-p)$. Using these values, the humidity equation for water in air becomes:

$$H = \frac{p}{(P-p)} \frac{28.9}{18} \tag{5}$$

To calculate the humidity for other systems, we simply use the partial pressure of the liquid, the total pressure, and the molecular weights of the actual liquid and purge gas used. To get the saturation humidity, we simply use the liquid vapor pressure for p in Eq. 5. Furthermore, the percent relative humidity is simply the partial pressure of the vapor in the purge gas over the vapor pressure at the same temperature times 100. That is, the percentage of the saturation value provided by the actual value.

The humidity can be used along with temperatures in a heat balancc to determine how much moisture will be evaporated if all the heat is used to evaporate the moisture, that is, if the process is adiabatic and no other heat losses are encountered. This heat balance is illustrated by Eq. 6.

$$H_s - H = \frac{C(T-T_s)}{HV} \tag{6}$$

where:

H_s = saturation humidity at the surface temperature
H = humidity of the purge gas
C = heat capacity of gas per weight of dry gas
T = temperature of gas
T_s = temperature of solid
HV = heat of vaporization at T_s

The left side of the equation is the amount of liquid evaporated per unit weight of dry gas based on gas humidity change. The right side is the same value based on heat input and heat of vaporization of the liquid.

Humidity or psychrometric charts giving the properties of some vapor mixtures in air versus temperature are given by Schurr (1984). These charts are useful because they give the relationships among saturated humidities, relative humidities, heat capacities, and temperatures for water, benzene, toluene, and carbon tetrachloride using air as a purge gas.

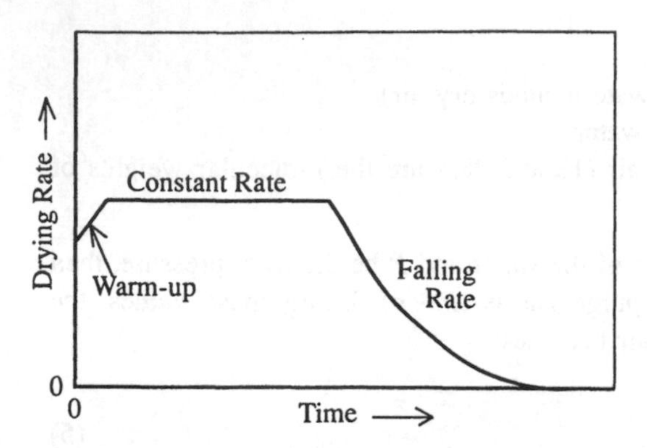

Figure 6.21 Drying rate versus time.

Drying Rates

Moisture removal does not take place at a constant rate throughout the drying process. This is because the mechanism of drying changes as the drying progresses. Figure 6.21 shows a typical plot of drying rate versus time. In the beginning, the rate increases as the wet solids warm up, then the rate remains constant as moisture evaporates from the saturated surface of the solid, and finally the rate falls as moisture evaporates from the interior of the solids. These three different regions are shown in Figure 6.21. The point where the constant rate region changes to the falling rate region is called the critical moisture content.

Constant Rate Region: In the constant rate region of drying, the moisture movement in the solid is fast enough to keep the surface saturated. Thus, the drying rate is controlled by the constant rate of heat transfer to the solid surface. The vapors evaporate from the solid surface and diffuse to the surrounding environment where they may be swept away by a purge gas. The value for the constant rate depends on the heat transfer rate, the area in contact with the purge gas, and the driving forces of differences in temperature and humidity between the gas stream and the wet solid. None of these factors depends on internal flow inside the solid particle and thus the constant rate period is independent of internal flow mechanism.

Falling Rate Region: The falling rate period begins when part of the solid surface is no longer saturated. At this point the drying rate begins to depend on the mechanism, and thus the rate, of internal moisture movement inside the solid. This can be by diffusion in homogeneous solids or capillary flow in porous solids, for example. Under these conditions the effect of heat transfer rate is greatly reduced and evaporation rate is almost completely controlled by internal flow rates. If the solids contain internal soluble liquid and must be dried to low levels of moisture content, the falling rate region controls overall drying time.

The entire drying time can be in the falling rate region if all the moisture is dissolved in the solids as, for example, with solvents dissolved in plastics. On the other hand, for non-porous solids in which there is no moisture solubility, the entire drying region can be in the constant rate region. There is no reliable method for calculating the critical moisture content at which the drying rates change from constant rate to falling rate since it is so dependent on the specific material. However, some approximate values for air drying of various materials are given by Schurr (1984). These values vary from gelatin at 300% on a dry basis to various inorganic crystals and powders below 10%.

6.3.2 Specific Applications

Drying is widely used in industry for the preparation of solid powders, granules, sheets, and large ceramic or painted objects. The specific industries involved are food, pharmaceuticals, paper, lumber, minerals, plastics, pottery, leather, rubber, chemicals, and others. The drying operation can be carried out either to remove the liquid to enhance the quality of the solid or to recover the liquid for its own value, and frequently both. For example, volatile organic solvents used in the processing of chemicals or pharmaceuticals need to be removed to eliminate contamination of the solid product. However, these solvents also need to be recovered both to prevent air pollution and for their value as recycled products. Because of the very large number of drying applications in existence, only a few related to the environment will be discussed here.

Cleaning Contaminated Soil

Soils contaminated with volatile organic contaminants (VOCs) such as solvents, oils, and fuels can be cleaned by drying. This is done by passing the wet solids either through a totally enclosed dryer operating under vacuum or through a rotary dryer-calciner using combustion gases at atmospheric pressure. The use of the vacuum dryer permits removal of higher boiling materials and also allows recovery of the VOCs removed. On the other hand, if the combustibles in the contaminated soil are expendable, they can be burned in the dryer-calciner to provide part of the heat needed for drying, while at the same time removing them from the vent gas.

Drying Waste Sludges

Sludges recovered from sedimentation, filtration, or centrifugation can be further dried using flash drying with disintegration. In this type of operation a flash dryer of the type sketched in Figure 6.22 is equipped with a cage mill to break up the wet sludge. Wet feed and recycled dry powder, as needed, are premixed

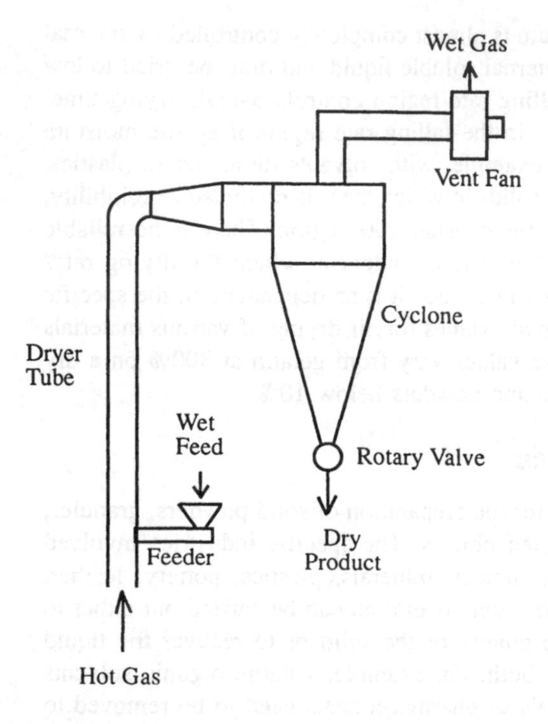

Figure 6.22 Sketch of a simple flash dryer.

in the feeder and then fed into the hot gases in the cage mill. They then pass into the drying column and up to the cyclone. Typical feedstocks might be sewage sludge, paper mill sludge, and wet filter or centrifuge cakes from pretreating industrial wastes.

Recovering Solvents

Vacuum dryers are commonly used to recover solvents from solids because of their effectiveness for complete removal of solvent and the ease of collecting the recovered solvent by condensation after compression by the vacuum pump. A typical batch operation would use a vacuum chamber in which the wet solid is spread in a series of shallow trays spaced vertically in the chamber. When drying is complete, the solids are recovered by manually emptying the trays. For larger batches, a vacuum rotary drum dryer can be used with indirect heating. Again the liquid can be recovered by condensation.

For continuous operation, screw-type indirectly heated units can be used either under vacuum or at atmospheric pressure. These units feed the wet solids at one end and propel them through the totally enclosed trough-like vessel by means of a heated screw conveyer. The solvent is recovered by condensation after compression.

These techniques are used to regenerate adsorbents used to capture organic solvents from exhaust streams after they can no longer be regenerated by their normal in situ techniques. Such exhausts are generated in paint shops, print shops, dry cleaning establishments, and chemical manufacturing plants.

Pretreating Dryer Feeds

Flash drying is the introduction of wet solids into a hot flowing carrier gas so that surface moisture is rapidly removed in a short retention time prior to collecting the solids in a cyclone. A typical flash drying layout is sketched in Figure 6.22.

Flash dryers are often used as pretreaters integrated with other types of dryers or calciners. When used with either direct- or indirect-heated rotary calciners, the flash dryer can improve overall thermal efficiency by using the exhaust gases from the calciner as the hot gas in the flash dryer. Furthermore, flash drying can be used as a feed pretreater to a fluid bed drying operation. This is especially helpful when the feed contains a lot of moisture but would be damaged by the high temperature necessary to remove all the moisture in the short time available in a flash dryer. By using a flash dryer to remove a lot of the moisture without overheating, followed by the fluid bed with its better temperature control, the efficiency of the overall drying operation can be increased without risk of product damage.

6.3.3 Equipment

Drying equipment can be classed as direct or indirect depending on how the heat is supplied to the wet solids. Subclassifications then include batch or continuous operation and specific mechanical design features. Such designs are classified as tray, tunnel, rotary, spray, flash, through-circulation, and fluid bed units. Each of these mechanical designs has been generated to handle a specific type of feed. However, the merits of the different dryers when applied to drying of various types of materials are given by Schurr (1984) and by Wentz and Thygeson (1979) and will be touched on only briefly here.

Direct Dryers

The direct dryers use direct contact between the wet solids and the hot gases to transfer heat. Temperatures can be up to about 1300°F, limited by the materials of construction and the thermal stability of the liquid being removed. The vaporized liquid is carried away by the hot gases, which also serve as the purge gas. These dryers are also frequently called *convection dryers*.

The moisture content of the feed gas is very important when drying solids at gas temperatures below the boiling point of the liquid being removed. To get

Table 6.2 Vendors of Drying Equipment

ABB Raymond 650 Warrenville Rd. Lisle, IL 60532-0434 (708) 971-2500	Andritz Ruthner 1010 Commercial Blvd. S. Arlington, TX 76017-7130 (817) 465-5611
APV Crepaco, Dryer Division 165 John L. Dietch Sq. Attleboro Falls, MA 02763 (508) 695-7014	Bepex Corporation Three Crossroads of Commerce Rolling Meadows, IL 60008 (708) 506-0100
Bethlehem Corp. PO Box 338 Roosevelt, NJ 08555-0338 (609) 443-4545	Fuller Company 2040 Avenue C, LVIP Box 2040 Lehigh Valley, PA 18001-2040 (215) 264-6765
Komline Sanderson Holland Ave. Peapack, NJ 07977 (908) 234-9487	Littleford Bros., Inc. 7451 Empire Drive Florence, KY 41042-2985 (606) 525-7600
MEC Company 1400 W. Main PO Box 330 Neodesha, KS 66757 (316) 325-2673	Processall, Inc. 10596 Springfield Pike Cincinnati, OH 45215 (513) 771-2266
Stokes Vacuum, Inc. 5500 Tabor Rd. Philadelphia, PA 19120 (215) 831-5400	Swenson Process Equip. Inc. 15700 S. Lathrop Ave. Harvey, IL 60426 (708) 331-5500
Universal Process Equip. Inc. Box 338 Roosevelt, NJ 08555 (800) 243-6228	Wyssmont Co. P.O. Box 1397 Fort Lee, NJ 07024 (201) 947-4600

rapid drying, it may be necessary to dehumidify the feed gas or use a dry inert gas as a heating medium. At temperatures above the boiling point of the liquid, superheated vapors of the liquid being removed can be used effectively for drying.

Because of the direct contact between the gas and the solids, as well as the large amounts of gas needed, it may be necessary to provide dust recovery equipment when drying fine particles. This greatly increases the cost of the installation.

Indirect Dryers

With indirect dryers the heat is transferred to the wet solids through a hot surface, usually made of metal. The vaporized liquid does not come in contact with the heating medium and is removed by other means, such as vacuum or inert stripping gas. The rate of drying depends on the contact between the wet solids and the hot surface. These dryers are often called *conduction* or *contact dryers*.

The surface temperatures of indirect dryers can vary from below freezing for freeze-drying of foods to above 1000°F. They use either condensing vapors or hot combustion products to provide this heat. The use of condensing vapors is usually less expensive in heat consumption because they provide only the heat required to vaporize the liquid. Indirect dryers are also effective when using vacuum or inert atmospheres in the recovery of solvents.

Table 6.2 gives the names of some vendors for drying equipment. A vendor should be contacted for help in choosing the right dryer for a given application and in getting reliable cost estimates.

6.3.4 Common Problems and Troubleshooting

Drying is generally the last wet processing step in the processing of solids. Therefore, the dryer must consistently produce a product that is conveyable, storable, and meets the necessary quality standards. To do this, the dryer must be appropriate for the task, it must be satisfactorily controlled, and adequate heat and mass transfer must be achievable.

The problems most frequently encountered with dryers are reliability, capacity, and product quality. These are discussed in order below, along with a few brief comments on safety.

Reliability

Reliability is the ability to consistently produce a satisfactory product at the required feed rate. When this is not achieved, the first place to look is the basic drying characteristics of the feed. Is the feed changing from time to time, and if so, how? Are upstream operations capable of making a consistent feed? If not, can they be economically modified to make a more consistent feed? If the answer is no, can better dryer control handle the variation in feed properties? And finally, if the answer is again no, it may be necessary to use a different type of dryer more suited to the variation in feed quality.

Capacity

Inadequate capacity results from an inadequate heat supply to the solids. This is caused by too low a temperature, too small a heat transfer area, or too low a heat transfer coefficient. Fouling of the heat transfer surface due to caking of the solids on the surface can cause loss of effective heat transfer area as well as reduction in the heat transfer coefficient. Furthermore, lumps of cake are slower to dry than free flowing particles. It may be necessary to use some mechanical agitation during drying to break up cakes and lumps. It may also be possible to modify caking character by recycling some dried product to the feed.

Product Quality

Product quality, as far as moisture content is concerned, can be assured by adequate residence time and temperature in the dryer. However, many solids are temperature sensitive and can be scorched, melted, or otherwise degraded if the dryer is pushed too hard and the solids are overheated. To assure adequate moisture removal without product degradation, a good control sytem is necessary. With the advent of good moisture detectors and inexpensive computers, it is now possible to achieve good control. Combined with the choice of a proper dryer, such control now makes it much easier to dry sensitive materials.

Good dryer control also makes it possible to dry closer to the final moisture content desired without having to overdry to assure reaching this level. This can have a major impact on operating costs by both shortening drying time and decreasing the amount of moisture removed.

Safety

Care must be taken to avoid fires and explosions when drying materials containing flammable solvents. This could include the use of an inert gas in direct-contact dryers and explosion-proof electrical fittings, as well as control of ignition sources by proper grounding. Furthermore, adequate prevention of fugitive toxic dusts and vapors is a must.

6.3.5 Economics

As with most other processes, the cost of the equipment for drying depends on the size and complexity of the operation. However, the operating cost for drying also strongly depends on how much moisture must be removed as well as its heat of vaporization. For example, it takes seven times as much heat to evaporate a pound of water as it does a pound of a hydrocarbon like gasoline. Also, the way the heat is generated and used greatly affects costs. Because the theoretical treatment of the heat and mass transfer balance in dryers has not been reliably worked out, the only way to get reliable cost data for various drying options is from the drying equipment vendors.

6.4 SUMMARY

Filtration and drying are related processes in that filtration removes solids from liquids and gases while drying removes liquids from solids. Filtration is primarily a sieving process where the solids are mechanically removed from the fluid by capturing them on a filter medium while allowing the filtrate to flow through. Drying, on the other hand, is a thermal process where the liquid is removed by vaporization. Thus, it is limited by the heat and mass transfer characteristics of the drying equipment. The two processes are also complementary in that filtration is commonly used as a pretreating step to concentrate solids in slurries before feeding them to a dryer.

Filtration is commonly used for treating process wastes to remove solid contaminants from liquids or exhaust gases. The clarified liquids can then be either recycled or disposed of safely and the gases can be vented to the atmosphere. Filtration is also widely used in the manufacture of solid products such as chemicals and pharmaceuticals to recover them from process solvents or dryer gases. These process fluids can also often be recycled.

Drying is frequently used for waste minimization to treat sludges and filter cakes. This is done to reduce their volume, and thus reduce the amount needing disposal, or to remove and recover volatile organic solvents. Drying is also an important process step in the manufacture of a wide variety of solid products, from food and drugs to construction materials. In these cases, drying is used to improve the quality of the product as well as to recover valuable solvents, when applicable.

The theory of filtration and drying has not yet been developed to the point of reliable design for all feedstocks. Therefore, the advice of experienced practitioners or vendors should be sought in planning equipment purchases in these areas.

REFERENCES

Cain, C. W. (1979a). Filter Aid Filtration, *Handbook of Separation Techniques for Chemical Engineers* (P. A. Schweitzer, ed.), McGraw-Hill, New York, pp. 4-9 to 4-14.

Cain, C. W. (1979b). Filtration Theory, *Handbook of Separation Techniques for Chemical Engineers* (P. A. Schweitzer, ed.), McGraw-Hill, New York, pp. 4-3 to 4-8.

Carman, P. C. (1938). Fundamental Principles of Industrial Filtration. *Trans. Inst. Chem. Eng. (London)*, *16*:171.

Emmett, R. C., Jr. (1979). Continuous Filtration, *Handbook of Separation Techniques for Chemical Engineers* (P. A. Schweitzer, ed.), McGraw-Hill, New York, pp. 4-35 to 4-54.

Hall, R. S. and Associates (1982). Current Costs of Process Equipment, *Chemical Engineering, 89(7)*:108–115.

Jacobs, L. J. (1984). Filtration, *Perry's Chemical Engineers' Handbook*, 6th edition, (R. H. Perry and D. W. Green, eds.), McGraw-Hill, New York, pp. 19-65 to 19-89.

Nickolaus, N. (1979). Cartridge Filtration, *Handbook of Separation Techniques for Chemical Engineers* (P. A. Schweitzer, ed.), McGraw-Hill, New York, pp. 4-85 to 4-93.

Perry, R. H. and Green D. W. (eds.) (1984). *Perry's Chemical Engineers Handbook*, 6th ed., McGraw-Hill, New York.

Schurr, G. A. (1984), Solids Drying, *Perry's Chemical Engineers' Handbook*, 6th edition, (R. H. Perry and D. W. Green, eds.) McGraw-Hill, New York.

Schweitzer, P. A. (ed.) (1979). *Handbook of Separation Techniques for Chemical Engineers*, McGraw-Hill, New York.

Tiller, F. M., and Crump, J. R. (1977), Solid–Liquid Separation, *Chemical Engineering Progress, 73(10)*:65.

Wentz, T. H. and Thygeson J. R. (1979). Drying of wet solids, *Handbook of Separation Techniques for Chemical Engineers* (P. A. Schweitzer, ed.), McGraw-Hill, New York, pp. 4-144 to 4-184.

Wrotnowski, A. C. (1979). Felt strainer bags, *Handbook of Separation Techniques for Chemical Engineers* (P. A. Schweitzer, ed.), McGraw-Hill, New York, pp. 4-95 to 4-111.

Zievers, J. F. (1979). Batch filtration, *Handbook of Separation Techniques for Chemical Engineers* (P. A. Schweitzer, ed.), McGraw-Hill, New York, pp. 4-15 to 4-34.

7

Centrifuging and Cycloning

7.1 INTRODUCTION

Centrifuging and cycloning are in principle very much like gravity settling, but with the force of gravity increased many times using centrifugal force. They can be used to separate solid and liquid particles from gases, immiscible liquids from each other, and solid particles from liquids. Thus, they find frequent use in applications where gravity settling would be considered but would take too long. Such applications involve either low density difference between phases to be separated, small particle size of the dispersed phase, or high viscosity of the continuous phase. The multiplication of gravity by centrifugal force makes these applications possible in a reasonable length of time.

Centrifuges are mechanical devices that generate centrifugal force by rapid spinning about either a vertical or horizontal axis. They can be operated either batchwise or continuously, with batch units usually spinning around the vertical axis. Small laboratory amounts of materials are usually separated in batch units, while both batch and continuous units are used for production-scale separations. High-speed centrifuges are capable of generating centrifugal forces greater than 20,000 times the force of gravity (g) although a level of 1000 to 10,000 g is more common for general applications. Thus, the force driving the separation is 1000 to 10,000 times that of gravity settling. This should theoretically increase the rate of the separation by the same magnitude.

In contrast to centrifuges, which generate the centrifugal force mechanically, cyclones convert the kinetic energy of fluid flow into centrifugal force. They do this by diverting the straight-line flow path of the fluid into rotating motion in a cylindrical vessel. This can generate a separating force up to about 2500 g, although much lower values are more common. This centrifugal force causes the higher density fluid or particles to migrate toward the outer wall of the vessel and the lighter fluid or particles to migrate toward the center of the vessel. Suitable take-off devices are then provided to collect the separated streams. Because fluid flow is used to generate the centrifugal force, cyclones are always operated with continuous flow. If the continuous phase is a liquid, the cyclone may be called a hydrocyclone or hydroclone.

7.2 CENTRIFUGING

This discussion on centrifuging is limited to the removal of particles from a carrier liquid. The particles may be liquid or solid or both. The type of machine chosen for the separation depends on the number and kinds of phases in the feed to be separated and on the physical properties of these phases. That is, density, viscosity, and particle size as well as total particulate content of the feed are all factors. There are two major classes of centrifuges: *sedimentation*, or solid bowl units, and *filtration*, or perforated bowl units. The sedimentation centrifuges use the centrifugal force to speed up the settling of particulates from a liquid carrier stream. The filtration units use the centrifugal force to drive the carrier liquid through a filter in the centrifuge and thus increase filtration rate. Ambler (1979) has presented an excellent discussion of the various types of centrifuges along with their application to a variety of separations.

Gas centrifuges, used in the nuclear industry for separating gaseous isotope compounds, will not be discussed in this chapter. However, the separation of liquid or solid particles from gaseous streams will be discussed in Section 7.3, "Cycloning."

7.2.1 Fundamentals of Centrifuging

A sedimentation centrifuge is basically a gravity settler with an extremely large gravitational force imposed, instead of the value of 1 provided by the earth. Figure 7.1 illustrates a simple batch sedimentation centrifuge. A slurry of higher density solids in a liquid is placed inside the solid centrifuge bowl, and the centrifuge is started. The large centrifugal force generated by the spinning

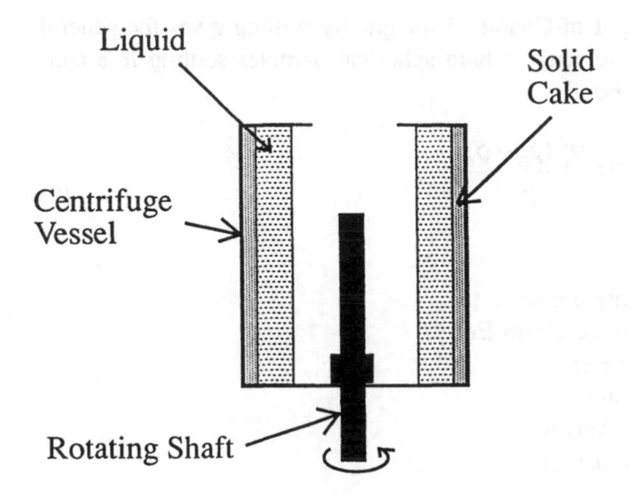

Figure 7.1 Sketch of simple batch sedimentation centrifuge.

centrifuge moves the denser solid particles to the outer walls of the bowl where they collect as a cake. The less dense liquid phase moves to the center of the bowl. After the separation has been made the centrifuge is stopped, the liquid phase is drained off, and the solid cake is removed manually. Continuous feed centrifuges, which are much more complex than this, are available and are discussed in Section 7.2.2, "Equipment."

The large gravitational force generated by the rotating centrifuge is proportional to the radius of the centrifuge bowl and the square of the revolutions per minute. Thus, the force of gravity in a centrifuge is shown in Eq. 1 below:

$$G = 1.43 \times 10^{-5} n^2 D \tag{1}$$

where:

G = centrifugal force in multiples of gravity
n = centrifuge rpm
D = diameter of centrifuge bowl in inches

Using this equation, a centrifuge with a 6-in. bowl running at 5000 rpm would theoretically generate a centrifugal force, G, of:

$$G = 1.43 \times 10^{-5} \times 25 \times 10^6 \times 6$$

$$= 2145$$

Substituting Eq. 1 into Eq. 1 of Chapter 5 for gravity settling gives the general equation for the terminal velocity of hard spherical particles settling in a centrifuge. This is shown as Eq. 2.

$$\text{Terminal velocity} = K'G\,\frac{D_p^2\,(\rho_p - \rho)}{\mu} \tag{2}$$

where:

K' = proportionality constant
G = centrifugal force (from Eq. 1)
D_p = particle diameter
ρ_p = particle density
ρ = carrier fluid density
μ = carrier fluid viscosity

The value of K' includes conversion factors for units and efficiency factors dependent on centrifuge design variables.

Comparison of Eq. 2 and Eq. 1 of Chapter 5 shows that the terminal velocity for centrifugation is G times the terminal velocity for gravity settling with some adjustment to K' for the differences in equipment characteristics. However, unlike gravity settling, the sedimenting particles in a centrifuge almost never reach the exact terminal velocity. This is because the particles are moving from the center of the centrifuge bowl to the outside wall and are in a region of increasing centrifugal force, i.e., increasing terminal velocity, throughout the entire settling time. Moir (1988) has noted that actual settling velocities are often only 50% of the calculated theoretical values.

A filtration centrifuge, as its name implies, uses centrifugal force to generate flow of a liquid through a bed of filtered solids. The solids are supported by a filtration barrier, which may be wire mesh or a supported cloth or paper membrane. Just as a sedimentation centrifuge is an enhanced gravity settler, a filtration centrifuge is an enhanced gravity filter, using centrifugal force in the place of gravity. The pressure drop across the filtration membrane is greatly increased by the centrifugal force, thereby increasing the filtration rate. A sketch of a basket-type filtration centrifuge is shown in Figure 7.2.

The slurry feed is introduced inside a porous basket spinning on either a vertical or horizontal axis. The filter bed is on the inside walls of the basket. The particles in the feed are filtered out as the feed passes through the filter bed under the influence of the centrifugal force. If necessary, filter aids can be added to the feed or precoated on the filter to improve the rate of filtration. A filtration centrifuge is very effective when the recovered solids require washing during recovery. In such cases, the slurry is fed continuously until the filter cake is fully developed; feed is then stopped and the cake is spun dry. Then

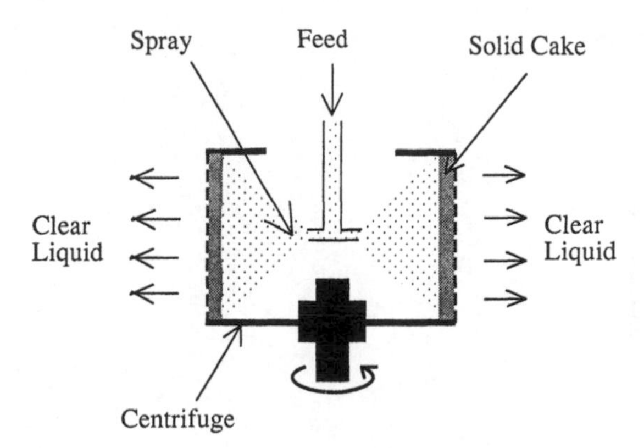

Figure 7.2 Sketch of basket-type filtration centrifuge.

the cake is washed, spun dry again, and removed before starting a new cycle. With a sedimentation centrifuge it is usually very difficult or impossible to wash the sedimented solid cake.

The operation of a filtration centrifuge is usually continuous for liquid filtrate, and either batch or continuous for solid filter cake, depending on the solids content of the feed slurry. The different designs are discussed in the following section.

7.2.2 Equipment

Sedimentation Centrifuges

There are a variety of centrifuge designs in both the sedimentation and filtration categories. They are distinguished by the manner in which the solids are removed from the centrifuge bowl. In the sedimentation or solid-bowl category the types are:

- Solids-retaining bowl
- Solids-ejecting bowl
- Nozzle bowl
- Decanter bowl

The *solids-retaining* solid-bowl centrifuges are illustrated in Figure 7.3. To shorten the settling distance as well as to increase the settling area, inserts have been installed in the bowls. The bowl shown in Figure 7.3a uses a series of conical disks to provide surface for the solids to collect underneath, and flow is toward the outer surface of the bowl. (The liquid flow direction is indicated by the arrows.) Figure 7.3b shows a bowl where the inserts are concentric cylinders where the solids collect on the inside vertical walls of the inserts as the slurry

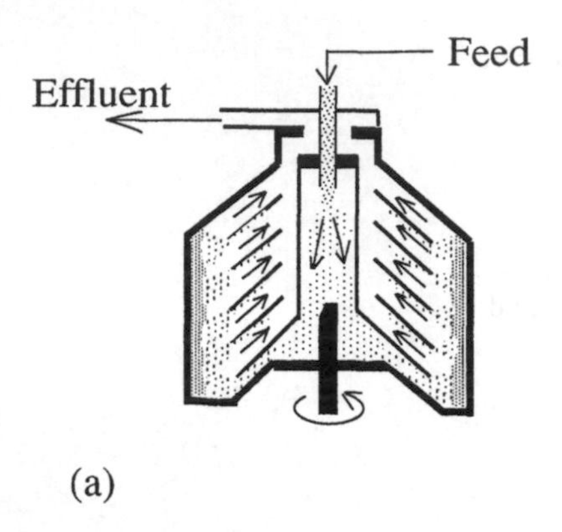

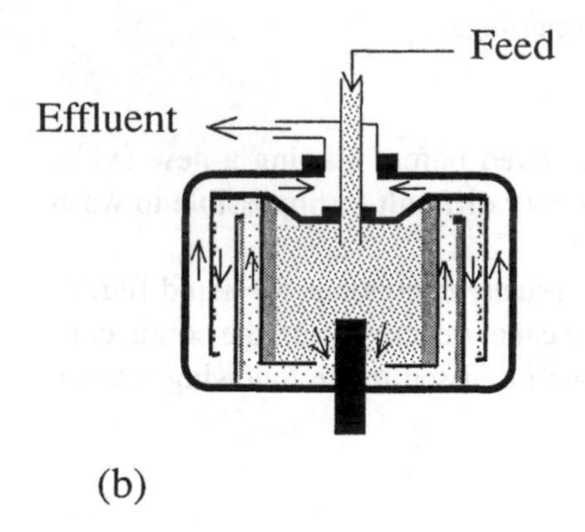

Figure 7.3 Example of solids-retaining solid bowl centrifuge. (a) Conical disk bowl, (b) cylinder insert bowl.

flows through the inserts in series. In both these units the centrifuge must be stopped for solids removal, performed manually. Clearly, the cycle time depends on the solids content of the feed slurry and is longer for more dilute slurries.

The *solids-ejecting* solid-bowl centrifuge is illustrated in Figure 7.4. The bowl is much like a clam shell, with the solids collecting at the outermost periphery where the upper and lower halves meet. Conical disks are also provided to improve the separation. The solids collect at the periphery of the bowl and are released periodically by a very brief lowering of the lower half, which opens an outlet at the periphery. The outlet is only open for a fraction of a second. The frequency of opening depends on the solids content of the feed slurry. Figure 7.4a shows the solids outlet open; while Figure 7.4b shows it closed.

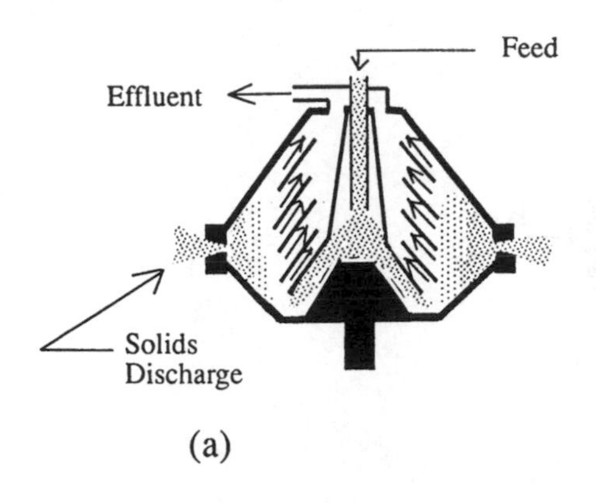

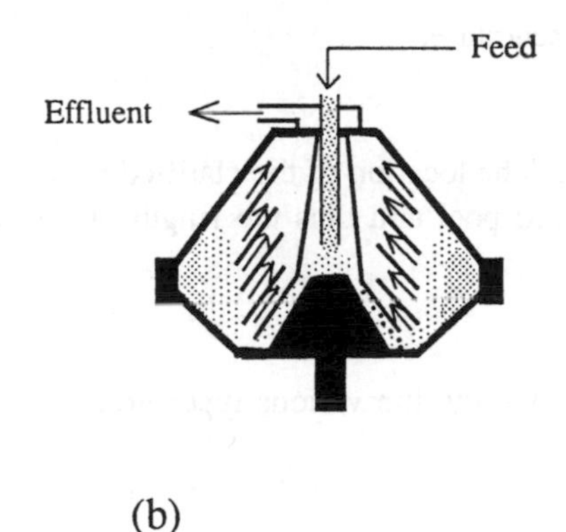

Figure 7.4 Example of solids-ejecting solid bowl centrifuge. (a) Solids outlet open, (b) solids outlet closed.

The *nozzle bowl* centrifuge is much like the solids-ejecting centrifuge. The major difference is that the nozzles around the periphery are small enough to remain open continuously. The flow is kept in balance with the feed rate by recycling solids as necessary.

The *decanter bowl* centrifuge rotates on a horizontal axis and is equipped with a concentric helical auger rotating at a slightly lower speed than the centrifuge bowl. This unit is illustrated in Figure 7.5. The bowl consists of a straight cylindrical section leading to a conical section, or beach. A liquid pool is mantained in the centrifuge with its surface falling part way up the beach. The centrifugal force of the rotation drives the solids to the outer wall of the centrifuge, and the auger transports the settled solids up the beach amd out of the liquid pool, where it is drained before being transported out of the centrifuge

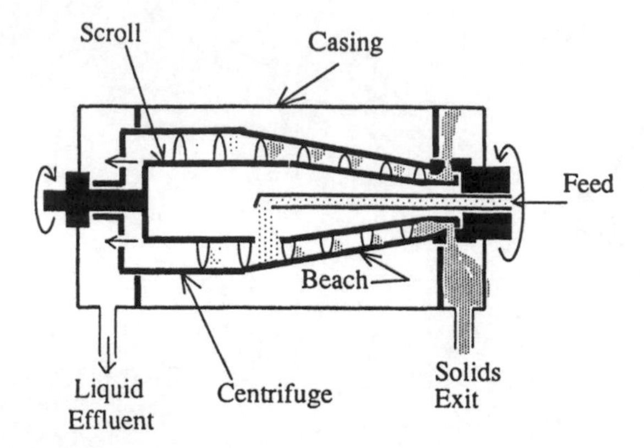

Figure 7.5 Sketch of decanter-bowl centrifuge.

at the small end of the conical section. The location of the clarified liquid outlet ports determines the depth of the liquid pool and thus the length of the beach for drainage.

Filtration Centrifuges

In the filtration, or perforated bowl category, the various types are:

- Basket
- Peeler
- Pusher

The basket centrifuge needs to be stopped for product removal, while the peeler and pusher centrifuges run at constant speed throughout the operating cycle.

The *basket centrifuge* is illustrated in Figure 7.6 rotating about a vertical axis. The intermittent operating cycle consists of a period onstream to build up the cake of solids, a spin dry period, washing of the cake, another spin dry period, and then removal of the dried cake. Figure 7.6a shows the unit onstream during the filtration operation, while Figure 7.6b shows the solids emptying part of the cycle. Solids can be emptied manually with the centrifuge stopped or by means of a scraper plough with the basket rotating at very low speed. Figure 7.6b shows solids removal with the scraper plough.

A *peeler centrifuge* rotating about a horizontal axis in the onstream part of the cycle in is illustrated in Figure 7.7a and in the solids discharge part of the cycle in Figure 7.7b. The peeler centrifuge goes through the same onstream, washing, and drying periods described for the basket centrifuge. The main difference is that the peeler does not have to be stopped for cake removal. The

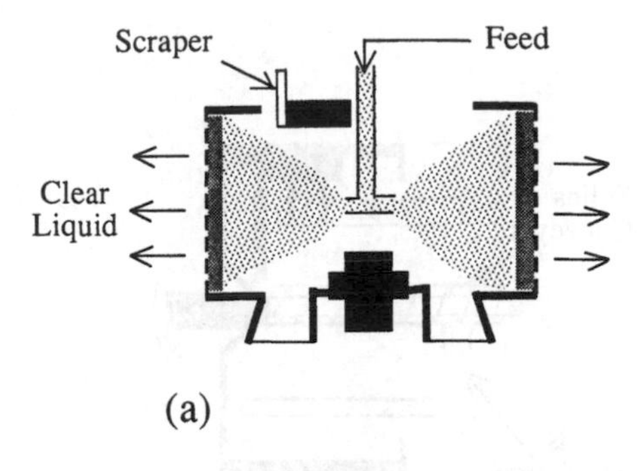

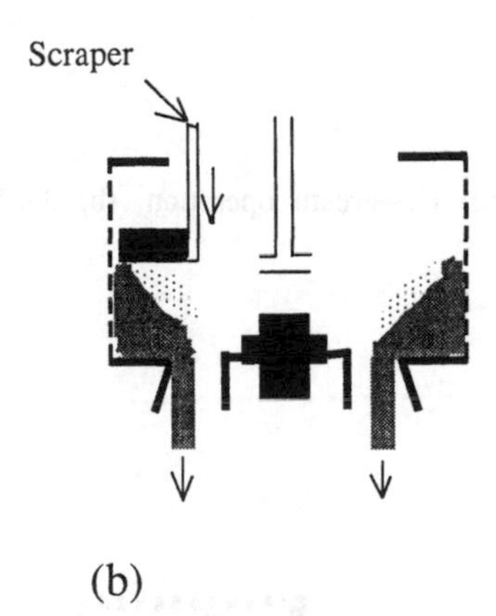

Figure 7.6 Sketch of basket centrifuge. (a) On-stream operation, (b) discharging cycle.

solids are removed by a scraper, as shown in the upper part of Figure 7.7b, as the centrifuge rotates. In Figure 7.7a the scraper is shown retracted.

The *pusher centrifuge* is illustrated in Figure 7.8 rotating around a horizontal axis. It operates with continuous feed and periodic but very frequent solids movement, caused by the reciprocating movement of a combination inner screen and pusher plate. A second, fixed position pusher plate is installed inside the inner screen to scrape the inner screen as it is retracted. On the forward stroke, the reciprocating scraper pushes the cake off the outer screen into the discharge shute, while on the reverse stroke the cake is pushed off the inner screen onto the outer screen. This type of operation provides continuous flow of both the solid and liquid phases.

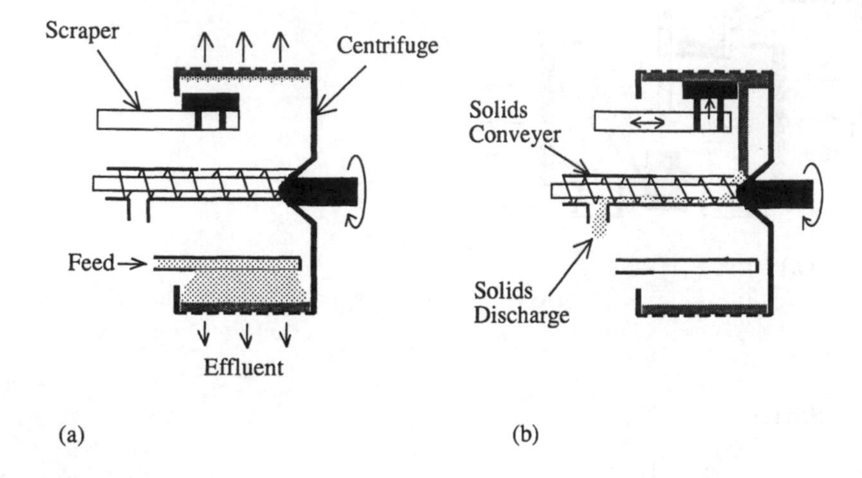

Figure 7.7 Sketch of peeler centrifuge. (a) On-stream operation, (b) discharging cycle.

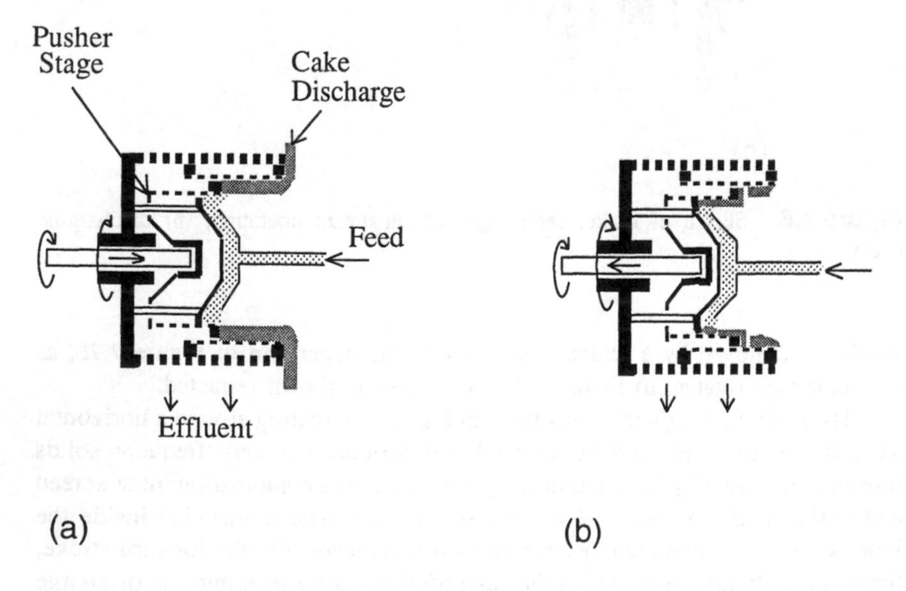

Figure 7.8 Sketch of pusher centrifuge. (a) Forward stroke, (b) backward stroke.

When considering the purchase of centrifugal separation equipment, it is a good idea to discuss the application with a centrifugal equipment supplier. Usually some pilot work is required for reliable choice of the proper equipment. Some suppliers of centrifuge equipment are listed in Table 7.1.

The choice of the type of machine for a particular application depends on the viscosity and density of the liquid and the amount, flowability, particle size,

Table 7.1 Vendors of Centrifuges

Alfa Laval Sharples 955 Mearns Road Warminster, PA 18974 (215) 443-4000	AML Industries 3500 Davisville Rd. Hatboro, PA 19040 (800) 258-4410
Barrett Centrifugals, Inc. Box 15059 Worcester, MA 01615-0059 (800) 228-6442	Bock Engineered Products, Inc. 3600 Summit, Box 5127 Toledo, OH 43611 (419) 726-2645
Broadbent, Inc, Box 185249 Fort Worth, TX 76118 (817) 595-2411	Centrisys Corporation 1931 Industrial Drive Libertyville, IL 60048-9738 (708) 816-8210
Flo Trend Systems, Inc. 707 Lehman Houston, TX 77018 (800) 762-9893	Flottweg 7095 Industrial Rd. P.O. Box 6270 Florence, KY 41042 (806) 283-0200
Humboldt Decanter, Inc. 3200 Pointe Parkway Norcross, GA 30092 (404) 448-4748	Niro, Inc. 9165 Rumsey Rd. Columbia, MD 21045-1991 (410) 997-8700
Sanborn Environmental Systems 25 Commercial Drive Wrentham, MA 02093 (800) 343-3381	Tabco Products, Inc. 1123 York Ave. Pawtucket, RI 02861 (401) 722-8888
Universal Process Equipment Box 338 Roosevelt, NJ 08555 (800) 243-6228	Western States Machine Co. Box 327 Hamilton, OH 45012 (513) 863-4758

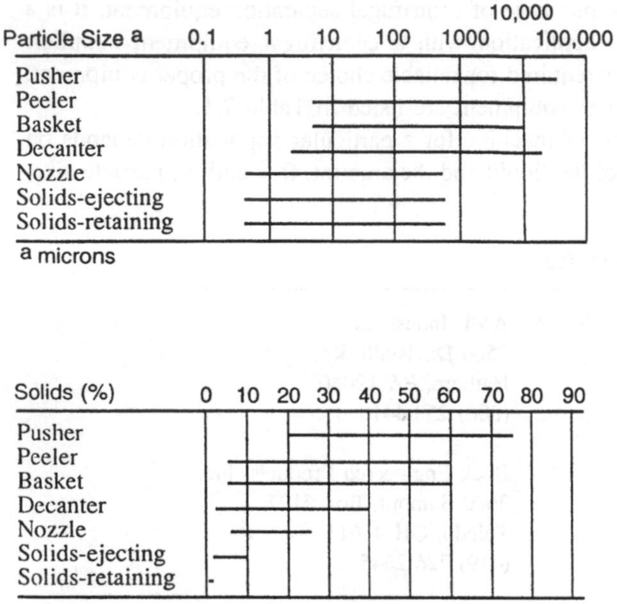

Particle Size a 0.1 1 10 100 1000 10,000 100,000

Pusher
Peeler
Basket
Decanter
Nozzle
Solids-ejecting
Solids-retaining

a microns

Solids (%) 0 10 20 30 40 50 60 70 80 90

Pusher
Peeler
Basket
Decanter
Nozzle
Solids-ejecting
Solids-retaining

Figure 7.9 Application ranges for various types of centrifuges.

particle density, and particle shape of the solids. Figure 7.9 gives a general range of applications by particle size and slurry solids content for the various types of centrifuges. In general, the pusher, peeler, and basket centrifuges handle thicker slurries of larger particles, while the solids-ejecting and solids-retaining centrifuges are applied to smaller particles and more dilute slurries. The decanter and nozzle centrifuges fall into a somewhat intermediate position. Some specific applications of the various machines are given in the following section.

7.2.3 Specific Applications

In a recent paper, De Loggio and Letki (1994) classified the applications of centrifuges by the type of separation desired. This section follows their pattern.

The applications of centrifuges include a variety of separation functions among which are:

- Clarification
- Purification
- Dewatering
- Classification
- Washing

- Extraction

In *clarification* the primary emphasis is on generating a clear effluent from which any haze caused by either liquid or solid particles of a second phase has been removed. Such applications include wastewater and intake water treating; clarification of paints, varnishes, juices, and beer; and recovery of valuable chemicals by removal of contaminants. An example of the latter is cleaning of transformer oil for recycling. For applications where the major requirement is liquid clarification, the solid-bowl sedimentation machines are preferred over filtration machines. On the other hand, if the major objective is a dry solids cake, the filtering machines are preferred, even though they do not produce as clear an effluent as the sedimentation machines. The filtering machines produce a cake containing only one-fifth as much liquid as the sedimentation machines.

According to Jacobs and Penney (1987), the order of effectiveness of producing a clear effluent for solid-bowl machines from most to least effective, is solids-retaining, nozzle, solids-ejecting, decanter.

Purification is the separation of two immiscible liquid phases having different specific gravities. A solid phase may also be present. Examples of this type of separation include (1) recovering liquid catalyst solutions and solvents for recycling, (2) breaking emulsions and separating cream from milk, (3) separating oil from water in petroleum production, and (4) recovering liquid chemicals for marketing. Disk stack centrifuges, with their high gravitational forces, are usually recommended for this service.

Dewatering has been defined simply as the concentrating of a slurry. The increasing cost of disposal of slurries and sludges as well as the environmental restrictions on such disposal have made the reduction in volume achievable by centrifuging very desirable. The goal is to remove as much water as possible so that as few pounds as possible need to be landfilled. The favorite centrifuges for this purpose are disk stack and decanter types. Scroll decanters are also widely used for dewatering of municipal sludge, coal and other minerals, heavy organic and inorganic chemicals, and plastic pellets. Pusher units are also used for dewatering crystals, particulate solids, and short fibers greater than 150 μm in size.

Classification is the separation of a suspension of particles into two streams with different particle size distributions. The separation is monitored by measuring the particle size distribution of both overflow and underflow product streams and determining the particle size that splits 50/50 between the effluent streams. This particle size can be decreased by increasing either the rotation speed or the number of disks in a disk stack centrifuge. The preferred centrifuge types for classification service are continuous solids-discharging centrifuges. These can be disk stack or decanter units. The typical applications are clay sizing, pigment sizing, and carbon black sizing.

Washing is the countercurrent washing of, or dissolving of impurities from, the solids carried in suspensions, either as crystals or amorphous solids. In a typical washing operation several centrifuges may be set up in series, with the solids separated from the primary feed in the first machine. The solids concentrate is then pumped to a second machine where the wash stream is injected and the solids are re-centrifuged. As many stages as needed can be used to get the desired product purity.

The food and biotechnology industries have particularly important applications of centrifuge washing. Here, ease of cleaning and continuous processing are important features of the equipment used. For these reasons, continuous solids-discharging disk stack centrifuges have become the most popular. Pusher units are also commonly used for washing crystals, particulate solids, and short fibers.

Extraction is the treating of a liquid solution with a second immiscible liquid, separating the phases, and then recovering the dissolved solute from the solvent. The major applications are in petroleum refining, the production of chemicals and pharmaceuticals, and in hydrometallurgy. The machines used in extraction processes can be any of the disk stack units or horizontal decanters.

7.2.4 Common Problems and Troubleshooting

The common problems encountered with centrifuges include wear, corrosion, and abrasion. These can usually be avoided by selecting the proper choice of the centrifuge. The materials used in the construction of the centrifuge can minimize corrosion and abrasion. For example, the use of special alloys or ceramics can eliminate corrosion. Abrasion can be minimized by either removing the abrasive components from the feed or, if that is impossible, by using hardfacing alloys, tungsten carbide, or ceramics for parts that are subject to abrasion. Proper balancing of the centrifuge to minimize vibration can help keep wear to a minimum.

Filtration centrifuges also have the additional problem of potential blinding of the filter medium. This can usually be remedied by using either an appropriate filter aid as a precoat or by changing the material used as the filter medium.

7.2.5 Economics

According to Ambler (1984), neither the capital nor the operating cost of a centrifuge can be correlated with any single parameter. These costs depend on:

- Physical nature of materials being separated
- Chemical nature of materials being separated
- Environment of the centrifuge
- Auxiliary equipment needed
- Difficulty of making the separation

Because of these factors, the cost of a centrifuge installation can vary over a wide range. Installation costs may vary from one to four times the purchase price of the centrifuge alone. Some of the factors contributing to higher costs are the need for:

- Abrasion protection (materials of construction)
- Corrosion protection (materials of construction)
- Toxicity protection (auxiliaries)
- Explosion protection (auxiliaries)
- Multiple or very large size centrifuges (capacity)
- High or low temperature operation (auxiliaries)
- Pressurized operation (auxiliaries)

Due to this wide variation in costs, it is best to consult with equipment manufacturers or centrifugation consultants when planning a centrifuge installation.

7.3 CYCLONING

The term *cyclone* is applied to mechanical collectors having no moving parts that use centrifugal force to separate particles from a fluid stream. Figure 7.10 shows a sketch of a typical cyclone. The centrifugal force is generated by forcing the flowing fluid to swirl in a tight spiral as it passes through a separation vessel. This swirling motion is achieved either by introducing the feed tangentially along the inner wall or by passing it axially through vanes that cause it to swirl. The centrifugal force generated by the swirling motion can vary from about 5 g for very large diameter cyclones up to 2500 g in small high pressure drop units. The centrifugal force causes the heavier phase to move to the vessel walls and the light phase to move to the center of the vessel, where it exits through an outlet pipe or vortex finder. The heavy phase slides down the outer wall of the vessel and is removed at the bottom into a collection hopper which is periodically emptied.

The carrier fluid can be either a gas or a liquid, and the particles can be either liquid or solid. Because the cyclone can operate at high ambient temperatures and pressures (up to 1800°F and 7500 psig) the gas-carrier version has found extensive use as a dust collector for catalysts in the process industries and for pollution control in size reduction and combustion processes. It is also frequently used to collect mists generated by either vapor condensation or vapor/liquid contacting.

The liquid-carrier version is often called a hydrocyclone and is used to separate either liquid or solid particles from a carrier liquid. It finds its major uses in (1) removing sand, grit, and other solids from intake, waste, or recycle waters and (2) recovering solids or liquids from process streams.

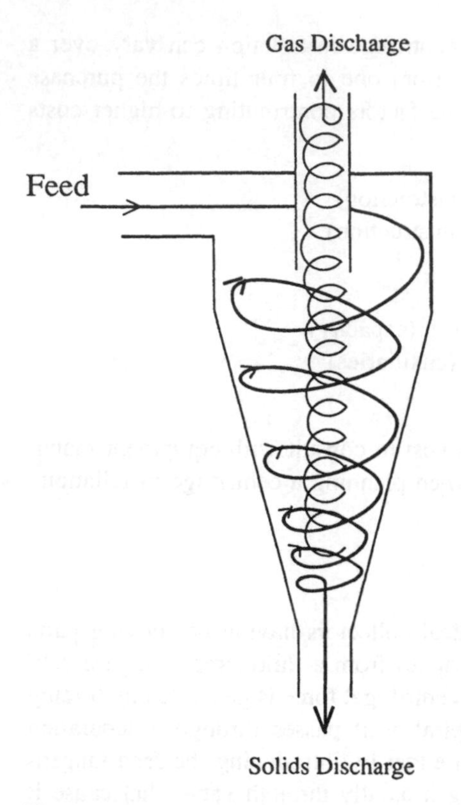

Figure 7.10 Sketch of typical reverse flow cyclone. (Courtesy Alfa Laval Sharples, Warminster, PA.)

7.3.1 Fundamentals of Cycloning

Cyclones should really be considered as classifiers because they separate a feed containing a distribution of particle sizes into two fractions, each of which also has a distribution of particle sizes. The smaller and less dense particles remain in the carrier fluid while the larger and more dense particles are removed. For particles of a homogeneous material, the particle size that distributes half to the overflow and half to the underflow is called the critical particle size. Larger particles go primarily to the underflow and smaller particles are essentially not collected. However, there is some overlap of particle size, the magnitude of which depends on the efficiency of the cyclone. If all the particles in the feed are larger than the critical size, the cyclone can remove essentially all of the particles in the feed. On the other hand, if they are all smaller than the critical size, the cyclone will be ineffectual.

Gas Flow Cyclones

The most common type of gas flow cyclone is the reverse flow design illustrated in Figure 7.10. In this design, the feed gas stream enters the top cylindrical section of the cyclone either tangentially or axially. If the entrance is axial, the swirling motion is obtained by a set of vanes in the inlet region. As the particle-laden gas swirls in the cylindrical portion of the cyclone, the particles are centrifuged by the spinning flow to the walls of the cyclone. The particles then flow down the conical section of the cyclone to the heavy phase outlet. The spinning gas goes part way toward the bottom outlet, but is not allowed to exit from the bottom. It then reverses direction and passes axially up the center of the cyclone as a vortex to the gas outlet at the top, which is called a vortex finder.

As with gravity settling and centrifuging, the separation achieved by cyclones depends on particle settling velocities. These are defined by the feed particle size and shape as well as the density difference between the phases. However, theoretical predictions of cyclone performance are not very reliable as yet. They will predict the direction of trends but not the absolute values of performance for specific cyclone designs. Coker (1993) and Doerschlag and Miczek (1977) have discussed cyclone design and presented the effects of a variety of variables on cyclone performance. Some generalizations are summarized below:

- Collection efficiency increases as particle size increases.
- Collection efficiency increases with gas pressure drop.
- Collection efficiency decreases with solids loading of feed.
- Limiting particle diameter collected increases with cyclone diameter.
- Limiting particle diameter decreases with cyclone length.
- Smaller diameter cyclones are more effective than larger diameter cyclones.
- Single units are more efficient than parallel units of the same size.
- Large single units are less effective than a bank of smaller parallel units of the same total capacity.

The general limit in particle size that can be collected is 5 μm. However, if the particles flocculate very readily, this limit can extend down to 0.1 to 0.2 μm.

7.3.2 Specific Applications

Dust Collection

Doerschlag and Miczek (1977) have listed the conditions for dust collection applications that favor the use of cyclones and those that mitigate against their use. Conditions favoring successful applications include:

- Dust must be collected in dry form.
- Temperatures are high.

- Dust concentrations are high.
- Gas is under high pressure.
- Dust or gas become corrosive when wet.

Conditions for which cyclones should not be specified are:

- Dust sticks to the equipment because of condensation or the surface properties of the dust.
- Dust particles are very small, below 1 to 5 μm.

Cyclones are particularly well suited when the collected dust is the valuable product, and the particle size is sufficiently large so that a high collection efficiency can be obtained. Furthermore, the ease of keeping the equipment clean and the lack of contamination from the collection device make cyclones particularly valuable in the collection of spray-dried foods.

The cyclone is also especially effective as a preconcentration device, used before final removal of the last traces of solids by other dust collection techniques, such as bag filters, wet scrubbers, or electrostatic precipitators. This application is used extensively in petroleum refining for fluid bed catalytic processes, both in reactors and regenerators. For example, in a fluid catalytic cracker, which operates the reactor at 900 to 1000°F and the regenerator at a couple of hundred degrees hotter, the vessels are so large that the cyclones are actually inside the vessels above the surface of the fluid bed. The gases entering the cyclones are so heavily laden with dust that two or three stages of cyclones in series are used before the exhaust gas is fed to a wet scrubber or electrostatic precipitator.

Some other common dust-collecting applications of cyclones are:

- High pressure removal of sand and grit from wellheads in natural gas production
- Coal gasification systems
- Fluid bed incinerators
- Ore roasting and cement production
- Sawdust recovery in lumber mills
- Solids recovery when classifyfing refractories

The purpose of the cyclone can be to recover the valuable product, to remove a potential pollutant from exhaust gases, or to provide a preconcentration before a final cleanup. They can be installed either internally in the large process vessels or externally in the exhaust gas line.

Mist Collection

Mist collection is much like dust collection except that the mist droplets coalesce into a bulk liquid phase when they strike the wall of the cyclone. This liquid

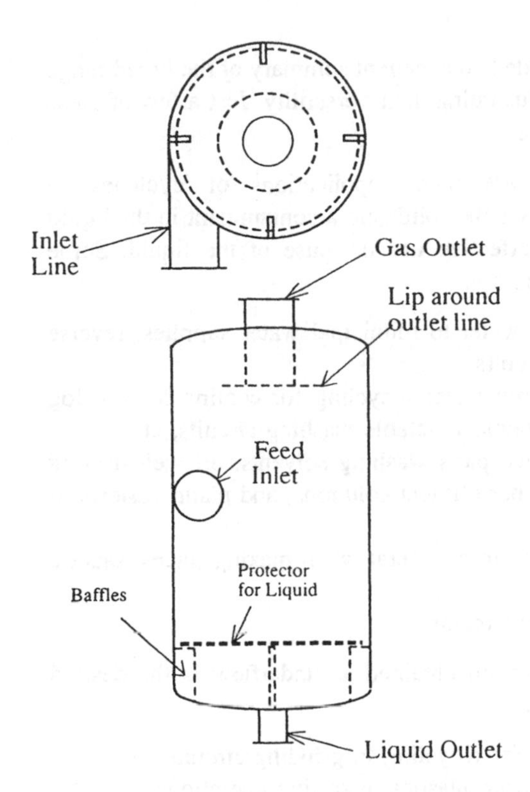

Figure 7.11 Sketch of mist collection cyclone.

then runs down to the underflow of the cyclone. The valuable product can be either the liquid phase, the gas phase, or both. Two applications of mist collection are:

- Removal of fogs from fixed gases
- Removal of entrainment from vapor/liquid contacting

Cyclones for collecting mists can be of the same design as for collecting dusts. However, some units are designed with a dished head bottom rather than the tapered cone of the reversed flow cyclone. Other common modifications are (1) lowering the inlet to near the middle of the cyclone, (2) supplying a baffle over the bottom outlet to prevent re-entrainment from the liquid pool above the liquid outlet, and (3) providing a lip around the gas outlet to prevent liquid from being re-entrained there. A sketch of the design recommended by Jacobs and Penney (1987) is given in Figure 7.11.

Liquid Phase Applications

Day and Grichar (1979) have provided an excellent summary of the broad range of application of hydrocyclones, illustrating their versatility. Just a few of these applications will be mentioned here.

Solid/Liquid Separations: There are many applications of cyclones to solid/liquid separations. In most cases the solids are a contaminant in the liquid and must be removed to permit effective use or reuse of the liquid. Some examples of this type of application are:

- Removal of sand from intake water to municipal water supplies, reverse osmosis units, and distillation units
- Degritting wash waters to permit water recycling for cooling towers, log flumes, steel plant quench systems, vegetable washing circuits, etc.
- Cleaning metalworking coolants, parts washing solvents, oil well drilling muds, limestone flue gas scrubber effluent solutions, and plant wastewater effluents
- Removal of seeds, pits, sand, or tramp metal when making juices, sauces, pastes, and purees
- Removal of coke from refinery streams

In other applications the coarser stream obtained as underflow is the desired product. Examples of this type are:

- Recovery of oversize particles for recycling in grinding circuits
- Recovery of metals and glass from plastics in sorting operations
- Recovery of liquid or solid catalysts from oils in refining

A third type of application is to improve the operation of downstream equipment. Examples of this type are:

- Reducing solids loading to increase capacity, minimize wear, reduce plugging, and reduce maintenance
- Pretreating feeds to filters and centrifuges
- Removing shot from fiberglass or grit from clays and classifying abrasives and pulp and paper stocks

Liquid/Liquid Separations: Separating oil and water is a major application of cyclones for liquid/liquid service. They are used for removing oil from produced water in petroleum production, which permits either disposal of the water or recycling it to waterflood operations. Furthermore, cyclones can be used to remove oil from runoff water from cleaning operations, from ship's bilge or ballast water, and from industrial wastewaters.

A second major use of liquid/liquid cyclones is to recover solvents or reagents used in chemical processing. Here, the aqueous product of the process

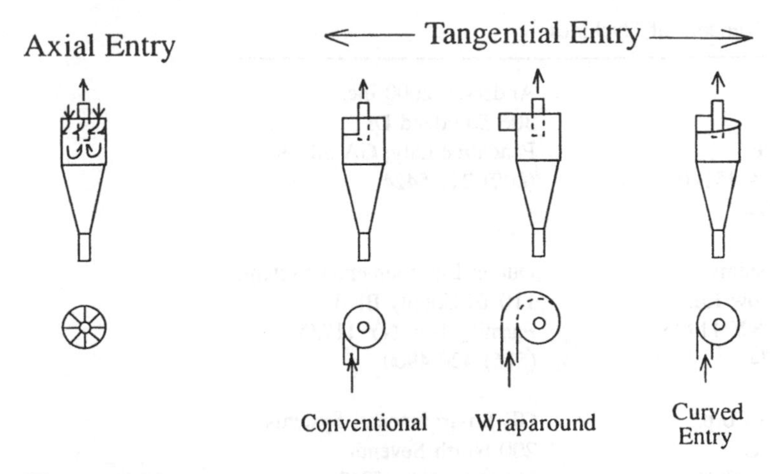

Figure 7.12 Various entry designs for cyclones.

is separated from the immiscible process solvent or reagent, which is recycled to the process. Immiscible non-aqueous liquids such as are encountered in solvent extraction operations in petroleum refining, can also be separated using cyclones.

7.3.3 Equipment

Depending on the size of the application, the cyclone installation can be either a single unit or a battery of units in parallel. Furthermore, either single stages or a series of stages can be used. In any case, the cyclones can have tangential or axial entry. For axial entry units, the inlet is equipped with vanes to provide the necessary swirling motion. These vanes are usually cast into the inlet of the cyclone. The tangential entry cyclones are made both with simple tangent entries and with involuted feed entries. These different design features are illustrated in Figure 7.12. Generally speaking, axial entry is used with smaller units and tangential entry is used for larger units. Some of the major suppliers of cyclones are listed in Table 7.2.

7.3.4 Common Problems and Troubleshooting

When a cyclone fails to operate properly, the first indication is a change, usually an increase, in the amount of heavy particulates in, and thus in the opacity of, the overhead stream. If the overhead is exhausted to the atmosphere or to a visible outfall, this can be readily seen. However, if the overhead stream feeds further processing units, it will only show up as a change in the loading on those units or contamination of products.

Table 7.2 Vendors of Cyclones

Airotech, Inc. Boyle Center 120 Ninth Ave. Homestead, PA 15120-1600 (412) 462-4404	Andersen 2000 Inc. 306 Dividend Dr. Peachtree City, GA 30269 (800) 241-5424
Clean Gas Systems 707 Broadhollow Rd. Farmingdale, NY 11735 (516) 756-2474	Ducon Environmental Systems 110 Bi-County Blvd. Farmingdale, NY 11735 (516) 420-4900
Fuller Kovako Corp. 2158 Avenue C Bethlehem, PA 18017 (800) 523-9480	GE Environmental Systems 200 North Seventh Lebanon, PA 17042 (717) 274-7000
Krebs Engineers 1205 Chrysler Drive Menlo Park, CA 94025 (415) 325-0751	M & W Industries Box 952 Rural Hall, NC 27045-0952 (919) 969-9526
Meyer, W.W. & Sons 861 Elmwood Ave. Skokie, IL 60076-2972 (708) 673-0312	Polymer Fabrication Co. Box 628 Arab, AL 35016 (800) 264-7659
C. A. Litzler, Quickdraft Div. Box 80659 Canton, OH 44708 (216) 477-4574	Rolfes Co., George A. Box 458 Boone, IA 50036-0458 (800) 247-8372
Ultra Industries, Inc. 145 Eastern Ave. Bellwood, IL 60104-1220 (800) 358-5872	

A simple but effective diagnostic tool for the detection of cyclone malfunction is a pressure drop manometer, which is connected across the cyclone from inlet to outlet. It responds to changes in flow rate, bypassing of gas flow, buildup of solids on pipe walls, and holes in the vessel, among other things.

Control of the dew point of gases is very important to prevent plugging by agglomeration of solids due to condensation of vapors on the solids and the resultant sticking together of the particles. Horzella (1978) has suggested the following precautions in solids drying operations.

1. Preheat the system before feed is started.
2. Continue to supply hot gas after shutdown until system is free of dust and water vapor.
3. Insulate the ductwork, cyclone, and dust hopper.
4. Provide artificial heating of the hopper.

The first three steps usually avoid the danger of corrosion and the fourth minimizes the chance of material buildup in the hopper as well as eliminates corrosion.

Erosion is one of the few maintenance problems encountered by cyclones. When abrasive materials are to be processed, the wear areas of the cyclone need to be either hard faced, lined with ceramic, or made from special alloys. These areas are the feed inlet and any vanes used to provide the initial swirling motion.

Re-entrainment can occur in all types of cyclones if the velocities get too high. The obvious remedy is to decrease the feed rate. However, in mist collecting cyclones liquid can sometimes run down the outside of the vortex finder to where the gas exits and be re-entrained by the high velocity of the exit gas. This can be avoided by putting a skirt around the outside of the vortex finder to keep the dripping liquid away from the bottom edge of the outlet pipe.

Short circuiting can occur in a cyclone when the vortex finder does not extend far enough below the feed inlet to send the flow spiralling down the cyclone. It can also happen if the solids outlet is plugged and the cyclone body fills with solids or liquid.

Loss of vortex occurs when the carrier fluid is allowed to flow out the apex of the cyclone in too large a quantity. This can occur when the dust pot is unsealed and has inadequate back pressure. It can also occur if erosion has worn a hole in the cyclone body.

7.3.5 Economics

Cyclones are generally the least expensive method for collecting dusts, provided the particle size of the dust is large enough. Cyclones can be fabricated in a wide range of sizes and from almost any material either by casting or fabrication from plate. They require little maintenance or operating labor, so their operating costs are essentially the cost of retiring the purchase price plus the cost of operating the blowers required to overcome pressure drop. However, the collection efficiency, solids loading, required materials of construction, and feed rate determine the size and number of cyclones needed, and thus the capital cost of the installation. To get good cost estimates it is best to check with vendors.

7.4 SUMMARY

Centrifuging and cycloning are techniques for increasing the force available for separating phases by utilizing their difference in density. Gravity settling uses the force of natural gravity and is the simplest separation based on density difference. Cycloning is next in degree of complexity, cost, and effectiveness. Centrifuging is the most effective due to the large separating force generated but is also the most expensive. Thus, gravity settling is the first choice where it will do the required job and flow rate is not an issue. If gravity settling is not satisfactory, cycloning would be the next choice, if suitable. It gives both lower holdup and better separation. Finally, if the separation is very difficult or high flow rate is essential, centrifuging may be the proper solution.

REFERENCES

Ambler, C. M. (1979), Centrifugation, *Handbook for Separation Techniques for Chemical Engineers* (P. A. Schweitzer, ed.), McGraw-Hill, New York, pp. 4-55 to 4-84.

Ambler, C. M. (1984), Centrifuges, *Chemical Engineers' Handbook* (R. H. Perry and D. W. Green eds.), McGraw-Hill, New York, pp. 19-89 to 19-103.

Coker, A. K. (1993), Understand cyclone design, *Chemical Engineering Progress*, December, 1993: 51.

Day, R. W. and Grichar C. N. (1979), Hydrocyclone Separation, *Handbook for Separation Techniques for Chemical Engineers* (P. A. Schweitzer, ed.), McGraw-Hill, New York, pp. 4-135 to 4-140.

De Loggio, T. and Letki A. (1994), New Directions in Centrifuging, *Chemical Engineering*, January, 1994: 70.

Doerschlag, C. and Miczek G. (1977), How to choose a cyclone dust collector, *Chemical Engineering*, February 14: 64.

Jacobs, L. J., Jr. and Penney W. R. (1987), Phase Segregation, *Handbook of Separation Process Technology* (R. W. Rousseau, ed.), Wiley & Sons, New York, pp. 129–196.

Horzella, T. I. (1978), Selecting, installing, and maintaining cyclone dust collectors, *Chemical Engineering*, January 30: 85.

Moir, D. M. (1988), Sedimentation centrifuges, *Chemical Engineering*, March 28: 42.

8

Chemical Precipitation and Sedimentation

8.1 INTRODUCTION

Chemical precipitation uses the addition of chemicals to convert a soluble component in a liquid feed into an insoluble solid material. The precipitate can then be removed from the solution by filtration, flotation, or settling. The process is widely used for the removal of dissolved toxic metals as insoluble metal salts. It is effective for both waste automotive lubricating oils and for aqueous wastes from the metalworking, metal finishing, and electronics industries. Chemical precipitation is also used as a pretreating technique for aqueous wastes destined for biological treatment. Such treatment can be done in either an on-site plant or a publicly owned treatment works (POTW). It can remove heavy metals and toxic materials which are above the limits allowed for feed to the biological treating plant. The commonest forms of anions used to form the insoluble metal salts are hydroxides from the addition of lime or caustic and sulfides from the addition of ferrous sulfide.

Sedimentation is the use of gravity settling for either concentrating solid particles or removing them from a liquid. It is commonly used with chemical precipitation in aqueous systems to remove the solid particles generated by the chemical addition. It can be subdivided into (1) thickening, where the primary purpose is concentrating the suspended solids in a feed stream for recovery of the solids, and (2) clarification, where the primary purpose is to remove the suspended solid particles and produce a clear effluent. The equipment used is very similar in design for both purposes. Both clarification and thickening are at least partially involved with each other in all applications to waste minimization. The details of equipment design determine whether the process unit is primarily a thickener, a clarifier, or both.

Flocculation and coagulation agents are commonly used in conjunction with sedimentation in aqueous systems. They improve the separation of the precipitated solids (or sludge) from the aqueous phase. Coagulation is the collection or aggregation of very small colloidal particles of precipitate to form larger aggregate particles. It is enhanced by the addition of strongly ionic chemicals such as acids, lime, alum, or ferrous sulfate in a region of intense mixing. Flocculation is the further growth of the coagulated aggregates by their coalescence as the aggregates contact one another. This is done under the influence of much milder agitation in the presence of polymeric additives. Flocculation is a much slower process than coagulation but produces larger, more easily settled particles. Special designs of sedimentation equipment have been developed to optimize the mixing time/intensity requirements for high throughput and optimum quality of the effluent streams. This chapter discusses the fundamentals of chemical precipitation and sedimentation and their applications in waste minimization.

8.2 PRETREATMENT OF WASTE

Dilute and non-toxic wastewaters are normally treated in a biological treating plant prior to disposal. However, when aqueous wastes are received from a variety of sources in a large industrial organization and treated in a common biological waste treating plant, some of these waste streams may be too toxic or too reactive to be acceptable as feeds to the biological treating plant. Furthermore, different waste streams may be incompatible due to potential chemical reactions between them which could release toxic gases to the atmosphere. This is particularly true of process wastewaters from the chemical process and metal finishing industries and for chemical wastes from laboratories. Such wastes must be pretreated to remove toxic metals or other toxic ions. They must also be brought to near neutral pH before they can be sent to either an on-site or a publicly owned water treatment works. The incompatibility of such waste streams is usually due to one stream containing a potentially toxic-forming material such as sulfide or cyanide anions and the other containing a reactive material, like strong acids, that can release toxic HCN or H_2S to the atmosphere. Examples of this type of incompatibility are (1) chromium- and cyanide-containing aqueous waste streams from metal finishing, where the acid pH needed to reduce

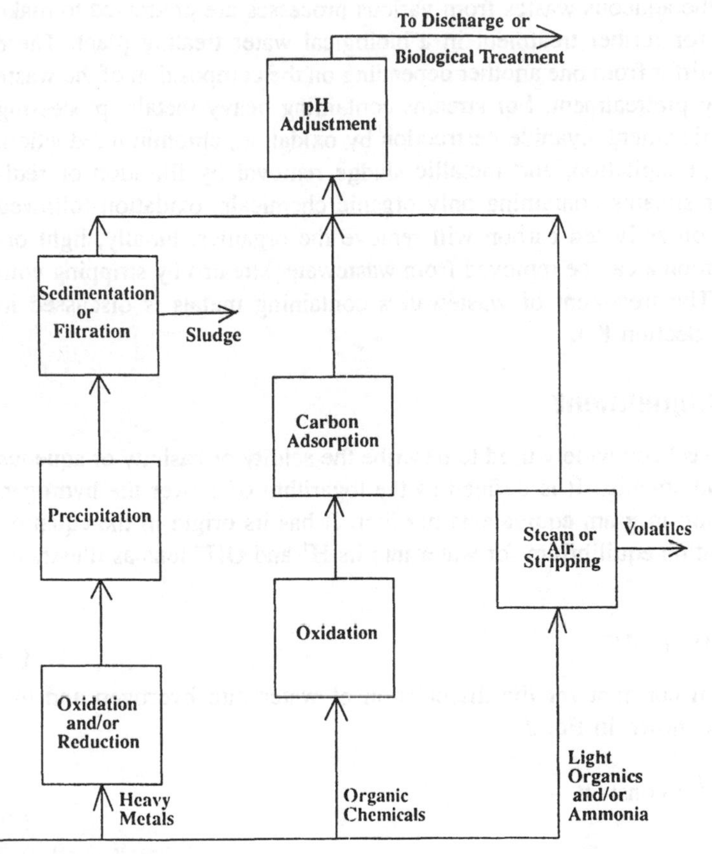

Figure 8.1 Flow sheet for pretreatment of industrial wastes. The pretreatment needed depends on the impurities in the wastewater. It differs for heavy metals, organic chemicals, and low molecular weight compounds.

chromium from the six-valence state to the three-valence state would release toxic HCN gas, and (2) highly basic streams and streams capable of releasing toxic vapors such as ammonia in the presence of bases. A more complete listing of incompatible materials is given in Appendix E of a National Research Council report (1983) on handling small quantities of laboratory wastes. In Chapter 6 of this same reference, procedures for laboratory destruction of hazardous chemicals are given including the chemistry involved.

A typical flowsheet for pretreating aqueous wastes prior to either disposal to a water body or feeding to a biological water treating plant is given in Figure 8.1.

In this figure the aqueous wastes from various processes are pretreated to make them suitable for further treatment in a biological water treating plant. These pretreatments differ from one another depending on the composition of the waste stream needing pretreatment. For streams containing heavy metals, processing includes pH adjustment, cyanide destruction by oxidation, chromium reduction, heavy metals precipitation, and metallic sludge removal by filtration or sedimentation. For streams containing only organic chemicals, oxidation followed by adsorption on activated carbon will remove the organics. Finally, light organics and ammonia can be removed from wastewater streams by stripping with air or steam. The treatment of wastewaters containing metals is discussed in more detail in Section 8.3.

8.2.1 pH Adjustment

The pH scale has been widely used to describe the acidity or basicity of aqueous and other liquid streams. It is defined as the logarithm of 1 over the hydrogen ion concentration in gram equivalents per liter. It has its origin in the equation for the dissociation equilibrium for water into its H^+ and OH^- ions as illustrated in Eq. 1.

$$H_2O \leftrightarrows H^+ + OH^- \tag{1}$$

The equilibrium constant for the dissociation of water into hydrogen and hydroxide ions is shown in Eq. 2.

$$\frac{[H^+]\,[OH^-]}{[H_2O]} = \text{constant} \tag{2}$$

The concentration of ions is very small as compared to the concentration of undissociated water. Therefore, the concentration of water can be also be considered a constant at 55 g mol/L. This allows Eq. 2 to be rewritten as Eq. 3.

$$[H^+]\,[OH^-] = \text{constant} \times [H_2O] = K \tag{3}$$

The value of the constant K has been determined to be 10^{-14}. Therefore, for a neutral solution which has the same concentration of hydrogen and hydroxyl ions, the hydrogen ion concentration is 10^{-7} and the pH is 7, from $\log 1/10^{-7} = 7$. Because the product of hydrogen and hydroxyl ions is always 10^{-14}, if the hydrogen ion concentration increases, the hydroxyl ion concentration must decrease and vice versa. For example, if the hydrogen ion concentration is increased to 10^{-2} by adding acid, the hydroxide ion concentration decreases to 10^{-12} to maintain the product, K, at 10^{-14}. This gives a pH of 2 for the solution. Thus, adding acid decreases pH, and adding base increases it. A pH below 7 is associated with acid systems, and a pH above 7 is associated with basic systems. The pH scale runs from 0 to 14 and is measured electrochemically using a glass electrode.

Two important effects result from the logarithmic nature of the pH scale. First, small additions of acid or base cause large changes in pH near pH = 7. However, large amounts of acid or base are required to change pH near the ends of the scale. Second, the amount of heat of reaction given off by changing the pH one unit by neutralization is very low near pH = 7 and very large near the ends of the scale. This is illustrated in Tables 8.1 and 8.2. It is a consequence of a pH change of 1 at pH = 7 representing a hydrogen ion concentration change of 0.9×10^{-7} g mol/L while at pH = 1 it represents a change of 0.9×10^{-1} g mol/L. This change represents the addition or neutralization of 3.3 μg/L of HCl at pH = 7 and 3.3 g/L at pH = 1. At the other end of the scale, a pH change of 1 requires the addition or neutralization of 3.6 g/L of sodium hydroxide at pH = 13 versus 3.6 μg/L at pH = 7. Thus, the heats of reaction for neutralization at pH = 1 or pH = 13 are a million times as large as the corresponding values at pH = 7. This frequently leads to such difficulties as overshooting the endpoint in the approach to pH = 7 by neutralization and overheating of the mixture in the neutralization of very strong acidic or basic solutions.

Neutralization of acids with bases (and vice versa) is a very rapid reaction usually limited by the degree of mixing available in the neutralization apparatus. Either hydroxide ions can be added to reduce the hydrogen ion concentration, or hydrogen ions can be added to reduce hydroxide ion concentration. The products in either case are water and heat, as illustrated in Eq. 4.

Table 8.1 Amounts of Reagents Required for a Unit Change of pH at Various pH Levels

			Amount of reagent			
			HCl		NaOH	
pH	$[H^+]$ g/L	$[OH^-]$ g/L	g/L	lb/1000 gal	g/l	lb/1000 gal
0	1.0	17×10^{-14}	33	268	36	300
1	0.1	17×10^{-13}	3.3	27	3.6	30
3	0.001	17×10^{-11}	0.033	0.27	0.036	0.30
5	10^{-5}	17×10^{-9}	3.3×10^{-4}	0.0027	3.6×10^{-4}	0.0030
7	10^{-7}	17×10^{-7}	3.3×10^{-6}	2.7×10^{-5}	3.6×10^{-6}	3.0×10^{-5}
9	10^{-9}	17×10^{-5}	3.3×10^{-4}	0.0027	3.6×10^{-4}	0.0030
11	10^{-11}	17×10^{-3}	0.033	0.27	0.036	0.30
13	10^{-13}	1.7	3.3	27	3.6	30
14	10^{-14}	17	33	268	36	300

Table 8.2 Heat Released by a Unit Change of pH at Various pH Levels at 20°C (68°F)

	Heat released	
pH level	Cal/L	Btu/gal
0	12,310	102.6
1	1,231	10.3
3	12.3	0.10
5	0.12	0.0010
7	0.0012	0.000010
9	0.12	0.0010
11	12.3	0.10
13	1,231	10.3
14	12,310	102.6

$$H^+ + OH^- \rightarrow H_2O + \text{heat} \tag{4}$$

The amount of heat given off per gram mole of water formed at 20°C is 13,680 cal. This corresponds to 24,620 Btu/lb mol of water formed, or 1,368 Btu/lb of water formed. Strong acids at hydrogen ion concentrations of 10 g mol/L (pH = 0.1) and strong bases at hydroxide ion concentrations of 10 g mol/L (pH = 15) can go off the 0–14 pH scale. Neutralization of these two streams gives off 136,810 cal in 2L of solution and thus gives a temperature rise of about 100°C (180°F). This would boil the aqueous solution resulting from the neutralization. Clearly, heat removal is required during the neutralization of such strong solutions.

Neutralization or pH adjustment can be done either on a batch or continuous basis. Batch neutralization is preferred for small volumes and continuous neutralization is preferred for large volumes of relatively constant composition. In either case the acid or base used to change the pH is added by means of a pump controlled by a pH controller. Usually pure sulfuric acid and pure lime or sodium hydroxide are used for the neutralization to avoid running into incompatibility problems. However, if the compositions of acidic and basic (alkaline) plant waste streams are known and no incompatible components are present, savings can be made by using acidic and basic plant wastes to neutralize each other. This is often done by mixing compatible acid and alkaline wastes in an equalization tank or lagoon prior to final pH adjustment.

Batch pH Adjustment

The flow diagram for batch pH adjustment is illustrated in Figure 8.2. In this diagram, the reactor vessel is a stirred tank equipped with a pH sensor and

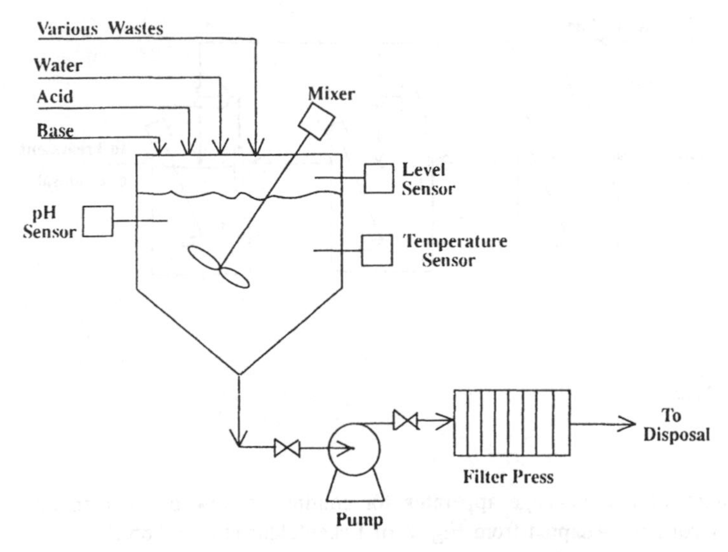

Figure 8.2 Sketch of a typical batch reactor used for pH adjustment.

temperature indicator, a mixer, waste and water fill lines, acid and base chemical feed inlets, and a high level alarm. Auxiliary equipment includes chemical metering systems for the acid and base as demanded by a pH controller. The pH controller operates from the signal of the pH sensor in the tank. If the wastes are very strongly acidic or basic, the feed may have to be diluted with water or the contents of the tank cooled to prevent temperature runaway. If solids are expected to be generated by the pH adjustment, a filter should be used in the drainage system to produce clear effluents when draining the tank after reaction.

Continuous pH Adjustment

The equipment for continuous pH adjustment is very similar to the batch equipment. However, usually two or three mixing vessels are used in series to make the approach to equilibrium at the desired pH easier to control. This is illustrated in the block diagram shown in Figure 8.3. An acidic stream is neutralized by preliminary reaction with alkaline plant wastes in a holding tank or lagoon. It is then brought to the desired final pH by adding lime or caustic incrementally in a series of three mixing tanks. If the major waste stream is alkaline rather than acid as in the above example, the preliminary treating in the lagoon would be with plant acid wastes. The final pH adjustment in the mixing tanks would be with sulfuric acid. Depending on the quality of the final effluents from pH adjustment, they can either go to further wastewater treatment or to disposal as a weak brine.

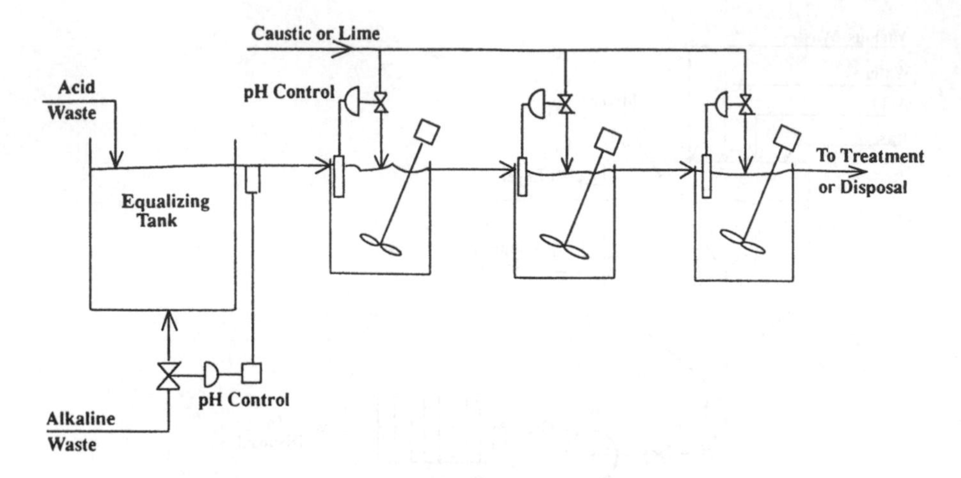

Figure 8.3 Sketch of a three-stage apparatus for continuous flow pH adjustment using automatic pH control. [Adapted from Fig. 2 of Eckenfelder et al. (1985).]

8.2.2 Chrome Reduction

Chrome reduction refers to the change in oxidation state of chromium and not to its concentration in wastewaters. It is usually required as a pretreatment for wastewaters from metal finishing plants. Chromium is a severely regulated toxic metal which must be removed from wastewaters before they can be discarded. It is removed along with other heavy metal ions by chemical precipitation as trivalent chromium hydroxide or sulfide. Unfortunately, in wastewaters from metal finishing the chromium is produced as hexavalent chromium in the chromate ion. This hexavalent chromium must be reduced electrochemically to trivalent chromium so that it can be precipitated later with other heavy metals. The reduction of chromium is carried out under acid conditions at a pH of 2.0–2.5. The reducing agent is usually sodium bisulfite, sulfur dioxide, or ferrous sulfate; sulfuric acid is used to maintain pH.

Sodium Bisulfite

Sodium bisulfite is the preferred reducing agent for small to medium-sized chromate waste streams, while sulfur dioxide is sometimes used for large installations. The chemistry of chromate reduction is given in Eq. 5. The reduction of the valence state of chromium from plus six in chromate ion to plus three in the chromic sulfate depends on sulfur being oxidized from the plus four valence state in the bisulfite to the plus six state in the sulfate.

$$2H_2CrO_4 + 3NaHSO_3 + 3H_2SO_4 \rightarrow Cr_2(SO_4)_3 + 5H_2O + 3NaHSO_4 \quad (5)$$

This equation shows that two molecules of chromic acid require three molecules of sodium bisulfite and three molecules of sulfuric acid. The products are one molecule of chromic sulfate, five molecules of water, and three molecules of sodium bisulfate. The pH must be maintained at 2.0–2.5 during the reduction.

Sulfur Dioxide

Sulfur dioxide can be used in place of sodium bisulfite because it forms sulfurous acid in water solution. Thus it provides bisulfite ions for reduction of chromium in the same way as sodium bisulfite. The sulfur is again oxidized from plus four to plus six valence while the chromium is reduced from the plus six state to the plus three state. This is illustrated in Eq. 6.

$$2H_2CrO_4 + 3H_2SO_3 \rightarrow Cr_2(SO_4)_3 + 5H_2O \tag{6}$$

In this reaction three molecules of sulfurous acid react with two molecules of chromic acid to form one molecule of chromic sulfate and five molecules of water. The reaction must be carried out in strong acid solution at a pH of 2.0–2.5, just as with sodium bisulfite reduction. However, the reaction does not consume sulfuric acid as is done in Eq. 5 because there are no sodium ions in the reactants to end up as sodium salts.

Ferrous Sulfate

Unlike the sulfur-based chromium reductions, use of ferrous sulfate as a reducing agent depends on the oxidation of iron from a plus two valence to a plus three valence. The chemical reaction is given in Eq. 7.

$$2H_2CrO_4 + 6FeSO_4 + 6H_2SO_4 \rightarrow Cr_2(SO_4)_3 + 8H_2O + 3Fe_2(SO_4)_3 \tag{7}$$

The amounts of chemicals required for the three reduction methods described are given in Table 8.3. The molecular weights of the various reagents are also given for reference purposes.

The quantities of chemicals needed to reduce 1 lb of chromium from a valence of plus six to a valence of plus three is given in a column for each of the three processes. In all of these reduction processes, the completeness of reaction is monitored by an instrument reading the oxidation reduction potential (ORP) of the solution. The endpoint reading of the ORP meter for each system is the sum of the potentials for all species and must be determined experimentally for each system.

The equipment used for chromium reduction is very similar to that shown in Figure 8.2 for pH adjustment. However, more chemical feed equipment, more chemical feed lines to the reaction vessels, and an ORP monitor need to be added. The soluble chromic sulfate produced by all three of the reduction methods can subsequently have its trivalent chromium ion precipitated as chro-

Table 8.3 Amounts of Reagent Needed to Reduce 1 lb of Chromium VI to Chromium III

		Pounds of reagents needed for reducing processes:		
Reagent	Atomic or molecular weight	$NaHSO_3$	SO_2	$FeSO_4$
Chromium	52	1.0	1.0	1.0
$NaHSO_3$	104	3.0	—	—
H_2SO_4	98	2.8	—	5.7
Sulfur dioxide	64	—	1.9	—
$FeSO_4$	152	—	—	8.8

Amounts are for the pure materials listed as reagent. Allowance must be made for waters of crystallization or diluents in reagents.

mium hydroxide. It can then be removed along with other metal hydroxides as described in Section 8.3.

8.2.3 Cyanide Destruction

Cyanide wastes from plating, heat treating, and other industrial processes are among the most toxic materials present in industrial wastewaters. They must be removed by pretreatment before wastes containing them can be processed in sewage treatment processes. This is because ordinary sewage treatment processes do not eliminate cyanides. Fortunately, cyanides can be destroyed efficiently and inexpensively by oxidation to less toxic cyanates and then to carbon dioxide and nitrogen. This is most frequently done by alkaline chlorination using strong sodium hypochlorite or chlorine as the oxidizing agent. The reaction proceeds in two steps. The first is oxidation of cyanides to cyanates at pH of 10.5 to 11.5. The second is destruction of cyanates by further oxidation to carbon dioxide and nitrogen at a pH of 8.0 to 8.5. The chemistry of these reactions is given in Eqs. 8 and 9.

Step 1 at pH = 10.5 to 1.5:

$$NaOCl + NaCN \rightarrow NaCNO + NaCl \quad (8)$$

In this reaction one molecule of sodium hypochlorite reacts with one molecule of sodium cyanide to oxidize the cyanide to cyanate and reduce the hypochlorite to sodium chloride. This step also breaks any metal-cyanide complexes that may be present in the solution, thus freeing the cyanide for oxidation to cyanate. The step 1 reaction is normally completed in a residence time of 10 to 15 min.

Step 2 at pH = 8 to 8.5:

$$2NaCNO + 3NaOCl + H_2O \rightarrow 2CO_2 + N_2 + 3NaCl + 3NaOH \tag{9}$$

In the second step, the cyanate is further oxidized by more hypochlorite to give carbon dioxide, nitrogen, sodium chloride, and sodium hydroxide. This reaction takes place very rapidly as the pH is lowered and is usually complete by the time the pH is lowered to 8.0–8.5 with sulfuric acid. The completeness of reaction is indicated by an ORP reading of greater than 600 mV. If the ORP reading is below 600 mV after pH has been adjusted to 8.0–8.5, more sodium hypochlorite must be added to get an ORP reading above 600 mV and insure complete destruction of cyanides. The pH must not be allowed to go below 8.0 to avoid release of toxic gases such as HCN or cyanogen chloride to the atmosphere. Sodium hypochlorite is available commercially as strong solutions in water. The amount needed per pound of cyanide will vary with the strength of the available solution. Every pound of cyanide in the waste stream requires 1.36 lbs of active chlorine in the first step and 2.0 lbs in the second step based on molecular weights of CN equal to 26 and Cl equal to 35.5.

The equipment for cyanide destruction is about the same as for chromium reduction and pH adjustment. Common equipment can be used for all three processes in blocked operation. A typical batch apparatus for doing this is sketched in Figure 8.4. *However, extreme care must be taken never to mix the acidic chromium wastes with the alkaline cyanide wastes because of the potential release of very toxic gaseous HCN to the atmosphere.* A variety of compact batch and continuous treating plants are available from commercial vendors for both pretreating and chemical precipitation. These will be discussed later in Section 8.6.

Excellent summaries of wastewater treating are given by Cheremisinoff (1990) and Eckenfelder et al. (1985). For groundwater treating, which is very similar, see Newton (1990). These references show how chemical precipitation and sedimentation fit into the overall water treatment technology.

8.3 CHEMICAL PRECIPITATION

Chemical precipitation is widely used in the treatment of hazardous wastes to remove toxic metals or other inorganic substances from aqueous waste streams. It is defined as a process which converts a soluble substance to an insoluble form by means of a chemical reaction. The insoluble product can then be removed by sedimentation or filtration. By far the most common precipitation agent is hydroxide ion from either lime (calcium hydroxide) or sodium hydroxide (caustic soda). The precipitation is carried out at alkaline pH. However, other precipi-

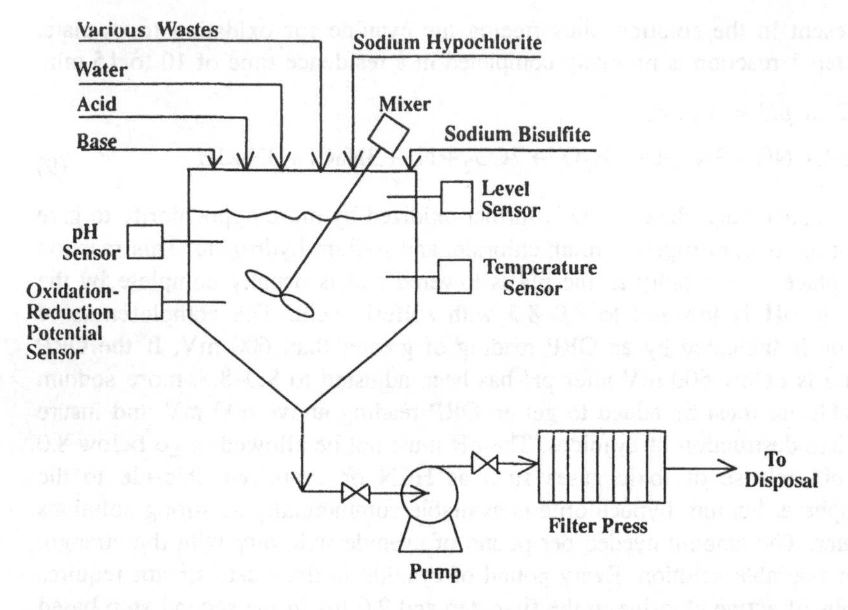

Figure 8.4 Sketch of batch apparatus for performing chromium reduction, cyanide destruction, and pH adjustment in a common unit with blocked operation.

tating agents such as sulfides, carbonates, phosphates, and sodium borohydride are also effective for a variety of metals. Very specific precipitation agents can be used when a waste stream contains only one metal or when recovery of a specific metal is required from a waste containing a variety of metals.

8.3.1 Chemistry of Precipitate Formation

The metals that can be precipitated from aqueous wastes in an insoluble form include arsenic, barium, cadmium, chromium, copper, lead, mercury, nickel, selenium, silver, thallium, and zinc. It is important to note that these metals are not present as elemental metal dissolved in the wastes, but are metallic ions from soluble dissociated metal salts. They can react with appropriate anions to generate insoluble particles of precipitate. The two most generic anions for metal precipitation are hydroxides and sulfides. However, as mentioned above, other specific anions are often used when metal recovery is the objective of treatment or only a single metal is present in the waste.

Hydroxide Precipitates

Most metals can be precipitated as hydroxides or oxides at high pH. The hydroxide precipitates are formed by adding hydroxide ions from lime or caustic soda to the waste stream. Insoluble metal hydroxides are formed as shown in Eqs. 10 and 11. The metal is shown as a divalent positive ion in Eq. 10 and as a trivalent positive ion in Eq. 11. In both cases, the insoluble metal hydroxides precipitate from solution as a rather gelatinous solid.

$$M^{++} + Ca(OH)_2 \rightarrow M(OH)_2\downarrow + Ca^{++} \quad (10)$$

$$2M^{+++} + 3Ca(OH)_2 \rightarrow 2M(OH)_3\downarrow + 3Ca^{++} \quad (11)$$

Among the divalent metal ions which precipitate as hydroxides are cadmium, copper, iron, lead, mercury, nickel, tin, and zinc. Ions which precipitate as trivalent hydroxides are aluminum, chromium, bismuth, gold, iron, antimony, and titanium. Silver I, copper I, and thallium IV can also be precipitated as hydroxides. A complete list of metal cations that can be precipitated as hydroxides along with the pH range for which they precipitate is given in Appendix J of National Research Council (1983).

The reason for giving the effective pH range for precipitation of different cations is that many of the metal hydroxides are amphoteric. That is, they can act as either bases or acids. The metal serves as cation in the bases and as part of the anion in the acids. In the presence of sufficiently strong bases, i.e., at high pH, the metal hydroxides may redissolve, thus eliminating the hydroxide precipitate. Figure 8.5 shows how the metal hydroxide solubility varies with pH for a variety of metals. Each metal has its own optimum pH for precipitation, i.e., the minimum in its solubility curve. However in cases where the waste contains a variety of metals, an optimum must be chosen with the help of experience and "jar tests" to find the best compromise. If a satisfactory compromise cannot be found, it may be necessary to run more than one precipitation stage with a pH change betweeen stages.

Sulfide Precipitates

Compared to hydroxides, sulfides have both advantages and disadvantages as a precipitant for metals. The major advantages are that sulfides have a lower solubility in water than hydroxides and do not have a pH sensitive minimum in the solubility/pH curve. This gives better metals removal and avoids redissolution of metals at high pH since the sulfides are not amphoteric. The solubilities of typical metal sulfides as a function of pH are illustrated in Figure 8.6.

The primary disadvantages to precipitation of metals as metal sulfides are (1) the potential for forming toxic hydrogen sulfide gas upon addition of sulfide chemicals to the metal-containing waste and (2) the possibility of leaving excessive sulfide in the effluent, which would then require posttreatment. As in all these chemical treatments, some excess reagent is required to guarantee completeness of the desired reaction. Neville K. Chung (1988a) has pointed out that the disadvantages of sulfide precipitation can be effectively avoided when

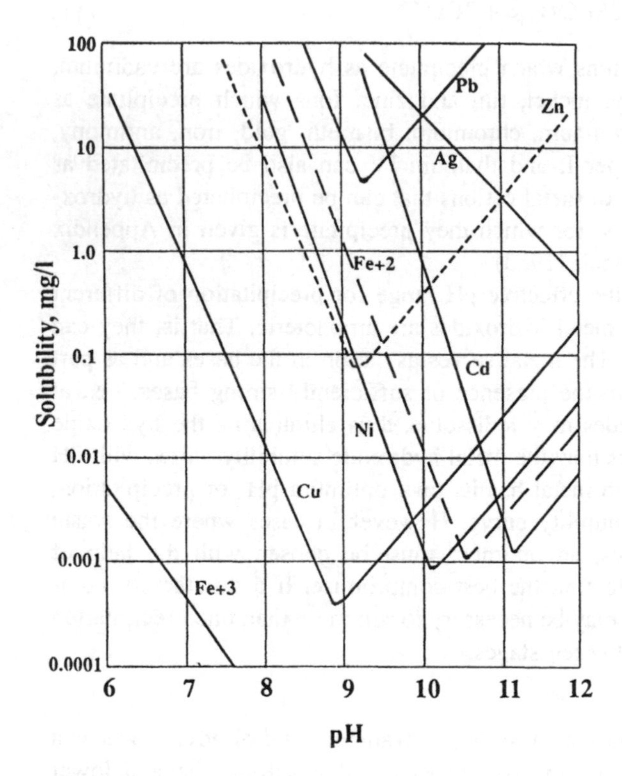

Figure 8.5 Solubilities of metal hydroxides as a function of pH. [From EPA (1983).]

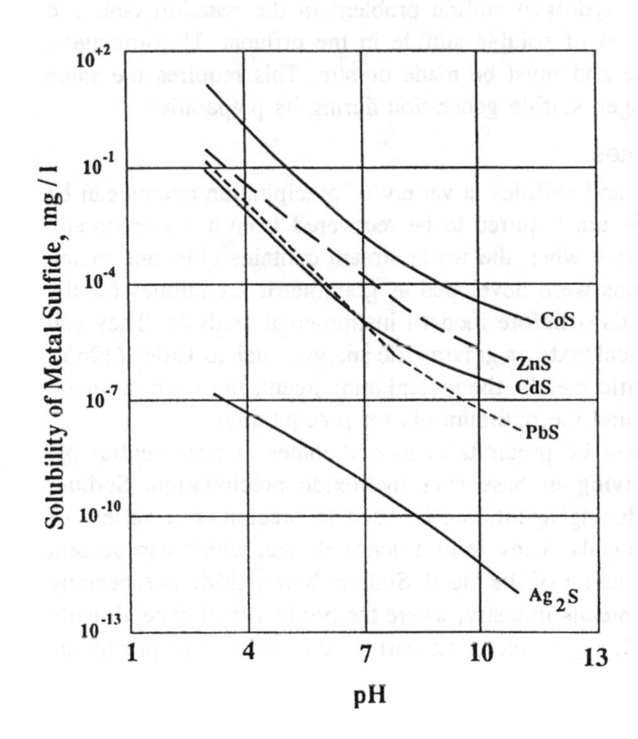

Figure 8.6 Solubilities of metal sulfides as a function of pH. [From EPA (1983).]

adding sulfide as soluble salts such as sodium sulfide or bisulfide. However, the addition must be done in a closed system with adequate ventilation, and the sulfide dosage must be carefully controlled by using a sulfide-specific ion electrode to measure free sulfide. This system is capable of maintaining a free sulfide concentration of 0.5 ppm. Chung (1988a) also pointed out that ferrous sulfide, which is only slightly soluble in aqueous streams, is an effective sulfide precipitant. It is soluble enough to precipitate other heavy metals, but its own solubility is low enough to maintain a very low free sulfide concentration. This essentially eliminates the hydrogen sulfide problem in the reaction tank and ensures a low concentration of soluble sulfide in the effluent. Unfortunately, ferrous sulfide is unstable and must be made on-site. This requires the same precautions against hydrogen sulfide generation during its preparation.

Specific Metal Precipitates

In addition to hydroxides and sulfides, a variety of precipitation agents can be used when specific metals are required to be recovered from a waste stream. They are especially effective when the waste stream contains only one metal. These precipitation methods were developed as gravimetric techniques for the analysis of metals in the days before modern instrumental analysis. They can usually be found in analytical texts on gravimetric analysis such as Erdey (1965). Table 8.4 lists some specific metals, the precipitating agent, the form in which the metal is precipitated, and the optimum pH for precipitation.

Cadmium and lead can be precipitated as carbonates at near neutral pH with a large attendant saving in base over hydroxide precipitation. Sodium borohydride, a strong reducing agent, can be used to precipitate a variety of metal ions as elemental metals. They form a dense sludge, which can be sent to metals recovery for recycling of the metal. Sodium borohydride is especially effective for the precious metals industry, where the product is of exceptionally high value. The use of sulfate to precipitate barium and chloride to precipitate

Table 8.4 Specific Metal Precipitating Agents

Metal	Precipitating agent	Form of precipitate	Optimum pH
Cd, Pb	Na_2CO_3	$CdCO_3$, $PbCO_3$	7.5–8.5
Pb, Hg, Ni, Cu, Cd, Au, Ag, Pt	$NaBH_4$	Elemental metals	8–11
Ba	Na_2SO_4	$BaSO_4$	4–5
Ag	NaCl	AgCl	7–10
Trivalent Fe, Al, Cr	Na_3PO_4	$FePO_4$, $AlPO_4$, $CrPO_4$	Acidic

silver are classical analytical methods. They are used in the gravimetric analysis of barium and silver as the sulfate and the chloride, respectively. Finally, the use of phosphate as a precipitant permits iron, chromium, and aluminum to be removed selectively from mixed wastes containing lower valence metal ions.

The preferred anions for precipitating all the metals are given in Table 6.1 from a report of the National Research Council (1983). They are primarily hydroxides, secondarily sulfides, and occasionally sulfates, carbonates, and chloride.

8.4 SEDIMENTATION

After a solid precipitate has been generated by addition of appropriate chemicals, the problem becomes one of separating the solids from the carrier liquid of the slurry. This must be done to recover the solids as a concentrated slurry or sludge and generate a clarified liquid for further treatment or disposal. This separation of solids from a liquid can be made by filtration or centrifugation, which are discussed in Chapters 6 and 7, respectively. However, in wastewater systems it is usually done by sedimentation, which is defined as the partial separation of suspended solid particles from a liquid by gravity settling.

Sedimentation is a physical process where particles suspended in a liquid are forced to settle by a combination of gravity and inertial forces on both the solid particles and the liquid carrier. The net forces on a particle in the desired direction of settling must be greater than the forces tending to move the particle in the opposite direction. This net force causes the particle to move in the desired direction. This direction would be upward for low density particles which float in water, or downward for high density particles such as the metal hydroxides and sulfides.

The basic elements of apparatus required for sedimentation processes are given by Cheremisinoff (1990) as:

- A basin or tank large enough to hold the wastewater being treated in a relatively quiescent state for long enough for the particles to settle
- A feed inlet into this reservoir which is conducive to settling and not remixing
- A means of removing settled particles or floating solids from the liquid

Sedimentation can be achieved in either batch or continuous equipment. However, continuous equipment is much more common, especially where large volumes of wastewater are to be treated. In recent years a number of manufacturers have developed compact, continuous units for the entire pretreatment of wastewater. These units provide first chemical precipitation and pH adjustment and then sedimentation and clarification of the resulting slurry. Sedimentation equipment and suppliers are discussed in Section 8.6 after further discussion of the process and its applications.

8.4.1 Fundamentals

The rate of sedimentation of a suspension of solid particles in a liquid depends on (1) the driving force for settling and (2) the resistance to settling. The driving force is a function of the solid particle size and the density difference between the solid and liquid phases. The resistance is largely a function of the viscosity of the liquid phase. However, the surface properties of the solid particles in the suspension and the ability to modify them by the use of small amounts of additives can make large differences in the rate at which suspensions settle. A generalized diagram relating settling regimes of suspensions of particles to particle stickiness (coherence) and solids concentration is reproduced from Perry (1984) in Figure 8.7.

"Totally discrete" particles shown at the left of the abscissa are usually larger than 20 μm in diameter and have little or no tendency to stick together. They are encountered in processing minerals, crystals of salts, and other such simple substances. At the other end of the abscissa scale, the "extremely flocculent particles" include (1) particles smaller than 20 μm unless they are stabilized as a dispersion due to repulsion of surface charges on the particles, (2) metal

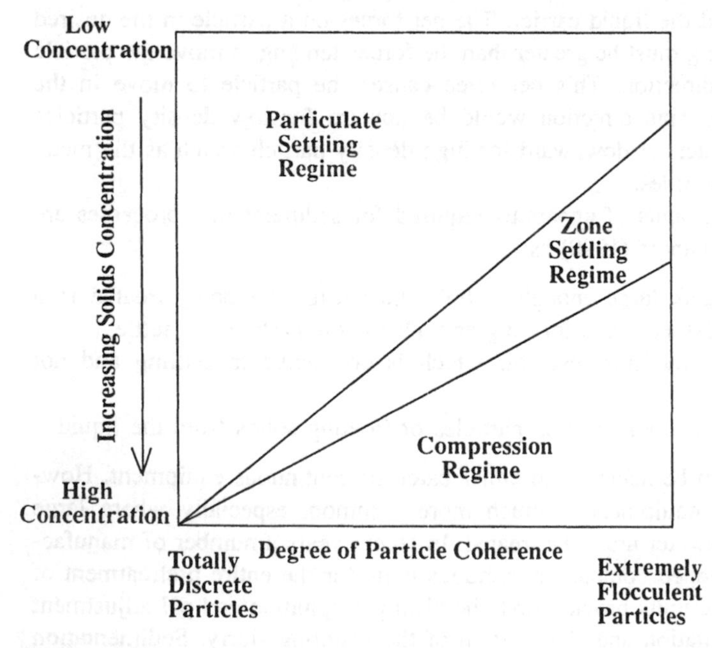

Figure 8.7 Interaction between particle coherence and solids concentration as it affects settling characteristics of a suspension. [From Perry (1984).]

hydroxides, and (3) many chemical precipitates. The materials normally encountered in chemical precipitation for wastewater treating are usually of the flocculent type.

At low concentration, solid particles are sufficiently far apart to settle freely regardless of their coherence or stickiness. Denser or larger particles, which settle faster, may collide with slower settling ones. If they do not stick on collision they will continue to settle at their own specific rates. Those that do stick on collision will form larger diameter floccules. These will settle at an even faster rate than the individual particles, but still as individual particles of floc.

In the zone settling regime, the particles are concentrated enough and sticky enough to form a mass of solid, which settles as a unit. The settling rate of this mass is a function of its solids concentration, other conditions being the same. As particle stickiness increases, the required particle concentration needed to form a mass of particles is decreased, making it easier to enter the zone settling regime.

As solids concentration is increased, it will finally reach a level where the particles interfere with each other during settling. Under these conditions, the weight of the particles in the upper portion of the settled floc bed presses on the settled floccules underneath, thus deforming them to fill the interfloc voids and squeeze out the interstitial liquid. This compacting of the settled floc bed by compression due to its own weight is called the compression regime of settling. In the compression regime, the rate of settling is a function of the depth of the pulp in the settled floc zone and the concentration of solids in the pulp, i.e., the weight of the floc bed.

Figure 8.7 illustrates that for particles which do not stick to one another, the effects of zone settling or compression are either totally absent or very minimal. Under such conditions simple settling devices are very satisfactory and can produce settled solids containing less than 20% liquid. However, for the flocculent metal hydroxides which have major zone settling and compression regimes, special thickening and clarification apparatus, as well as additives to modify the flocculent nature of the solids are needed to get satisfactory product streams. Starting with a feed containing a low concentration of solids, it may pass through all of the settling regimes shown in Figure 8.7 as it is thickened due to sedimentation. Any one of these regimes may be rate-determining and thus may fix the size of the required sedimentation equipment.

8.4.2 Coagulation

Historically, the terms coagulation and flocculation have been used interchangeably. Both terms were used to describe a process for treating suspensions of solids in liquids whereby particles too small to settle effectively can be made to agglomerate into larger, more settlable particles. However, the mixing regimes needed and the mechanisms of operation involved are different in the two cases. Therefore we shall discuss them separately even though some settling additives are capable of performing both functions.

Coagulation is the term usually applied to the neutralization of surface charges on particles of colloidal or near colloidal size. These charges are of the same sign for all the particles in a given suspension and can be either positive or negative. Thus they repel each other and hold the particles apart preventing their agglomeration. Destroying the repulsions between particles by neutralizing the charges that stabilize the suspension allows the particles to approach each other closely enough to touch. They can then collect into larger aggregates which can settle effectively.

The additives that are effective for neutralizing surface charge on solid particles are highly ionic electrolytes, preferably containing ions of multiple charge. The effectiveness of trivalent ions is 10 to 100 times that of monovalent or divalent ions, as measured by coagulation time. Typical coagulation additives are acids, lime (calcium hydroxide), alum (aluminum sulfate), and ferric chloride or sulfate. The additive is mixed into the wastewater with intense mixing to distribute it rapidly throughout the feed stream. This is in direct contrast with flocculation, which is enhanced by gentle mixing.

8.4.3 Flocculation

Flocculation is the further enlargement of agglomerated particles by either the presence of sticky metal hydroxides or the addition of an organic polymer. The suspension is gently mixed to allow bridges to form between particles. This binds them together into larger agglomerates called flocs. The formation of these larger flocs is enhanced by mild mixing, while intense mixing breaks the bridges and redisperses the particles. Some additives such as aluminum sulfate (alum), ferric salts, and certain ionic polymers (polyelectrolytes) can enhance both agglomeration and flocculation.

The aluminum and iron salts hydrolyze to form insoluble aluminum or iron hydroxide gels, which form flocs that collect and trap colloidal materials. The polyelectrolytes are polymers of water-soluble organic molecules. They are large organic molecules having polar groups spaced along their length. These polar groups can be cationic, anionic, or non-ionic. The ionic groups give the polymer water solubility as well as the ability to dissociate and form a polymer molecule with ionic charges spaced along its length.

These polymers work by dissociation of the ionic sites in the polymer, thus providing ionic charges on the polymer. The ionic charges then interact with the charge on the colloidal particles and form flocs by collecting a lot of individual particles to a single polymer molecule. They also bridge from one polymer molecule to another. Cationic polymers contain nitrogen in amine or quaternary ammonium groups as the active sites. Anionic polymers contain carboxylic, sulfate, sulfonate, phosphate, or any of the ionizable derivatives of mercaptan groups. The non-ionic polymers work by attracting the charges on the particles to the polar sites in the polymer, which normally contain oxygen as alcohol, ether, aldehyde, or ketone groups.

8.4.4 Jar Tests

For design purposes, the effectiveness of a given combination of pH, mixing intensity, and additive package on clarification of a specific wastewater feed can be determined in a pilot plant or bench scale treating unit. However, this is an expensive and slow way to follow the modifications, which need to be made in day-to-day operation as changes occur in the composition of the wastewater feed to a given treating plant. To overcome this problem, plant operators and plant designers have developed a technique called *jar tests*. These give a quick reading on the effect of operating changes on operating results. The name apparently comes from the fact that these tests were developed using vertically calibrated round battery jars. They are still used today along with tapered conical Imhoff tubes and graduated cylinders to measure settling rates and settleability of suspensions and sludges.

In the jar tests the required precipitant dosages, optimum pH, settling aid requirements, and effluent concentrations achievable by chemical precipitation and sedimentation can be determined. For batch treating, each batch of wastewater can be jar tested to determine the optimum recipe for effective settling. For continuous units, the jar test can predict the solids content of the clarified overhead product as well as the water content of the underflow-settled pulp or floc. Design overflow and underflow rates for thickeners and clarifiers should not exceed half of those indicated by the jar tests. A separate jar test is required for each combination of pH, chemical dosage, and additive dosage. As experience is gained with various wastewater compositions, only a few tests are required to optimize settling conditions.

Clarification

The jar test for clarification consists of taking an approximately 2- to 3-qt sample of the wastewater to be treated in a calibrated glass or plastic battery jar, adjusting pH, adding any necessary chemicals, and mixing thoroughly. Then a specific dose of settling additives is added with mild stirring to disperse the additives, and then the settling of the floc formed is timed. Analytical samples for measurement of solids content can be taken from immediately below the vertical midpoint of the settling jar contents, taking care not to resuspend solids that have settled. The time intervals for sampling should increase consecutively, such as 5, 10, 20, 40, and 80 min. The suspended solids concentration plotted on a log-log scale against total detention time usually gives a straight line. The static detention time required to achieve a specific solids concentration in the overflow from an ideal settling basin can be picked directly off the line, which has the form given in Eq. 12.

$$C = K t^m \tag{12}$$

where:

C = solids concentration
K = a constant coefficient
t = time
m = a constant exponent

The values of K and m are characteristic of the particular suspension being tested. To get the settling time/effluent solids concentration relationship for an actual clarifier, compared to an ideal settling basin (or the jar test), some sort of efficiency must be included. This is to account for the effects of turbulence and non-uniform flow in the actual clarifier. Dahlstrom et al. (1984a) give such a detention efficiency as a function of the diameter-to-depth ratio in a circular continuous clarifier. To get the desired solids concentration in the effluent from the continuous clarifier, the time obtained in the jar test must be divided by the efficiency to obtain the retention time required in the continuous clarifier. This efficiency usually is in the range of 35 to 50%, or 0.35 to 0.50, for diameter-to-depth ratios varying from 4 to 2. Higher efficiencies are obtained at lower diameter-to-depth ratios.

The detention time required to achieve a given effluent solids concentration from an actual clarifier is determined from the jar test retention time and the clarifier efficiency. Then the volume of the clarifier is determined from the overflow rate times the time required for settling. This volume is the volume in the clarifier above the level of the settled pulp interface.

Thickening

In contrast with clarification, which makes a clear effluent from a dilute suspension of solids in liquids, thickening has as its objective the production of a concentrated solid slurry or sludge from a relatively concentrated suspension of solids in a liquid feed. Whereas clarification specializes in getting solid particles out of the liquid, thickening is aimed at removing liquid from the solid particles. Thus, where settling of particles from the liquid determines clarifier size, collapse of floc beds to release water determines thickener size for aqueous suspensions.

A method developed by Kynch (1952) makes use of a plot of pulp settling level versus time to determine the thickener area required to obtain the desired underflow solids concentration. This test is normally run in a graduated cylinder

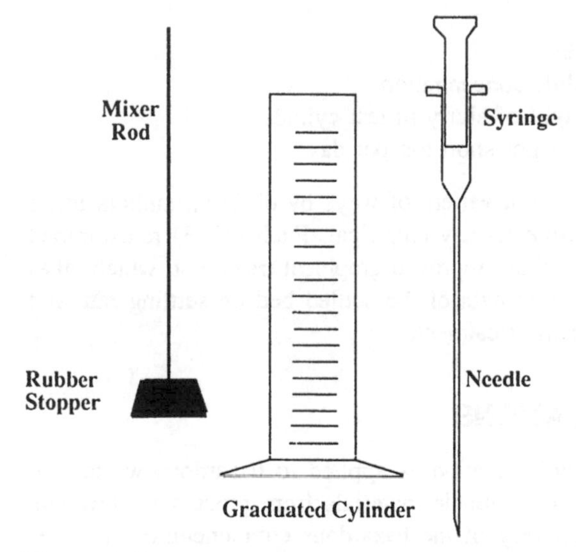

Figure 8.8 Sketch of apparatus for determining settling rate of pulp in a thickener.

and is the preferred test when flocculation with polymer additives is utilized. To run the test, apparatus of the type shown in Figure 8.8 is collected.

An inverted rubber stopper with a diameter about three-quarters that of the graduated cylinder is equipped with a rod for gently raising and lowering it through the sample. This provides mild agitation during the addition of flocculating agents to the sample at the beginning of the test. The syringe and a long needle are supplied to provide a means of adding the flocculating agents. To start the test, a sample of pulp to be settled is placed in the graduated cylinder. The flocculating agent solution is then placed in the syringe, and the flocculating agent is added to the pulp over a period of 3 to 15 s. Mild agitation of the pulp is provided during the addition by raising and lowering the stopper through the cylinder from top to bottom. The stopper is then removed from the cylinder and the position of the top of the settled pulp bed is recorded versus time until the the desired underflow concentration is reached.

The results of the settling time data can be converted to a term called *unit area* which is simply the plan area of the thickener required to settle a certain mass of solids per unit time. Any consistent set of units can be used, a typical one being square feet per short ton of solids per day. The unit area is defined by Eq. 13.

$$\text{Unit area} = \frac{t_u}{C_o H_o} \tag{13}$$

where:

t_u	=	time in days
C_o	=	initial solids concentration
H_o	=	initial height of slurry in test cylinder
Unit area	=	square feet per short ton per day

The value of t_u was estimated in a variety of ways by different authors using graphical procedures which were largely empirical. Fitch (1983) re-examined the Kynch method and proposed an improved graphical treatment, which takes into consideration the effect of growth of the settled bed on settling rate and thereby improves the rigor of the treatment.

8.5 SPECIFIC APPLICATIONS

Chemical precipitation and sedimentation as applied to hazardous wastes are usually used to remove toxic or valuable materials from process wastewaters or used oils. This permits recovery of the hazardous component(s) for either recycle or disposal of a greatly reduced volume. This also renders the non-hazardous component(s) suitable for recycling or disposal in the case of aqueous systems and for recycling or re-refining in the case of used oils.

8.5.1 Aqueous Systems

The aqueous systems that can be rendered non-hazardous by chemical precipitation and sedimentation usually contain toxic metals. These streams include process wastewaters and contaminated groundwaters recovered from cleanup of uncontrolled hazardous waste dump sites, such as the Superfund sites. The

Table 8.5 Metal-Containing Hazardous Wastes Identified by EPA

EPA identifying number	Metal contaminant
D002	Corrosion products
D004	Arsenic
D005	Barium
D006	Cadmium
D007	Chromium
D008	Lead
D009	Mercury
D010	Selenium
D011	Silver
D062	Iron, nickel, chromium

specific hazardous wastes in this class that have been identified by RCRA and the EPA are given in Table 8.5. These wastes usually come from (1) the ferrous and non-ferrous metal finishing industries (as pickling, plating, and electropolishing solutions), (2) the electronics industry (as cleaning and plating solutions), and (3) the mining and inorganic pigments industries. As discussed in Section 8.3, these metals are usually precipitated as hydroxides or sulfides and settle with extensive flocculation.

8.5.2 Metals from Waste Oils

The metals in waste automotive lubricating oils are either (1) wear products from metal parts in internal combustion engines, (2) lead from leaded gasoline which gets into the oil as piston blowby from the combustion, or (3) metals contained in the commercial additive packages blended into the oil by the manufacturers to improve performance. Typical metals found in waste automotive oils are given in Table 8.6 along with their quantities, sources, and allowable levels under EPA regulations.

In preparing these used oils for recycling or re-refining, the metals are removed by either (1) chemical precipitation as an insoluble solid sludge, (2) physical solvent precipitation as a dense liquid phase or asphalt, or (3) vacuum distillation as a bottoms pitch or residue. In the case of chemical precipitation, a suitable precipitating agent is added to the oil as an aqueous solution and the mixture is cooked at elevated temperature and pressure to generate solid reaction products containing the metals. The solid reaction products are removed from the oil by filtration with the help of solid filter aids. However, this presents a

Table 8.6 Typical Metal Concentrations in Waste Motor Oils and Maximum Levels for Unrestricted Burning

	Typical level, ppm		Maximum level allowed	
Metal	Mean	Median	California CCR Title 22	U.S. EPA
Arsenic	17	5	5	5
Barium	132	48	100	NA
Cadmium	3	3	2	2
Chromium	28	7	10	10
Lead	665	240	50	100
Zinc	580	480	250	NA

NA, not applicable

disposal problem for the metal-contaminated solids generated. The chemicals used for precipitation of metals from oils have included diammonium phosphate, ammonium sulfate, aluminum sulfate, and sulfuric acid and clay. The trend is away from chemical precipitation and toward vacuum distillation. This is because of easier disposal of the metal-containing bottoms from vacuum distillation as asphalt or roofing tar.

Waste industrial hydrocarbon oils such as hydraulic oils are treated in much the same way as automotive lubricants to prepare them for reuse. Waste oil-containing coolants and water emulsions from the machine shops of the metal shaping industries are treated like aqueous wastewaters, as described in Section 8.3, prior to recycle or disposal.

8.6 EQUIPMENT AND VENDORS

The process apparatus for chemical precipitation and sedimentation comes either as batch or continuous units and range in size from processing a few gallons per minute up to over 2,000 g/min. Batch processing is generally preferred for volumes under 30,000–40,000 g/day (21–28 g/min). However, continuous flow package units for integrated chemical precipitation and sedimentation are available for volumes as low as 5 g/min. The components of a complete wastewater treatment plant may be separate units, such as a tank, a mixing vessel, and a settling pond, or combined in a highly complex integrated package plant that

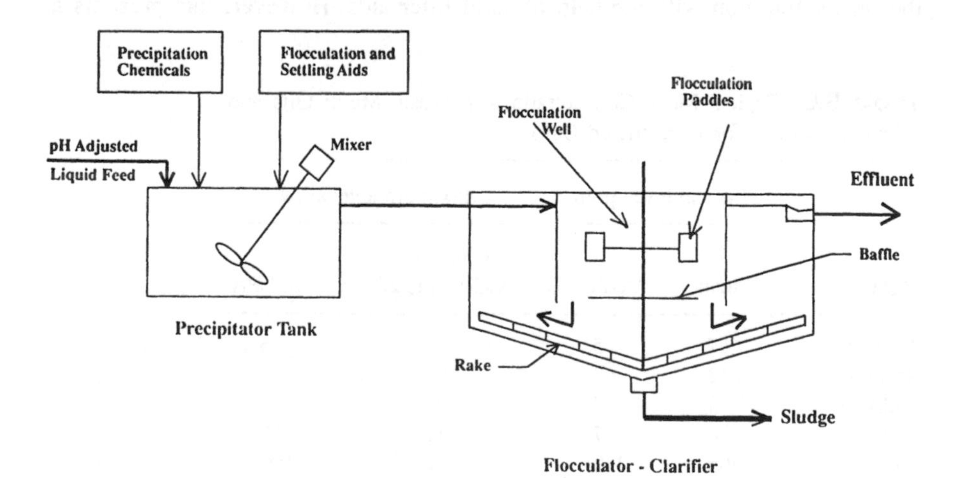

Figure 8.9 Schematic of components needed in a chemical precipitation treating plant.

is factory assembled. In either case, the functions carried out in the flow sheet given in Figure 8.9 are probably needed.

Feed systems to add the chemicals needed for precipitation, coagulation, and flocculation must be provided. Furthermore, the design must also provide the mixing volume necessary to thoroughly mix the chemicals with the pH-adjusted wastewater feed. The effluent suspension from the precipitation tank is held in a relatively quiescent holding volume to allow time for the suspension to settle. Very mild mixing may be required to speed up flocculation and break up the settled floc layer, helping to increase its solids content. The types of equipment used as clarifiers, thickeners, and package plants, as well as the vendors listed for them in the Pollution Engineering Buyer's Guide (1991), are discussed below.

8.6.1 Clarifiers and Thickeners

The clarifiers and thickeners used with chemical precipitation have essentially the same design and are used to provide a clear overhead effluent and a concentrated sludge underflow. They are called *clarifiers* for dilute suspension feeds and *thickeners* for more concentrated suspension feeds. They can be either rectangular or circular, but the types used with chemical precipitations which require flocculation for adequate settling are usually round. They may have built into them separate zones for chemical mixing, precipitation, flocculation, and settling, in which case they may be called reactor-clarifiers. These are essentially large-scale versions of integrated package treating plants. A schematic drawing of a reactor-clarifier is given as Figure 8.10. This unit is of the type given in Eckenfelder et al. (1985).

These units vary in size from about 30 to 200 ft in diameter. They include a wide variety of design details for the various zones and mixing devices in the units provided by different vendors. However, they all include the zones shown in Figure 8.10. The vendors for sedimentation equipment are given in Table 8.7.

Inclined Plate Clarifiers

A popular settling device which is increasing in use for clarification of solid suspensions is the tilted plate separator commonly used for settling of immiscible liquid suspensions discussed in Chapter 5. To be effective for solid suspensions, the bundle of plates (or tubes) must be inclined at an angle greater than the angle of repose of the settled sludge so that the settled sludge will slide down the slanted plates in counterflow to the upward-flowing clarified liquid. The purpose of the bundle of plates is to provide narrow settling channels that minimize the particles' settling distance, and thus the time required for settling. The steep angle of 55–60 degrees from the horizontal required for solid suspensions provides a lot of settling area in a given horizontal cross-section. This makes this device very useful in combination package units where area is an

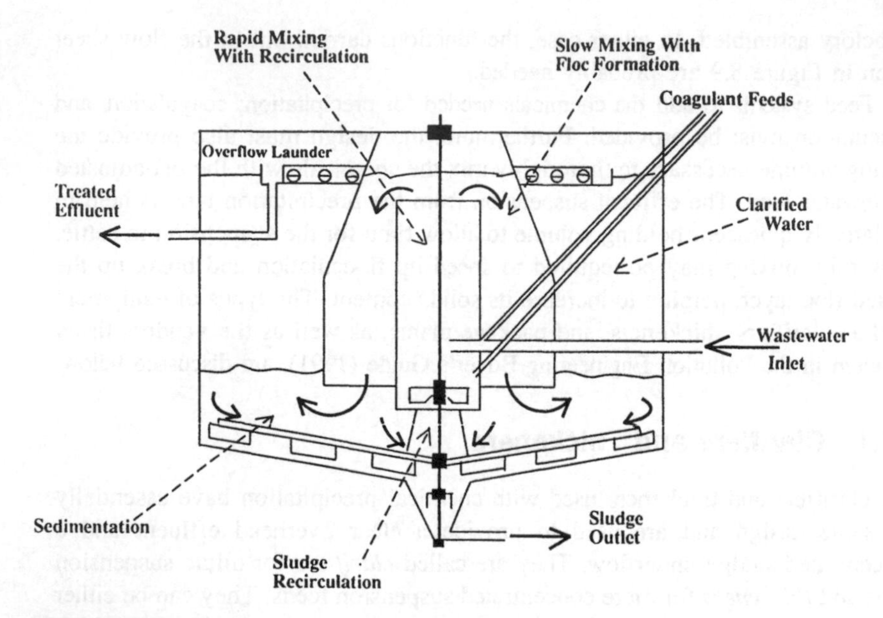

Figure 8.10 Schematic of a reactor-clarifier designed for both coagulation and settling. [Adapted from Fig. 6 of Eckenfelder et al. (1985).]

Table 8.7 Vendors of Sedimentation Equipment

CBI Walker, Inc. 1245 Corporate Blvd. Suite 102 Aurora, IL 60504 (708) 851-7500	Dorr Oliver, Inc. 612 Wheelers Farm Road RD Box 3819 Milford, CT 06460 (203) 876-5500
Eimco Process Equipment Co. P.O. Box 300 Salt Lake City, UT 84110 (801) 526-2000	Filtronics, Inc. 1157 N. Grove Street Anaheim, CA 92806 (714) 630-5040
Great Lakes Environmental 463 Vista Addison, IL 60101 (708) 543-9444	Permutit Co. 30 Technology Drive Warren, NJ 07059 (908) 668-1700
Schreiber Corporation, Inc. 100 Schreiber Drive Trussville, AL 35173 (205) 655-7466	WesTech Engineering, Inc. P.O. Box 65068 Salt Lake City, UT 84165 (801) 265-1000

important consideration. However, this design does not permit internal flocculation, and it results in a variable and less concentrated sludge than the other thickeners. Thus, it requires separate flocculation equipment and tankage as well as final sludge thickening before disposal. However, its area efficiency makes it a popular component of packaged combination treating plants. Its major drawbacks are a lower solids underflow concentration than other gravity settlers and the difficulty of cleaning if scaling or deposition of solids occurs.

8.6.2 Packaged Combination Units

The main differences between a reactor-clarifier and a packaged combination unit are the use of a slanted-plate settler and a sludge-aging chamber in the latter. This gives a more compact unit and still provides adequate solids content in the final sludge underflow. The elements in a typical packaged unit are illustrated in Figure 8.11. These elements perform the same functions as those shown in Figure 8.9, but use separate chambers for flocculation, sedimentation, and sludge handling. The total package is very compact; a photograph of a typical unit is given in Figure 8.12.

The packaged plants differ in design details but contain apparatus to perform the same basic functions. These are pH adjustment, precipitation, flocculation, settling, and sludge concentration. A list of vendors for packaged units for chemical precipitation and sedimentation is given in Table 8.8.

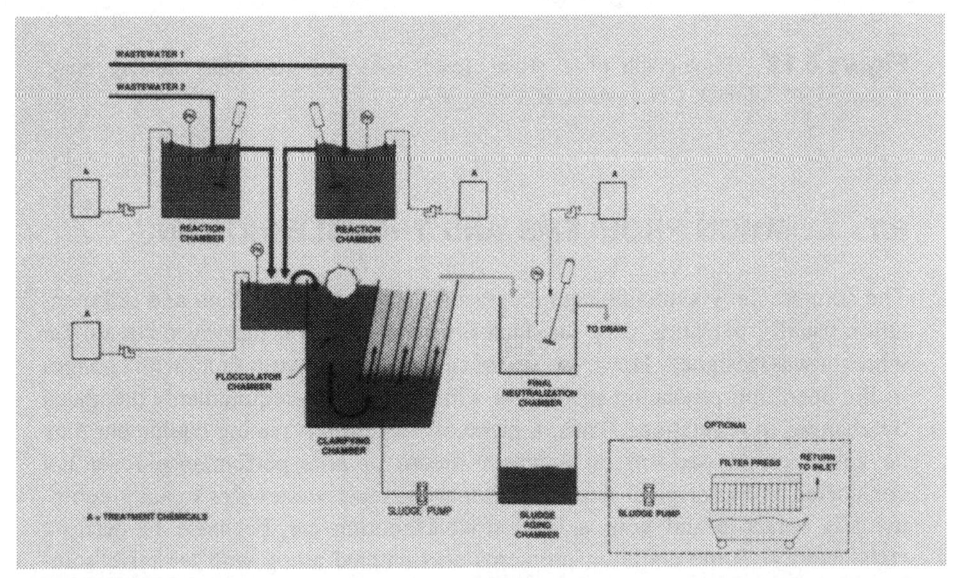

Figure 8.11 Typical elements in a packaged water treating plant. (Courtesy of Memtek Corporation.)

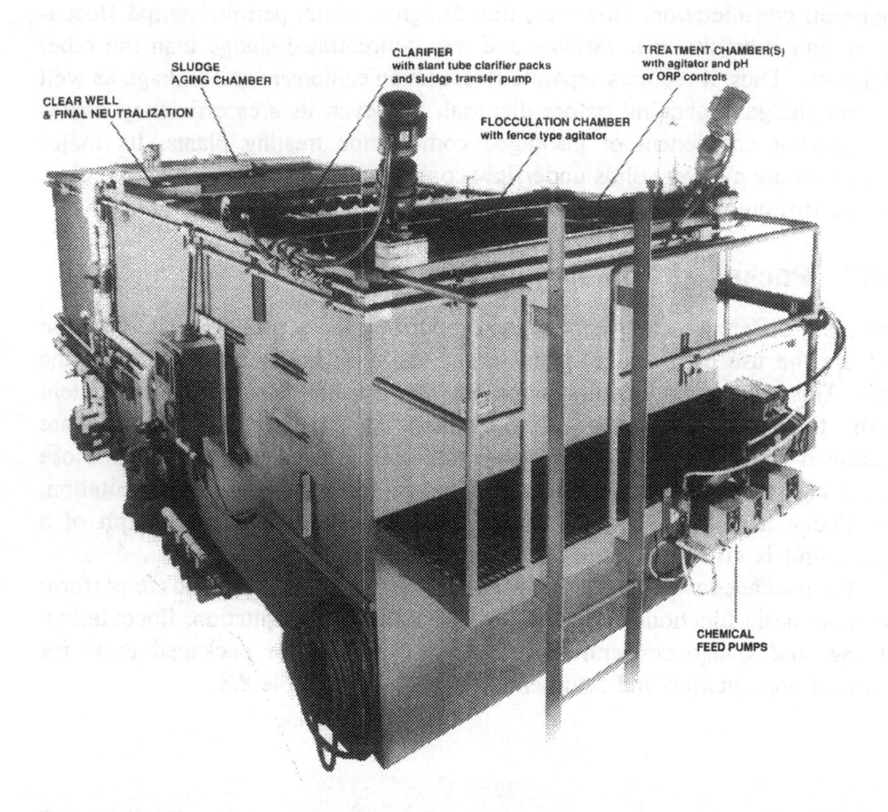

Figure 8.12 Photograph of a "Total Treat" packaged wastewater treating plant. (Courtesy of Memtek Corporation, Billerica, MA.)

8.7 COMMON PROBLEMS AND TROUBLESHOOTING

The commercially available equipment for chemical precipitation and sedimentation usually performs very satisfactorily when used on the waste stream for which it was designed. However, waste streams change with time, with changes in the operation producing them, and with the effluent requirements mandated by changes in regulations. Thus, a piece of wastewater-treating equipment may be called on to perform on a waste stream or at a performance level not considered in the original design. Fortunately, most wastewater treating equipment is versatile and with a little troubleshooting can be made to perform satisfactorily. Some common problems encountered along with tentative solutions are addressed in the following sections.

Table 8.8 Vendors of Packaged Units for Chemical Precipitation and Sedimentation

Andco Environmental Processes, Inc. 595 Commerce Drive Amherst, NY 14150 (716) 691-2100	CPC Microfloc Route 20, Box 36 Sturbridge, MA 01506 (508) 347-7344
Duriron Company, Inc. P.O. Box 1145 Dayton, Ohio 45401 (513) 226-4000	Filtronics, Inc. 1157 N. Grove St. Anaheim, CA 92806 (714) 630-5040
Great Lakes Environmental, Inc. 463 Vista Addison, IL 60101 (708) 543-9444	Lancy International, Inc. 181 Thorn Hill Rd. Warrendale, PA 15086 (412) 772-0044
Memtek Corp. 28 Cook Street Billerica, MA 01821 (508) 667-2828	Met-Pro Corp, Systems Div. 238 Cassell Rd, Box 144 Harleysville, PA 19438 (215) 723-6751
Niagara Environmental Assoc., Inc. 88 Okell St. Buffalo, NY 14220 (716) 822-3921	Pasco Water Pollution Control, Inc. P.O. Box 401 Devault, PA 19432 (800) 548-2545
Pollution Technology Systems 2922 Benton St. Garland, TX 75042 (214) 272-4010	Quantum Technologies, Inc. Jones Mill Industrial Park, Box 223 Jones Mill, AR 72105 (501) 844-4700
Smith and Loveless, Inc. 14040 Santa Fe Trail Lenexa, KS 66215 (913) 888-5201	Tenco Hydro, Inc. 4620 Forest Ave. Brookfield, IL 60513 (708) 387-0700
Unocal Chemicals Division P.O. Box 7376 Fullerton, CA 92634 (714) 525-9225	Water Technology, Inc. 564 Industrial Way East Macon, GA 31201 (912) 743-3050
Western Filter Co. P.O. Box 16323 Denver, CO 80216 (303) 288-2617	Zimpro/Passavant, Inc. 301 W. Military Rd. Rothschild, WI 54474 (715) 359-7211

8.7.1 pH Adjustment

The most common problem with pH adjustment is overshoot of the endpoint. This is primarily due to the logarithmic nature of the pH scale which requires a lot of acid or base to change the pH by one unit near pH = 1 or 14 at the ends of the pH scale, but very little at pH = 7, the neutral point. This was illustrated in Table 8.1. To get reasonably rapid neutralization at the ends of the pH scale requires relatively concentrated acid or base while to get sensitivity near the endpoint requires relatively weaker acid or base and less of it. For batch equipment it is often recommended that the final endpoint be approached manually rather than automatically to give a more sensitive approach to the endpoint. This permits addition of small amounts of dilute acid or base on an intermittent basis with stirring time allowed between additions. The entire reactor contents then equilibrate without continuous addition of reagent.

For continuous units, it is recommended that addition of acid or base be made in two or more stages with pH control in each stage, as was illustrated in Figure 8.3. This is done to spread out the approach to the endpoint in steps. A further refinement would be to use less concentrated reagent in each stage as the endpoint is approached. This would increase the complexity of the chemical feed storage but would make the approach to the endpoint much easier to control.

A second problem common to pH adjustment for concentated acid or basic waste streams is the heat generated on neutralization. Such acid streams are found in metal-containing wastes from pickling, plating, and metal finishing operations. With such concentrated streams, the use of concentrated neutralization agents can release sufficient heat to cause localized boiling or splashing at the point of addition of neutralization reagents. Chung (1988b) has pointed out that excessive temperature rise can be avoided in a batch treating tank in a variety of ways:

- Use dilute concentrated wastes, preferably with dilute wastes of the same type.
- Put reagents and dilution water into the tank prior to adding concentrated waste.
- Add reagents slowly to permit cooling by heatloss from the reaction tank.
- Cool the reaction tank externally or by aeration.

For continuous units, feed dilution and external cooling are the preferred methods.

8.7.2 Precipitation

Precipitation of metals as hydroxides requires good control of pH during the precipitation. This is because each metal hydroxide has an optimum pH giving a minimum solubility of the precipitate. At lower and higher pH, the solubility of the precipitated metal hydroxide increases and the effluent contains more of the dissolved metal. This optimum pH is not the same for all metals in wastes containing mixtures of metals, and therefore a compromise pH may have to be determined by jar tests to get adequate effluent quality.

Precipitation reagents must be added to the wastewater in a region of intense mixing to get a uniform dispersion of precipitate and to maintain a uniform pH throughout the treated wastewater.

8.7.3 Flocculation and Settling

At the present state of surface chemistry, flocculation is more of an art than a science. While there are a lot of qualitative guiding principles, the quantitative aspects as applied to flocculation must be determined in jar tests. Thus, the optimum combination of pH, amount and kind of coagulation agents and settling aids, etc., must be determined for a given wastewater and precipitation agent to provide the best effluent clarification and sludge solids concentration. Since flocs are disrupted by strong agitation and enhanced by mild agitation, there also exists an optimum mixing rate in the flocculating zone of a reactor-clarifier, which can be determined only on the existing unit. Too intense mixing in the flocculation zone redisperses flocculated solids by breaking the weak bonds holding the particles together as a floc. This greatly inhibits settling and should be avoided.

If settling turns out to be inadequate due to changes in the wastewater composition, the remedy is probably to change the amount or type of flocculating agent used. Since there are many types of coagulating and flocculating agents available on the market, it is usually possible to find a satisfactory combination for proper settling of a given wastewater. The jar test can be used to find this combination. Polymeric flocculants come in a wide variety of molecular weights and functionalities. In general, anionic flocculants work well with neutral or alkaline waters, non-ionics work well with acid suspensions, and cationics are best for colloids and organic material.

8.7.4 Sludge Concentration

The desirable concentration of solids in the underflow from a clarifier-thickener is as high as can be reasonably pumped, i.e., about 50%. However the achievable level of solids in the underflow is usually below this, on the order of 10 to 20%. This achievable level is obtained only by (1) maintaining a depth of pulp in the bottom of the clarifier to increase its compression for dewatering the pulp and (2) raking the sludge to break up bridged particles and let them settle to a denser state. If inadequate underflow concentration is encountered, the remedy is once again adjusting the flocculation and settling aid dosage, as determined by jar tests, to improve settling.

8.8 ECONOMICS

Since the purpose of this chapter is not to aid design or permit cost estimating of wastewater treating plants, only a few rule-of-thumb generalizations about the cost of such plants will be given here. Eckenfelder et al. (1985) give cost estimates for a variety of wastewater treatment facilities varying both in complexity and size. The results show, as would be expected, that both capital and operating costs increase with both size and complexity. Capital costs increase as the 0.6 power of size and operating expense increases as the 0.85 power of size. Dahlstrom et al. (1984b) show a range of installed costs for thickeners from about $20/ft^2$ for a 400-ft diameter unit to about $315 to $510/ft^2$ for a 10-ft diameter unit. These figures are in 1981 dollars. Costs can vary widely with different types of construction. Furthermore, if chemicals are needed for flocculation, their cost can be the major operating cost. Equipment vendors should be contacted for more detailed cost estimates for specific applications.

8.9 SUMMARY

Chemical precipitation is widely used in both aqueous and non-aqueous systems to remove toxic metals from waste streams. The precipitate is concentrated by various sedimentation techniques, such as coagulation and flocculation. The process can be used for aqueous systems as a pretreating step prior to treating wastewaters in a publicly owned biological treating plant. In certain cases, it can be the complete treatment required. Package plants are available from many vendors in small to intermediate size. They include pH adjustment, chemical precipitation, flocculation, clarification, and sludge concentration into a single package. Large installations offering the same options can be custom designed for specific applications.

REFERENCES

Cheremisinoff, P. N. (1990). *Pollution Engineering*, September, 1990:68.

Chung, N. K. (1988a). *Standard Handbook of Hazardous Waste Treatment and Disposal* (H. M. Freeman, ed.), McGraw-Hill, New York, p. 7.24.

Chung, N. K. (1988b). *Standard Handbook of Hazardous Waste Treatment and Disposal* (H. M. Freeman, Ed.), McGraw-Hill, New York, p. 7.29.

Dahlstrom, D. A., Emmett, R. C., and Klepper, R. P. (1984a). *Perry's Chemical Engineer's Handbook*, 6th ed. (R. H. Perry and D. W. Green, eds.), McGraw-Hill, New York, pp. 19-54.

Dahlstrom, D. A., Emmett, R. C., and Klepper, R. P. (1984b). *Perry's Chemical Engineer's Handbook*, 6th ed., (R. H. Perry and D. W. Green, eds.), McGraw-Hill, New York, pp. 19-65.

Eckenfelder, W. W., Jr., Pataczka, J., and Watkin, A. T. (1985). *Chemical Engineering*, September 2, 1985:60.

Environmental Protection Agency (1983). Development Document for Effluent Limitations, Guidelines, and Standards for the Metal Finishing Point Source Category, EPA 440/1-83/091, June, 1983.

Erdey, L. (1965). Gravimetric Analysis, Part II, Pergamon Press, New York.

Fitch, B. (1983). Kynch Theory and Compression Zones, *AIChE Journal*, *29*:940, November, 1983.

Kynch, G. J. (1952). A Theory of Sedimentation, *Transactions of the Faraday Society*, *48*:166.

National Research Council (1983). Prudent Practices for Disposal of Chemicals from Laboratories, National Academy Press.

Newton, J. (1990). *Pollution Engineering*, November, 1990:94.

Perry's Chemical Engineer's Handbook, 6th ed., (1984). R. H. Perry and D. W. Green, eds., McGraw-Hill, New York.

Pollution Engineering (1991), Buyer's Guide, Pudvan Publishing Co., Northbrook, IL.

9
Membrane Processes

9.1 INTRODUCTION

The membrane processes commonly used for waste minimization are reverse osmosis, ultrafiltration, and microfiltration. Permeation and electrodialysis are also finding increasing use. With the exception of permeation, which is still considered a developing technology, all the above are commercial processes. These processes make use of a thin membrane barrier which can be produced with properties that vary over a wide spectrum. The membrane is essentially solid for reverse osmosis, permeation, and electrodialysis; it is porous with continuous discrete holes for microfiltration. Ultrafiltration membranes can be of either type depending on the application. The processes are somewhat arbitrarily defined by the apparent pore size of the membrane barrier used. This apparent pore size is reported for measured experimental diffusion rates even when it is known that no continuous pores or holes through the membrane exist. This can be done because the form of the diffusion equations is the same for solid and porous membranes. The pore size reported for the solid barrier is that which will give the same flow rate for a given pressure drop through a membrane which does have continuous pores.

The boundary between solid and porous membranes is further clouded by the physical chemistry of solutions. Some dissolved components, such as ions from salts, don't exist as simple entities but are associated with many solvent molecules to form a *solvation cluster*. This cluster is much larger than the simple ion. Furthermore, ions of higher valence attract more solvent molecules and thus have larger cluster sizes. This larger size accounts for their lower permeation rates than monovalent ions. Many polar molecules, such as organic acids, exist in solution as dimers, trimers, etc. This produces effective sizes larger than expected from the molecular weights of the monomeric molecules and thus lower than expected permeation rates.

Allowance for these factors often explains unusual behavior in flow through membranes. For example, ions may be excluded from a membrane by their electrical charge or by the size of their solvation shell. For the above reasons it is usually necessary to get experimental data on the specific application of a given membrane to get reliable data for design purposes.

Solid barriers are used in reverse osmosis, permeation, electrodialysis, and some ultrafiltration membrane processes. They are made of polymeric materials and may contain (1) regions that appear crystalline because segments of the polymer chains are aligned closely to give apparent crystalline order in the aligned region, (2) regions that are amorphous because the polymer chains are not aligned and are much more loosely packed and relatively mobile in the solid, and (3) regions that are structurally defined by the number and type of interchain cross-links between segments of the polymer. These structures are illustrated in Figure 9.1.

The transport of material through these apparently solid barriers occurs through the amorphous region of the polymer by the creation and movement of free volume in the polymer. This free volume moves due to the thermal motion of the unrestrained segments of the polymer chains in the amorphous regions.

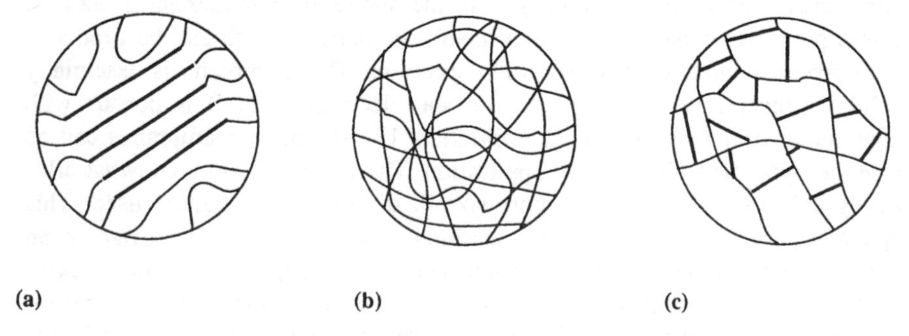

Figure 9.1 Structures found in polymeric membranes: (a) crystalline regions, (b) amorphous regions, and (c) cross-linked regions.

These segments flip from one position to another, opening free volume where they came from and filling free volume where they go. This causes movement of the free volume through the solid membrane. The free volume is statistically distributed in size and is in the range of 2 to 20 Å equivalent pore diameter. This polymer segment motion permits flow rates through the membrane equivalent to what would be obtained with continuous pores of about 5 Å in size even though no such continuous pores exist in the membrane. Because free volume motion is thermal, both apparent pore size and diffusion rate increase with temperature. Correspondingly, selectivity of the separation decreases with increased "looseness" of the membrane. It is a general rule that membranes give better selectivity and lower permeation rates with tight membranes as compared to looser ones. The degree of cross-linking in the polymer can be used to control the looseness of the polymer segments and thus both permeation rate and selectivity. Cross-linking is controlled during the polymer manufacturing process.

For solid membranes, part of the feed (usually the carrier solvent or water) diffuses through the membrane. This flow is driven by a concentration difference for the diffusing component between the upstream and downstream sides of the membrane. If more than one component of the feed dissolves in the barrier material, as is usual with permeation, the effectiveness of the separation is determined by the relative diffusion rates of these components through the barrier. However, if the barrier excludes all but one material, as in reverse osmosis for purifying water, the separation will be essentially complete.

The barrier is called semipermeable when either (1) it will not permit some materials to dissolve in the membrane and thus excludes them from transport through the membrane or (2) it gives markedly different diffusion rates for different components of the feed. For example, reverse osmosis membranes will dissolve water but will not dissolve salts. Thus, they permit transport of water through the membrane while excluding dissolved inorganic materials. They are much less effective for removing organic materials, which can also dissolve in the membrane. Permeation membranes usually dissolve non-polar molecules more readily than polar ones and can thus be effective for separating organic molecules from water by holding back the water. Ion exchange membranes used for electrodialysis can exclude ions of a given charge and thus prevent their transport through the membrane. At the same time, they allow oppositely charged ions to pass through with relative ease.

Porous barriers are made of either polymeric or ceramic materials and contain discrete holes, which can be characterized by a pore size distribution. Their action is much like a sieving action where molecules or particles of larger size than the holes are excluded from transport through the membrane. The flow of material through the membrane is caused by a pressure gradient through the barrier to give a purified downstream product. The rest of the feed is retained on the upstream side of the barrier as a concentrated residue.

In contrast to reverse osmosis and permeation, membranes which have an equivalent pore diameter of 2 to 20 Å, ultrafiltration membranes can have actual continuous pores through the membrane with diameters of 20 to 1000 Å. At the low end of the size range these membranes can also be of the solid polymer type described previously. The microfiltration membranes have even larger pore sizes from 1000 Å (0.1 µm) to 100,000 Å (10 µm). Thus, the microfiltration process represents basically an extension of conventional filtration to smaller particle sizes.

The U.S. Department of Energy (1990) has provided a useful comparison of these membrane processes in terms of the pore diameter of the membranes as compared to the diameters of typical molecules and particles encountered in practice. This comparison is reproduced as Figure 9.2. The figure shows that reverse osmosis is effective at separating water from salt and sugar in solution

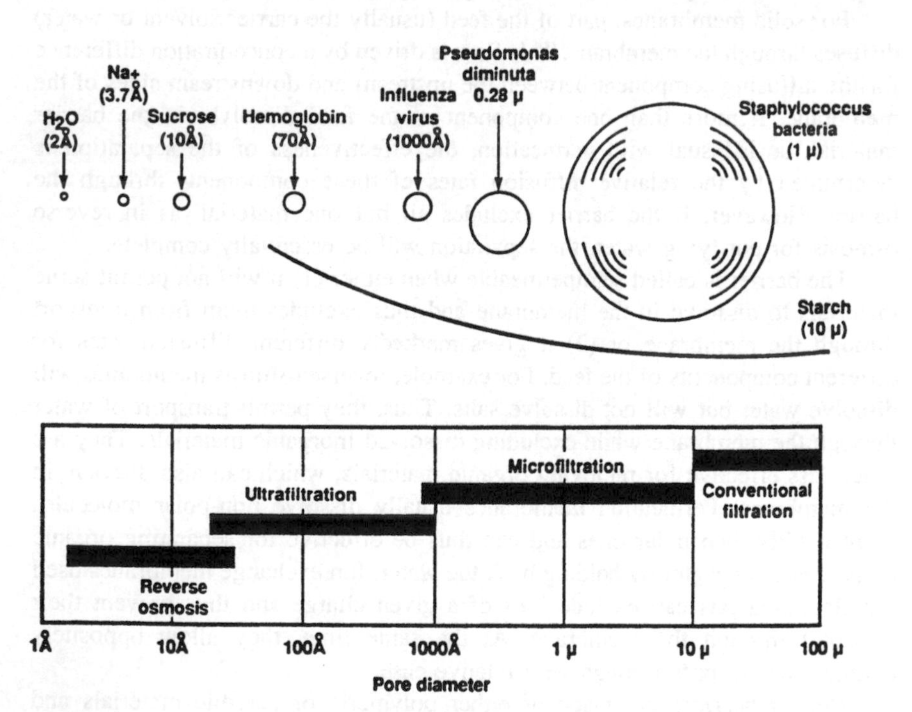

Figure 9.2 Reverse osmosis, ultrafiltration, microfiltration, and conventional filtration are all related processes differing principally in the average pore diameter of the membrane filter. Reverse osmosis membranes are so dense that discrete pores do not exist. Transport in this case occurs via statistically distributed free volume areas. The relative size of the different solutes removed by each class of membrane is illustrated in the schematic. [From Baker et al. (1990).]

as well as from any insoluble particles. This is because the membrane has no continuous pores and doesn't dissolve salt or sugar. Permeation membranes are like reverse osmosis membranes in that they also have no continuous pores. Ultrafiltration is effective for sterilizing aqueous streams containing particulate viruses and bacteria because the membrane pores are too small to pass these materials. It is also capable of recovering large molecules, such as polymers, from aqueous or non-aqueous waste streams by retaining them on the upstream side of the membrane. Microfiltration is effective for removing particles that are too small for effective removal by conventional filtration.

The barriers used in membrane processes are the separation media and can be in the form of sheets, coils, or fine tubes such as hollow fibers. The various processes are distinguished from each other by both the size of the pores in the barrier and the driving force used to get flow through the barrier.

Other membrane separation processes also exist in addition to those mentioned above. Examples are gaseous diffusion through porous barriers, conventional dialysis, and diffusion through liquid membranes. However, they do not at present find much use in waste minimization and are not included in this chapter.

9.2 REVERSE OSMOSIS

Reverse osmosis gets its name from the fact that it is based on reversing the solvent flow that would normally be caused by osmosis through a semipermeable membrane, i.e., one which would pass solvent but retain solute. This can be done by applying a sufficiently large pressure drop across the membrane so that flow occurs in a direction opposite to that expected for osmosis. This is illustrated in Figure 9.3, which shows the equilibrium osmotic pressure of 367 psi established for seawater by a semipermeable membrane.

If a semipermeable membrane that will pass water but not salt is placed at the bottom of a U-tube and the U-tube is filled to equal levels with seawater in the left arm and pure water in the right arm, the pure water will flow through the membrane from right to left. This will raise the liquid level in the left arm and thus the pressure on the left side of the membrane until an equilibrium is established. This equilibrium occurs when the additional pressure caused by the rising liquid level in the left arm is adequate to prevent any further flow of pure water through the membrane. This additional pressure is called the osmotic pressure of the salt solution. It results from the difference in chemical potential between seawater and pure water. If, on the other hand, a sufficiently long column of seawater is added to the left column at the beginning of the experiment to offset the osmotic pressure of the seawater, no flow of pure water in either direction will occur. The pressure differential across the barrier under these conditions is the osmotic pressure of seawater. Finally, to get reverse osmosis flow, the pressure gradient opposing the osmotic pressure must be

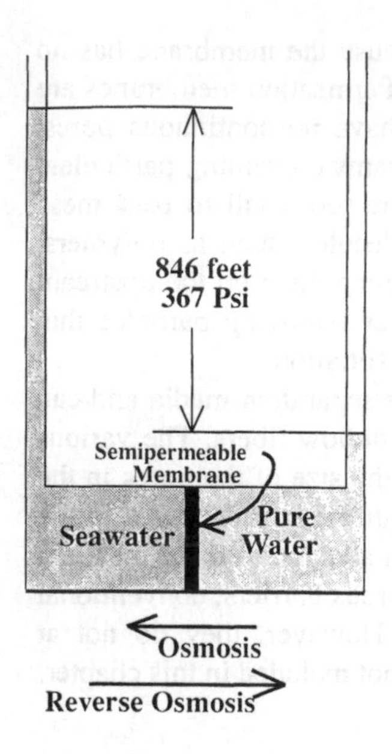

Figure 9.3 Equilibrium osmotic pressure established by a semipermeable membrane in seawater.

greater than the osmotic pressure so that the flow of pure water will actually be from left to right through the membrane (see Figure 9.3).

The required pressure drop across the membrane is the sum of the pressure needed to overcome osmotic pressure to prevent reverse flow and the pressure drop required to get reasonable flow rates. This depends on the concentration of impurities in the water phase and is on the order of 800 to 1000 psig for seawater, of which about 367 psig is required to overcome osmotic pressure and the rest to provide adequate flow rate through the membrane. The osmotic pressure in water is usually about 10 psi per 1000 ppm of soluble salts. For brackish water that is much less concentrated in salts, the osmotic pressure is much lower than for seawater and lower overall pressures can be used to drive reverse osmosis, i.e., 200–400 psig.

9.2.1 Fundamentals of Reverse Osmosis

The flow of solvent through a reverse osmosis membrane involves the dissolving of the solvent in the upstream side of the membrane and its diffusion through the membrane under the influence of the concentration gradient of solvent (strictly the chemical potential gradient) between the upstream and downstream sides of the membrane. For forward flow to occur, the concentration, or chemical potential, gradient on the upstream side must be greater than that on the downstream side of the membrane. However, in a system with pure solvent on one side of the membrane and a solution on the other side, the chemical potential and concentration of the solvent in the pure solvent is greater than that of the solvent in the solution. Thus, the normal tendency would be for solvent to diffuse from pure solvent through the membrane into the solution. One way to stop this reverse diffusion, or osmosis, as was illustrated in Figure 9.3 for pure water and seawater, is to raise the pressure on the solution until the solvent solubility on the solution side of the membrane is the same as the solvent solubility on the pure solvent side of the membrane. At this pressure no flow would occur; the equilibrium pressure required would be the osmotic pressure.

To achieve forward flow, i.e., reverse osmotic flow, the pressure on the solution side of the membrane must be further increased to generate a higher solubility of solvent on the solution side of the membrane than on the pure solvent side. Thus, the effective driving force for diffusion of solvent through the membrane is the actual pressure gradient across the membrane minus the osmotic pressure gradient. This is illustrated for water from seawater by Eq. 1. The corresponding equation for flow of salt is given in Eq. 2.

$$F_W = K_W (\Delta P - \Delta \Pi) \qquad (1)$$

$$F_S = K_S (C_1 - C_2) \qquad (2)$$

where:

F_W and F_S	=	water and salt fluxes through membrane
ΔP	=	total pressure drop across membrane
$\Delta \Pi$	=	osmotic pressure drop across membrane
C_1 and C_2	=	salt concentrations on the two sides of the membrane
K_W and K_S	=	empirical constants that depend on membrane structure, temperature, and composition of the salt

The osmotic pressure drop for water across the membrane is usually equal to Π because the downstream water is essentially pure and has no osmotic pressure. Furthermore, C_2 in Eq. 2 is negligible with respect to C_1 for the same reason. This means that the actual flow rate of solvent through the membrane will be proportional to $(\Delta P - \Pi)$ and can be increased by either increasing total pressure or decreasing osmotic pressure. Flow rate for salt is proportional to C_1 and is usually close to zero for effective membranes. Osmotic pressure is decreased by decreasing the concentration of solute in the feed stream. This means that more dilute wastewaters process more rapidly by reverse osmosis than do concentrated streams. It also means that flow rates in reverse osmosis are limited by the maximum differential pressure the membrane can stand.

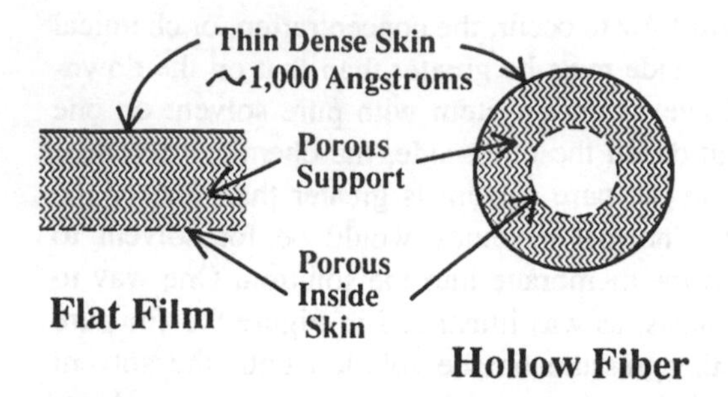

Figure 9.4 Schematic representation of hollow fiber and flat film asymmetric membranes.

Asymmetric Membranes

The development of reverse osmosis processes was greatly accelerated by the invention of the Loeb-Sourirajan (1962) asymmetric cellulose acetate membrane for purification of water. The cellulose acetate will dissolve water but not salt. The Loeb-Sourirajan membrane consists of a very thin layer of solid cellulose acetate supported by a much thicker porous cellulose acetate substrate. The structure of asymmetric membranes is illustrated in Figure 9.4 for both flat films and hollow fibers.

This asymmetric structure is produced by special solvent casting and membrane drying techniques. The thin active layer provides high water flow through the membrane and the integrity of this same layer prevents leakage of salt water feed through the membrane and thus gives excellent salt retention. The porous sublayer of the membrane provides low resistance to water flow and structural strength to withstand the large pressure gradients needed for reverse osmosis of seawater. This type of membrane is still in commercial use for recovering purified water from salt solutions or solutions of electrolytes.

Once the physical structure of the practical cellulose acetate reverse osmosis membrane had been determined along with practical ways of generating this structure, it was only natural that research turned to ways of generating this same asymmetric structure from other materials. The most notable success was with polyamides and polyamide hydrazides by Stannett et al. (1979). Cellulose acetate and the polyamides are the commonest materials in use today as membranes for reverse osmosis of aqueous solutions.

Composite Membranes

Composite membranes are the result of attempts to make the equivalent of an asymmetric membrane using different materials for the thin active layer and the thicker, porous support layer. The support layers are usually porous ultrafiltration membranes and the thin active layer is often polymerized in place on the surface of one side of the support membrane. In this way the properties of the active layer and the support layer can be separately optimized for their individual functions. For reverse osmosis, the most successful of these membranes to date uses a thin coating of aromatic amide on a polysulfone membrane substrate.

9.2.2 Specific Applications of Reverse Osmosis

Reverse osmosis was developed to produce potable water from saline waters varying from seawater to slightly brackish river and lake waters. This ability to recover pure waters from aqueous electrolyte solutions makes reverse osmosis a useful process for eliminating hazardous waste generated by the rinsing operations in the electroplating industry. A flow sheet for this application is given in Figure 9.5.

The plating bath chemicals carried to the rinse baths along with the plated parts are recovered from the rinse water by reverse osmosis. They can be further concentrated by evaporation, if necessary, before returning them to the plating bath. This configuration permits both the recycling of the recovered rinse water and the recovery and recycle of the plating bath chemicals, thus eliminating a major potential pollution source.

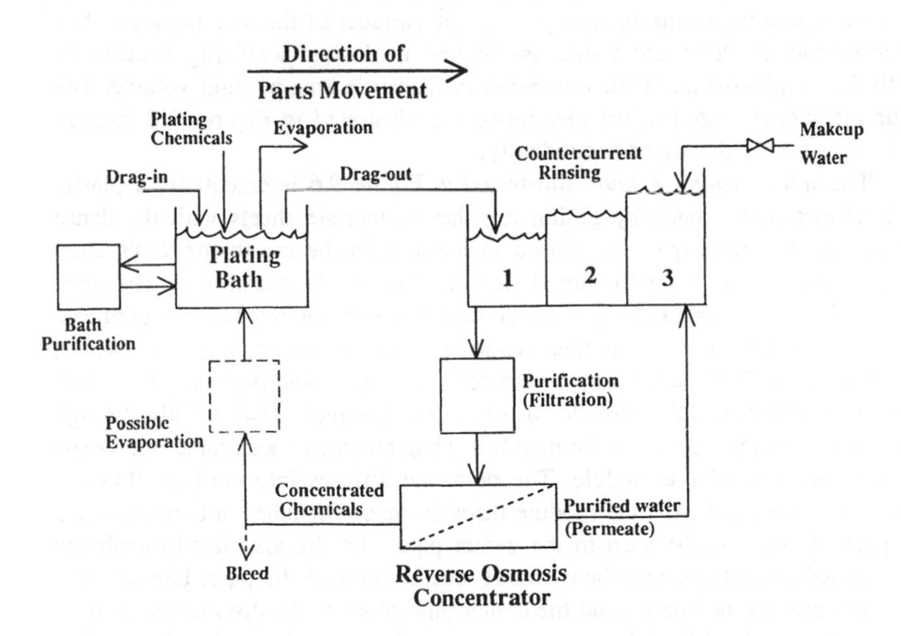

Figure 9.5 Flow diagram for a reverse osmosis system treating plating rinse water. [From U.S. EPA (1976).]

9.2.3 Equipment for Reverse Osmosis

The relatively high pressures required for reverse osmosis dictate to a large extent the mechanical design of reverse osmosis equipment. It is desirable to use cylindrical shells to contain the pressure and therefore the membranes are designed so that they can be inserted inside these pressure vessels. Another requirement is that a large membrane area be contained in a reasonably small volume. These considerations have led to the development of spiral wound membrane sheets and hollow membrane fibers as the most common forms for the reverse osmosis membranes. These membranes are contained in cylindrical modules and the desired total membrane area is obtained by using the appropriate number of modules. The popularity of these two membrane modules comes from the fact that they permit very large membrane areas to be achieved in a small volume and thus the equipment cost per square foot of membrane surface is relatively low. However, they are susceptible to fouling and plugging and may require pretreating of the feed by ultrafiltration before it goes to the reverse osmosis unit.

Two less frequently used membrane configurations are the plate and frame assemblies and tubular modules. The tubular modules still find specialized uses because of their larger clearances and thus their greater tolerance of suspended solids in the feed stream. Furthermore, they are easier to clean for sterile service. It is even possible to mechanically clean the surfaces of the membranes if they become fouled. Plate and frame assemblies are losing popularity because of both their high cost and their low membrane surface area per unit volume. The four membrane modules just mentioned are illustrated in Figures 9.6 through 9.9 and are now discussed individually.

The *spiral wound module* illustrated in Figure 9.6 is essentially a plastic bag of membrane made by gluing together membrane sheets with the dense sides out. A porous spacer is placed inside to hold the membrane walls apart and provide a flow path for liquid flowing through the membrane. The open end of the bag is sealed to a perforated center pipe through which the permeate exits the module. The bag is then wound around the center pipe along with a mesh spacer to hold succeeding layers of the bag apart and provide a flow path for the feed through the module. In operation, the feed flows axially through the module on the outside of the membrane bag and the concentrated brine exits at the other end of the module. The permeate diffuses into the bag, flows to the center tube, and exits the module through the center tube. One or more sets of perforations may be used in the center pipe with the associated membrane bag sealed to each set and then wound together around the pipe. Use of more than one membrane bag around the center pipe shortens the distance the permeate must travel (and thus the pressure drop on the permeate side) before exiting from the center pipe.

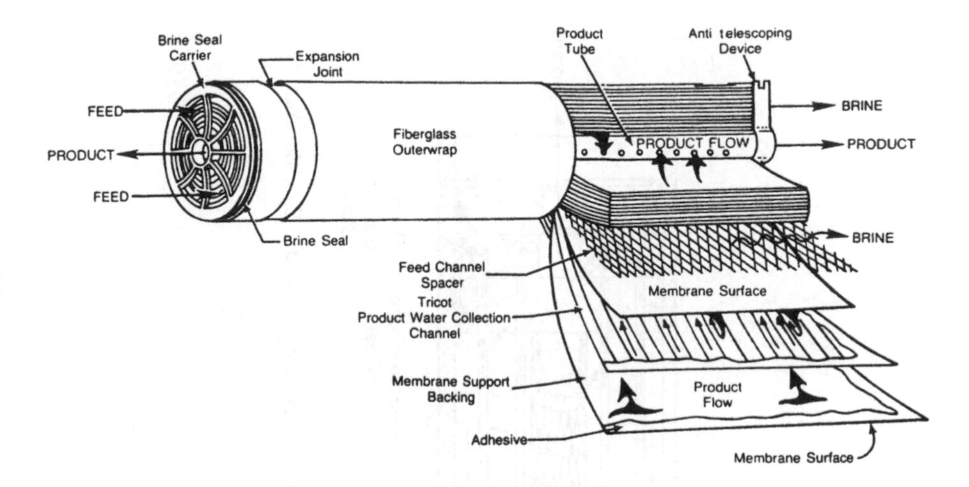

Figure 9.6 Sketch of a spiral wound membrane module. (Courtesy of Permasep Products, E. I. du Pont de Nemours & Co., Wilmington, DE.)

The *hollow fiber module* shown in Figure 9.7 is of the Du Pont Permasep design, but only a few fibers are shown to make visualization easier. A bundle of looped hollow fiber membranes has the open ends potted in a tube sheet of epoxy resin. The tube sheet can then be sealed with O-rings into one end of the containment vessel. The U-bend ends of the hollow fibers are also potted into epoxy resin. However, this plug is just to hold the membranes in place and guide the flow of the non-permeating liquid across the hollow fiber bundle. The Du Pont Permasep reverse osmosis module made in this way is sealed into a pressure vessel of 4- to 10-in. diameter and up to 4 ft long. Each module can contain up to 8,000 ft^2 of membrane area. The small diameter of the fibers (0.0016 in.) permits such very large surface areas to be obtained in a small volume. Other manufacturers use larger fibers (up to 0.045 in. inside diameter). These are still small enough to give large areas per unit volume.

The feed is introduced at the center of the fiber bundle shown in Figure 9.7 through a perforated pipe. It flows radially across the fibers until the non-permeant concentrated brine reaches the inside wall of the containment vessel and exits through the U-bend end of the module. The pure water diffuses through the hollow fiber walls to the center of the hollow fibers. It then leaves the fibers through the open ends which are sealed through the epoxy plug at the other end of the module.

Other manufacturers have somewhat different configurations for the hollow fiber bundle used for reverse osmosis. However, in all cases the high pressure

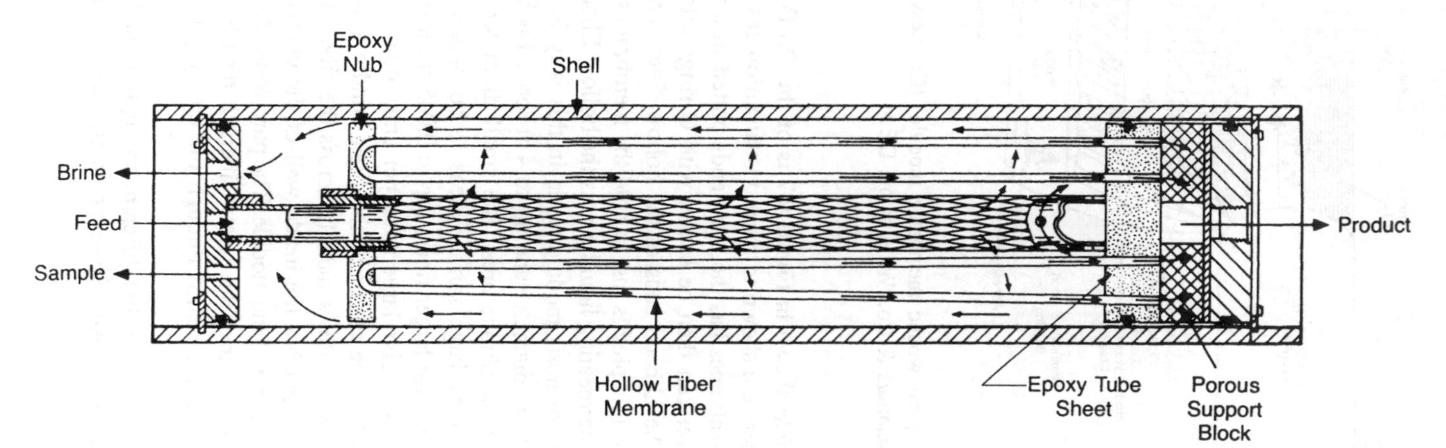

Figure 9.7 Illustration of a hollow fiber membrane reverse osmosis process. (Courtesy of Permasep Products, E. I. du Pont de Nemours & Co., Wilmington, DE.)

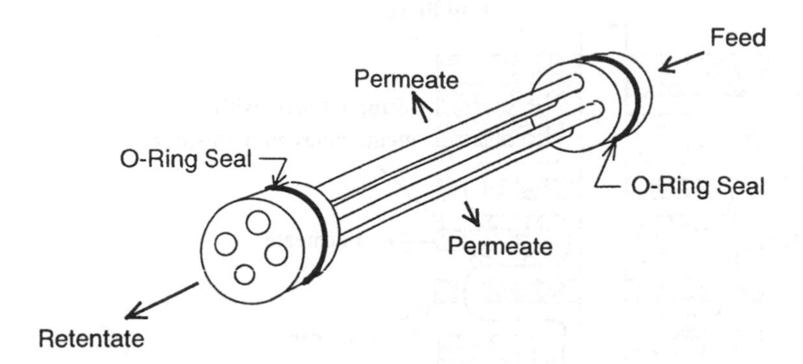

Figure 9.8 Sketch of a tubular membrane module.

is on the outside of the hollow fiber for increased fiber strength and the permeate exits from the center bore of the fibers.

The *tubular membrane module* shown in Figure 9.8 consists of a bundle of four porous tubes, each lined with a semipermeable membrane. The tube provides the structural support to withstand the pressure drop over the membrane and the membrane gives the selectivity for salt rejection. The inside diameters of the tubes vary from 0.25 to 1.5 in. and they are usually made of epoxy reinforced fiberglass. The number of tubes in a bundle is usually between one and ten. The membrane tube bundle is mounted in a closely fitting containment vessel and sealed at both ends, usually with O-rings. The whole unit is much like a shell and tube heat exchanger. Feed solution is fed inside the tube and pure water diffuses through the membrane and backup tube and is removed through an outlet in the side of the shell. The basic advantage of the tubular module is the larger clearances on the upstream side of the membrane as compared to the other module designs. This makes it less susceptible to plugging and easier to clean. The price paid for this benefit is that less membrane area can be packed into a given volume of module.

The *plate and frame* configuration shown in Figure 9.9 is much like the configuration of a plate and frame filter press except that the support plates, membranes, and spacers are usually round to permit insertion in a round pressure containment vessel. The assembly of membranes, support plates, and spacers is held together by a central bolt and inserted into a cylindrical vessel. The spacers guide the flow radially across the membrane surfaces and the porous support plates gather the pure permeated water from the downstream surfaces of the membranes. They also provide a flow path to the inside of the cylindrical shell, where the pure water is collected. Only one support plate with associated

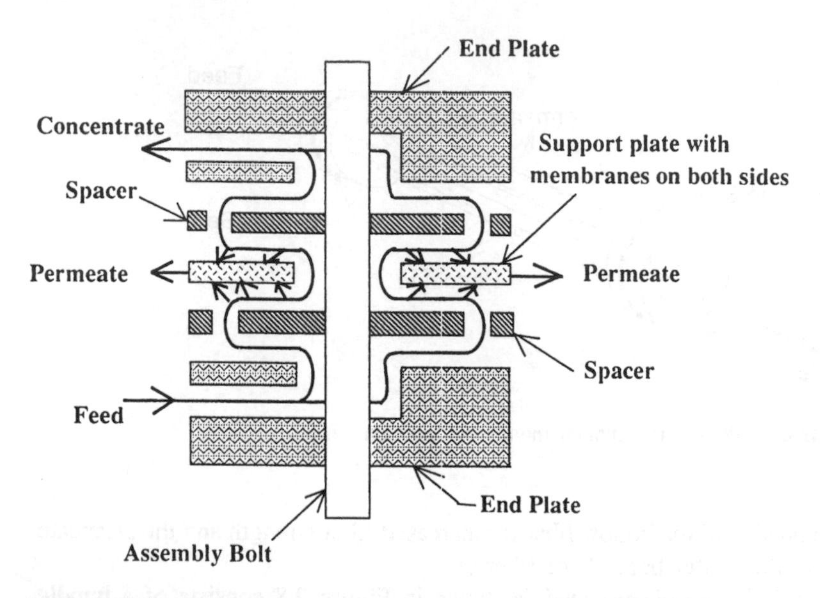

Figure 9.9 Sketch of plate and frame membrane configuration.

membranes is shown in Figure 9.9. However, by alternating more support plates and spacers in a stack, much more membrane area can be obtained.

The membrane equipment manufacturers for reverse osmosis are given in Table 9.1. The advantages and disadvantages as well as the problems with each type of membrane process will be discussed together later in this chapter after the descriptions of the individual processes.

9.3 MEMBRANE PERMEATION

Membrane permeation is the selective diffusion of molecules through a solid polymer film and is also called pervaporation or perstraction when applied to liquid feeds. It can be applied to either liquid or gaseous feed mixtures but has found most of its commercial application to date in gas separations. Like reverse osmosis, the feed stream is applied to one side of a solid polymer membrane. Flow of individual components of a feed through the solid polymer film is based on the concentration difference between the upstream and downstream sides of the film for each dissolved component.

For *gaseous feeds* the concentrations of the gaseous components of the feed dissolved in the two sides of the membrane are a function of the partial pressures of these components in the feed and product. Therefore, the driving force for flow through the membrane can be written in terms of partial pressures on the

Table 9.1 Vendors for Reverse Osmosis Systems

Aqua-Chem Inc. Water Technologies Div. P.O. Box 421 Milwaukee, WI 53201 (414) 961-2830	Arrowhead Industrial Water 300 Tri-State International Suite 320 Lincolnshire, IL 60069 (708) 940-1300
Cat Pumps Corporation 1681 94th Lane NE Minneapolis, MN 55434 (612) 780-5440	Desalination Systems, Inc. 1238 Simpson Way Escondisa, CA 92029 (619) 746-8141
E. I. Du Pont de Nemours & Co. 1007 Market Street Wilmington, DE 19898 (800) 441-7515	Filmtec Corporation 7200 Ohms Lane Minneapolis, MN 55435 (612) 835-5475
Illinois Water Treating Co. P.O. Box 560 Rockford, IL 61105-0560 (815) 877-3041	IonPure Technologies Corp. 10 Technology Drive Lowell, MA 01851 (800) 783-PURE
Koch Membrane Systems 850 Main Street Wilmington, MA 01887 (508) 657-4250	Matt-Son Inc. 28W005 Industrial Ave. Barrington, IL 60010 (708) 628-8766
Mobile Water Technology P.O. Box 14867 Memphis, TN 38114-0867 (800) 238-3028	Osmonics Inc. 5951 Clearwater Drive Minneapolis, MN 55343 (800) 328-0992
Permutit Company, Inc. E 49 Midland Avenue Paramus, NJ 07652 (800) 631-0878	Polymer Research Corp. 2186 Mill Avenue Brooklyn, NY 11234 (718) 444-4300
Sanborn Environmental Systems 25 Commercial Drive Wrentham, MA 02093 (800) 343-3381	Smith and Loveless, Inc. 14040 Santa Fe Trail Lenexa, KS 66215 (913) 888-5201

two sides of the membrane. The concentrations in the membrane are related to the partial pressures through the Henry's law constants for each component of the feed. Diffusion takes place through the space made available by the motion of segments of the polymer chains, just as in the active layer of the reverse osmosis membrane. The concentration gradient for permeation of gases is usually linear through the membrane. However, it can be markedly non-linear for some gases and particularly so for organic vapors. It is greatly influenced by the degree of swelling or plasticizing that occurs as gases or solvent vapors dissolve in the polymer. This interaction of permeating components with the polymer membrane causes the diffusion rate for each component of the feed to depend on the nature of the other components present in the feed through their effect on the steady-state polymer membrane structure. It also causes changes in selectivity and diffusion rate with time as the membrane comes to structural equilibrium with the specific feed being processed.

For *liquid feeds* the downstream concentration is maintained at a low value by either (1) applying a vacuum or a sweep gas to the downstream side of the membrane or (2) sweeping the downstream suface with a miscible, but easy to separate, solvent. The first configuration is called pervaporation and the second perstraction. Pervaporation is by far more common than perstraction. For liquid permeation the concentrations of the feed components in the upstream surface of the membrane are a function of their concentrations in the feed liquid and their solubility in the membrane polymer.

9.3.1 Fundamentals of Membrane Permeation

The membrane permeation process is a simple application of selective diffusion through a permeable solid barrier. In its simplest form the permeable solid barrier separates two chambers, one of which contains the feed stream at higher pressure and the other the product stream, which has diffused through the barrier to a lower pressure. This is illustrated in Figure 9.10 for a gaseous mixture of

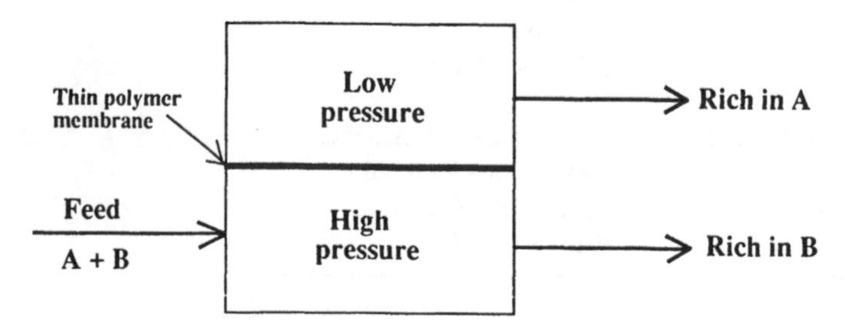

Figure 9.10 Schematic of a simple membrane process for separating gas mixtures.

A and B, where A diffuses through the barrier faster than B. However, the feed can be either a liquid or gaseous mixture and can be composed of more than the two components shown in Figure 9.10. The concentration of the individual feed components in the upstream surface of the membrane is in solubility equilibrium with the feed composition.

The flow of fluids through the membrane is induced by maintaining a solubility gradient through the membrane for the components of the feed. This is done by providing a low concentration of feed components on the downstream side of the membrane in equilibrium with the membrane surface and/or increasing the pressure on the upstream side of the membrane to increase the solubility on that side. The transport of the various feed components through the membrane occurs by diffusion under the influence of the concentration gradients established across the membrane. This is the solution-diffusion model of permeation which applies to both reverse osmosis and membrane permeation.

The flow equation for diffusion known as Fick's (1855) first law is given in Eq. 3 as it applies to membrane diffusion.

$$F = D\frac{C_1 - C_2}{l} \tag{3}$$

where:

F	=	flux through the membrane
D	=	diffusivity in the membrane
C_1 and C_2	=	concentrations in membrane at upstream and downstream surfaces, respectively
l	=	membrane thickness

This equation can be applied to either overall flow or to flow of the individual components of the feed. The selectivity of the separation is determined by the relative fluxes of the individual components of the feed. Eq. 3 also shows the importance of membrane thickness in achieving rapid permeation rates. This is the reason the Loeb-Sourirajan asymmetric membrane made such a difference in the development of reverse osmosis. It achieved a very thin active membrane for diffusion without sacrificing structural strength and the allowable operating pressure gradient.

Equation 3 can be rewritten as Eq. 4 by converting the concentrations to pressures through the use of Henry's law, which relates solubility to pressure. This form of the equation is more convenient for diffusion of gases through membranes.

$$F = P\frac{p_1 - p_2}{l} \tag{4}$$

where:

F	=	flux through membrane
P	=	permeability constant of the membrane
p_1 and p_2	=	pressures upstream and downstream of the membrane, respectively
l	=	membrane thickness

By manipulating these equations together, it can be shown that the permeability, P, is related to the diffusivity, D, and solubility, S, as shown in Eq. 5.

$$P = DS \tag{5}$$

If the downstream concentrations or pressures in Eqs. 3 and 4 are small with respect to the upstream values, which is the normal case, the separation factor for permeation of a binary feed becomes that shown in Eq. 6.

$$\text{Separation factor} = \alpha_{ab} = \frac{P_A}{P_B} = \frac{D_A S_A}{D_B S_B} \tag{6}$$

where:

P	=	permeability
D	=	diffusivity
S	=	solubility on the upstream side of membrane

Subscripts refer to components A and B of feed

The preceding equations all assume that the permeation or diffusion constants are independent values for each feed component. Further, they assume that the selectivity of the separation of a given binary mixture can be determined from the ratio of permeabilities or diffusivities of the pure components. This is usually a satisfactory assumption for diffusion of gas mixtures. However, it is often far from true for diffusion of liquids and vapors. With liquids and vapors, the polymer membrane may swell up to over 100% by volume by imbibing the feed liquid. This makes the membrane barrier into a gel-like solid whose diffusion resistance properties are more like those of a liquid than those of a solid. Such intensive swelling causes a major loss of both strength of the barrier and its selectivity as a separating medium. However, it does increase permeation rate. The selectivity decreases because the feed component which normally has the lower solubility in the solid membrane, and thus the lower diffusion rate, has its solubility and diffusion rate increased by the swelling. This means that at appreciable solubilities of feed in the membrane, the permeation rate for each component is affected by the other components in the feed. For this reason membrane permeation is normally carried out under conditions where such dramatic swelling of the membrane does not occur, i.e., moderate temperatures and membranes with low swelling properties.

In practical applications of permeation, it is convenient to include membrane area and thickness in the permeation constant and give the flux in terms of amount permeated per unit time per module. For liquids this is usually in gallons per day per module and for gases, cubic feet per hour per module. These values can be typically up to 125 gal/day and 1200 ft^3/hr, respectively, per module. Larger capacities are achieved by running modules in parallel and higher selectivities are achieved by running modules in series as a multistage cascade.

9.3.2 Specific Applications of Membrane Permeation

Membrane permeation is just beginning to make a place for itself in waste minimization although it shows a lot of promise as more selective membranes are becoming available, and confidence in their capabilities is growing. Gas permeation has taken over a large segment of the nitrogen market through the production of nitrogen from air. Many companies offer equipment for this service. It is also beginning to be used for purification of gases for recycle, removal of toxic impurities from commercial products, and dehydration of gas streams. Typical examples of these operations are given by Cooley and Dethloff (1985). They give results of field tests for (1) recovery of purified hydrogen for recycling to a petrochemical processing unit, (2) recovery of carbon dioxide for recycle to enhanced oil recovery, and (3) removal of carbon dioxide, hydrogen sulfide, and water from natural gas to meet quality requirements of the natural gas. It is expected that many more gas separation applications related to hazardous waste reduction will be introduced as the process becomes more familiar to environmental engineers.

Applications of liquid permeation are largely confined to pervaporation rather than perstraction because perstraction introduces an additional separation process into the separation scheme. That is, the liquid solvent used to generate the concentration difference across the membrane usually must be recovered for recycling to the membrane process. However, in the unusual case where both the perstraction solvent and the permeating component being recovered are to be feeds to processing units upstream of the permeation unit, perstraction may be the preferred membrane process. No applications of this type are yet known to the author.

The applications of pervaporation, where the concentration gradient is induced by maintaining a downstream vacuum while liquid feed is in contact with the upstream side of the membrane, are summarized for removal of volatile organic compounds from water by Lipski and Cote (1990) and by Wijmans et al. (1990). These references discuss the removal of oxygenated hydrocarbon, and halogenated solvents and conclude that pervaporation is technically feasible and economically attractive for removing volatile organic compounds from contaminated water. It is especially attractive for removal of relatively hydrophobic volatile compounds such as chlorinated solvents, naphthas, benzene, toluene, etc.

9.3.3 Equipment for Membrane Permeation

The equipment used for membrane permeation is very much like that used for reverse osmosis in that the membrane is assembled from flat sheets into spiral wound or plate and frame modules, or from hollow fibers or tubes into shell and tube modules. The major differences are (1) the units operate at much lower pressure because it is not necessary to overcome large osmotic pressures and (2) more flow area is provided by thicker spacers for gaseous flows in the spiral wound modules. The lower pressures also permit putting the upstream pressure on the inside of the fibers for hollow fiber modules. The most common units are the spiral wound and the hollow fiber units and both asymmetric and composite membranes are the norm because of their higher fluxes. Suppliers of permeation equipment are given in Table 9.2.

Table 9.2 Vendors for Permeation Systems

Gas Separations

Air Products and Chemicals, Inc.
Advanced Separations Dept.
7201 Hamilton Blvd.
Allentown, PA 18195-1501
(800) 426-3050

Generon Systems
515 West Greens Road
Suite 180
Houston, TX 77067
(713) 873-5100

Linde, Union Carbide
39 Old Ridgebury Road
Danbury, CT 06819-0001
(203) 794-2000

Permea, Inc.
11444 Lackland Road
St. Louis, MO 63146
(314) 995-3300

Pervaporation Systems

GFT, Represented by
Carbone of America Corp.
14 Eastman's Road
Parsippany, NJ 07054
(201) 503-0777

Membrane Technologies and Research
1360 Willow Road, Suite 103
Menlo Park, CA 94025
(415) 328-2228

Zenon Environmental, Inc.
845 Harrington Court
Burlington, Ontario
Canada L7N 3P3
(416) 639-6320

9.4 ULTRAFILTRATION AND MICROFILTRATION

Ultrafiltration and microfiltration are unlike the other membrane processes in that the membrane barrier may have actual holes through it. Membranes with holes do not have to depend on polymer segment motion in the membrane to provide a channel for flow through the membrane. As mentioned, this distinction is somewhat fuzzy for ultrafiltration. The distinction between ultrafiltration and microfiltration is made on the basis of the size of the materials screened out by the membrane and thus on the apparent size of the holes in the membrane. As shown in Figure 9.2, the range of pore sizes described for ultrafiltration membranes is generally accepted as 20 to 1000 Å, while microfiltration membranes have pores in the range of 1000 to 100,000 Å (0.1 to 10 μm). These pore size ranges describe what is meant by ultrafiltration and microfiltration membranes. However, the actual size range in a given membrane is usually much narrower than those which define the process classification. This provides a sharper actual particle size cutoff point for the separation. Most ultrafiltration membranes in use today are of the asymmetric type and are made from polymeric materials in the same way as the Loeb-Sourirajan reverse osmosis membranes. The main difference is that the apparent pore diameter is larger.

The membranes for microfiltration can be either of two main types having fundamentally different pore structures. One type consists of a relatively thick spongy structure of polymer much like the support layer of a reverse osmosis membrane. It has tortuous pores through the membrane with a relatively wide size distribution. Furthermore, the pore diameter can vary along the pore length. Thus, plugging of the pores can be a problem if the pores are too close to the size of the particles to be rejected. The other type of membrane, often called the Nuclepore or track-etch membrane, has a narrow size distribution of straight-through pores. It is usually much thinner than the spongy membrane. It is made by irradiating a non-conducting polymer with ionizing radiation. The linear tracks of the radiation are then chemically etched to form cylindrical pores through the membrane. The original Nuclepore membranes were made from polycarbonate films and they are still in wide use today.

A new development in membrane technology is the introduction of porous silica glass hollow fiber membranes by PPG (1990) and Schott Glaswerk (Mainz) (1990). These fibers can be made with small pores and narrow pore size distributions in sizes suitable for either ultrafiltration or microfiltration. Their temperature stability and corrosion resistance to acids and organic solvents should give them a good future in a variety of membrane separation processes.

Both ultrafiltration and microfiltration are pressure-driven filtration processes. However, the smaller pores of the ultrafiltration membranes make the retention of large (i.e., greater than 500 molecular weight) dissolved species possible along with retention of any particulate matter. This has provided the incentive for the application of ultrafiltration to a large number of waste treatment feeds.

In contrast to ultrafiltration, microfiltration is an extension of conventional filtration to smaller particles by using controlled pore size membrane filters. With its very real pores of relatively large size, the microfiltration membrane does not retain dissolved materials, but only particulate matter of sufficiently large size to be screened out.

9.4.1 Fundamentals of Ultrafiltration and Microfiltration

Because of the higher molecular weights of the solutes removed by ultrafiltration as compared to reverse osmosis, osmotic pressure is low enough not to be a consideration. Pressure gradients as low as 5 to 100 psig across the membrane are sufficient to achieve satisfactory flux rates. Furthermore, the larger pore size in ultrafiltration membranes, whether real or equivalent, leads to passage of ionic materials and small molecules, while retaining larger dissolved organic molecules. These large retained substances may be dissolved polymers and viruses or insoluble particulates such as complexed heavy metals, colloids, oil and grease, bacteria, and other suspended solids.

The flow rate through the membrane for both ultrafiltration and microfiltration is proportional to the pressure drop across the membrane. The proportionality constant is determined by the membrane structure. The rejection of solute or particulate matter is determined by size of solutes and suspended particles and is directly related to the equivalent or actual pore size through the membrane.

9.4.2 Specific Applications of Ultrafiltration and Microfiltration

The effectiveness of ultrafiltration at removing all sorts of troublesome materials from solutions and suspensions has led to many applications in waste minimization. Some of the more common uses as given by Parekh (1991), MacNeil and McCoy (1988), and by Eykamp and Steen (1987) are:

1. Pretreatment of dirty wastes to prepare them for reverse osmosis
2. Emulsion breaking and waste oil recovery from coolants used in metal-forming operations
3. Removal of complexed toxic metals from metal finishing rinse water
4. Concentrating oily wastes from a variety of dilute wastewater streams
5. Removing dyes and sizing chemicals from textile industry wastewaters
6. Removal of low molecular weight diluents and contaminants from paint baths to concentrate the paint for recycle

7. Concentration, clarification, and fractionation of various process streams and waste streams in the food, dairy, pharmaceutical, and wine industries
8. Concentration and pollution abatement in the paper and paint manufacturing industries
9. Lanolin recovery from wool scouring
10. Concentration of solids in the latex industries

In contrast to ultrafiltration, the applications of microfiltration are primarily aimed at particulate removal from a suspension or slurry. These particulates can be bacteria, chemical precipitates, fine floc particles, or contaminants picked up from process equipment. The ability to reject bacteria leads to a number of sterilization applications in the pharmaceutical and food industries. Some of these applications are listed, taken from Porter (1979):

1. Sterilization of fluids for medical use such as:
   a. Intravenous solutions
   b. Tissue culture media
   c. Vitamin solutions
   d. Bacteriological broth media
   e. Ophthalmic solutions
   f. Antibiotics and vaccines
   g. Vegetable or mineral oil
   h. Serum and plasma
   i. Phage and virus suspensions
   j. Albumin and allergins
2. Sterilization of gases used in fermentation processes in the pharmaceutical industry to prevent the presence of unwanted microorganisms during fermentation
3. Sterilization of wine beer and soft drinks to replace pasteurization and preserve flavor
4. Final purification of water used in the electronics industry to remove particulates
6. Sterilization of wastewater prior to disposal or recycle
7. Clarification of wastewater prior to disposal or recycle

The last two items in the list are particularly effective for small and medium quantity generators, where they do not have to compete with the economies of very large-scale wastewater plants.

9.4.3 Equipment for Ultrafiltration and Microfiltration

The equipment for ultrafiltration and microfiltration is much like that for reverse osmosis and permeation. The membrane modules for industrial operations are

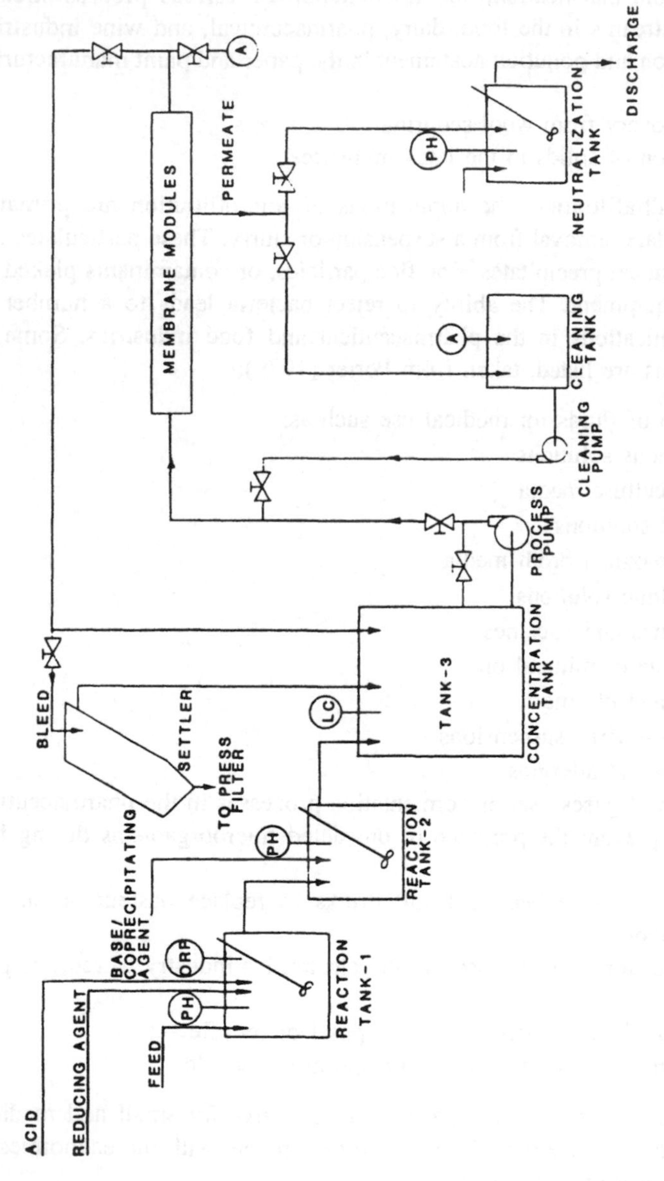

Figure 9.11 Typical flow diagram for membrane filtration application in treating wastewater from metal finishing. (Courtesy of MEMTEK Corporation, Billerica, MA.)

Figure 9.12 Photograph of a compact microfiltration unit for filtering solids generated by chemical precipitation. (Courtesy of MEMTEK Corporation, Billerica, MA)

usually either of the spiral wound, hollow fiber, or tubular type. The membranes can be made from a wide variety of polymers, glass, and ceramics. The pressure drop across the membrane is small compared to reverse osmosis so that hollow fiber membranes can be made for feed flow either inside or outside the hollow fibers. Typical operating pressure gradients over the membrane are 5 to 100 psi. Because of the lower pressure drop and the need for larger clearances in slurry and suspension handling service, tubular modules with their greater ease of cleaning find more use in ultrafiltration and microfiltration than they do in reverse osmosis.

A typical flow sheet for a membrane filtration application in wastewater treating is shown in Figure 9.11. In this application the membrane filter is used to concentrate solids generated by chemical precipitation from industrial wastewater. The solids can either be recycled to recover valuable components or disposed of. The clean water can be recycled or sent to the drain.

A photograph of a Memtek unit for the application shown in Figure 9.11 is shown in Figure 9.12. This photo illustrates the compact construction of a membrane filter containing 24 membrane modules. Most suppliers of ultrafiltration or microfiltration equipment can supply either individual components or complete package plants sized to fit the application. Some of the major suppliers of these units are given in Table 9.3.

Table 9.3 Vendors for Ultrafiltration and Microfiltration Systems

Amicon Division, W.R. Grace 72 Cherry Hill Drive Beverly, MA 01915 (800) 426-4266	Andco Environmental Processes 595 Commerce Drive Amherst, NY 14228 (716) 691-2100
Desalination Systems, Inc. 1238 Simpson Way Escondida, CA 92029 (619) 746-8141	Illinois Water Treatment 4669 Shepherd Trail Rockford, IL 61105-0560 (815) 877-3041
IonPure Technologies, Inc. 10 Technology Drive Lowell, MA 01851 (800) 783-PURE	Koch Membrane Systems, Inc. 850 Main Street Wilmington, MA 01887 (508) 657-4250
Lancy International, Inc. 181 Thorn Hill Road Warrendale, PA 15086 (412) 772-0044	Matt-Son Inc. 28W005 Industrial Avenue Barrington, IL 60010 (708) 628-8766
MEMTEK Corporation 28 Cook Street Billerica, MA 01821 (508) 667-2828	Sanborn Environmental Systems 25 Commercial Drive Wrentham, MA 02093 (800) 343-3381
Serfilco Ltd 1234 Depot Street Glenview, IL 60025 (800) 323-5431	Smith and Loveless 14040 Santa Fe Trail Lenexa, KS 66215 (913) 888-5201

9.5 ELECTRODIALYSIS

Unlike the membrane separation processes just mentioned, which use pressure or concentration gradients to cause flow through membranes, the driving force for electrodialysis is a direct current electrical field. Under the influence of the field, positive ionic species (cations) in solution move toward the cathode, or negative electrode. The negative ionic species (anions) move toward the anode or positive electrode. Obviously, anything that does not ionize in solution is not influenced by the electrical field or by the ion-repelling nature of the membranes through which the ions must pass. Therefore, electrodialysis is limited to de-

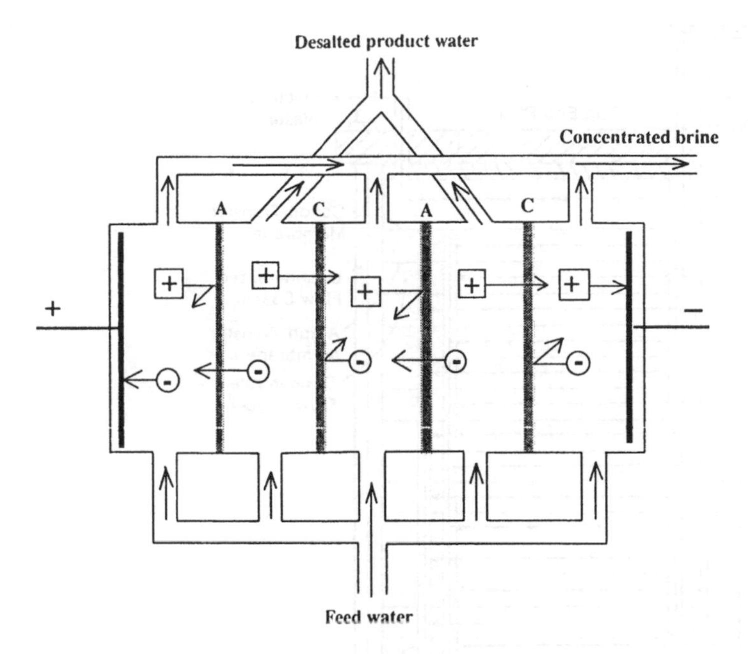

Figure 9.13 Simplified representation of the electrodialysis process. A, Anion permeable membrane; C, cation permeable membrane. [Adapted from Klein et al. (1987).]

mineralization of solutions containing solutes that can dissociate to form ions. Alternating ion selective membranes are placed between the electrodes to divide the electrodialysis apparatus into a group of chambers enriched in electrolyte and a group depleted in electrolyte, as shown in Figure 9.13.

The membranes are ion-exchange membranes which reject either cations or anions, but not both. The oppositely charged ion flows readily through the membrane. By dividing the apparatus into a group of flow chambers with alternating cationic and anionic membranes, the feed through the adjacent chambers is either enriched or depleted in electrolyte content. The flow chambers are manifolded so the concentrated streams and the depleted streams are collected separately.

In practice the membrane cells are combined into stacks, which may contain hundreds of individual cells in the form of alternating membranes and spacers. The spacers have channels to guide the flow of feed across the surface of the membranes and are equipped with ports to distribute feed to and collect product from the individual cells. These stacks may have their cells aligned in either a horizontal or vertical direction and may contain a variety of staging configurations in a single stack. The stack is clamped together in much the same way

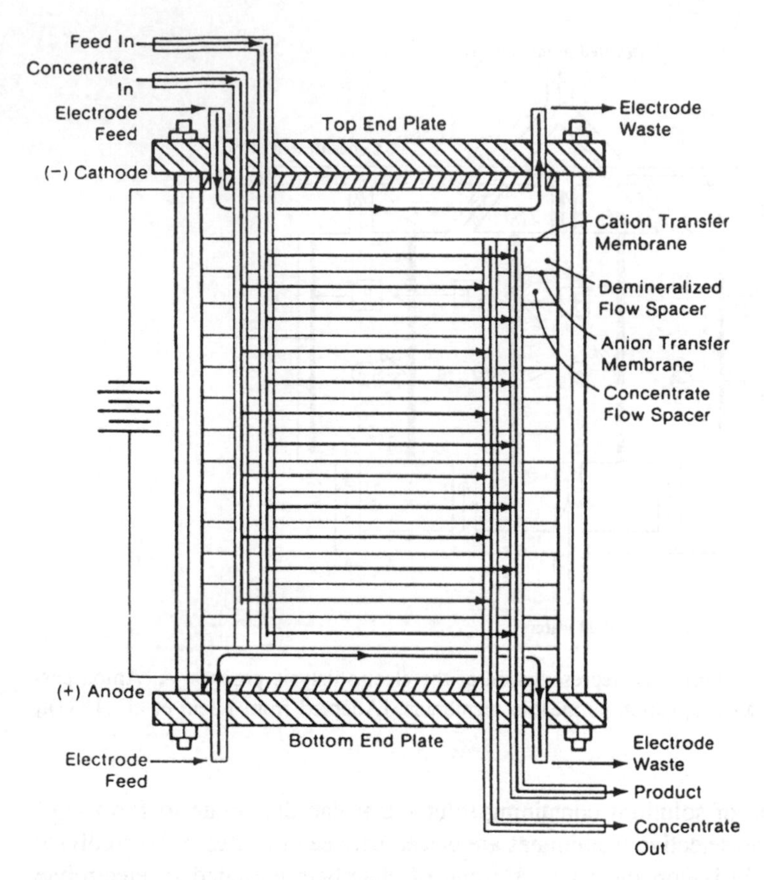

Figure 9.14 Sketch of a typical electrodialysis stack. (Courtesy of Ionics, Inc., Watertown, MA.)

as a plate and frame membrane module, which it really is. A simplified diagram of a typical horizontal electrodialysis stack is shown in Figure 9.14. The feed and concentrate pass through alternate flow chambers in the stack, and concentrate and product are collected at the downstream side of each chamber. The electrode feeds flush out of the stack the oxygen and hydrogen generated at the electrodes. A photograph of a 7-ft tall electrodialysis stack manufactured by Ionics, Inc. (Watertown, MA) is shown in Figure 9.15

A recent development in electrodialysis, called *reversible electrodialysis*, is the use of frequent reversal of polarity of the electric field to decrease the sensitivity to fouling of the membrane stack. This current reversal removes

Figure 9.15 Photograph of a seven foot high electrodialysis stack. (Courtesy of Ionics, Inc., Watertown, MA.)

surface films and precipitates from the membranes by redissolving them or flushing them off the membrane surfaces. However, the flow must also be shifted from one cell to another as the concentrating and depleting chambers are reversed with the current reversal. The flow pattern reversal is done by installing automatic valve systems to change flow manifolds as the electric field across the stack is reversed. Nearly all new electrodialysis installations use the reversal process.

9.5.1 Fundamentals of Electrodialysis

Electrodialysis membranes are ion exchange membranes with structures much like the cross-linked membranes shown in Figure 9.1. However, they also contain fixed ion exchange sites grafted onto the polymers used to form the

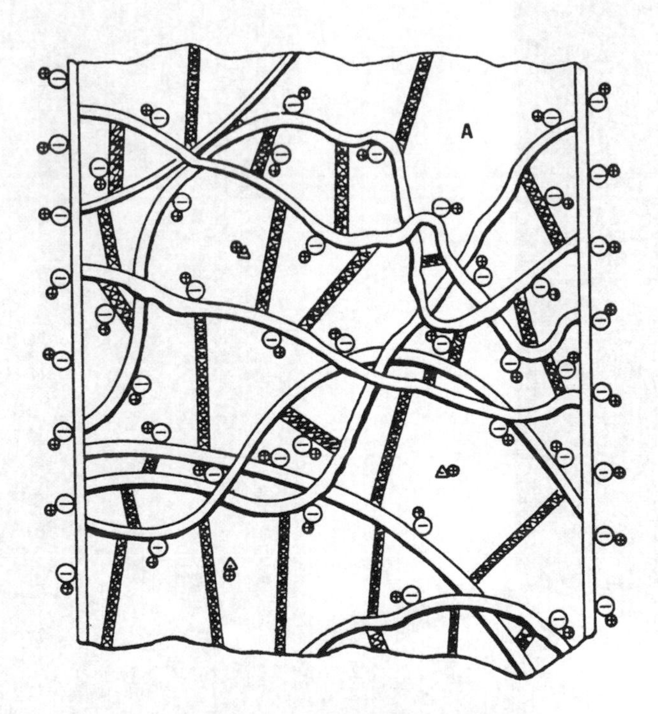

⊖ FIXED NEGATIVELY CHARGED EXCHANGE SITE

⊕ MOBILE POSITIVELY CHARGED EXCHANGEABLE CATION

═ POLYSTYRENE CHAIN

▩ DIVINYLBENZENE CROSSLINK

Figure 9.16 Representation of an ion exchange membrane. [From Klein et al. (1987).]

basic membrane. These fixed sites can be made anionic with sulfonates and carboxylates (or, less frequently, other anions) or cationic with quaternary ammonium ions and amine ions. The drawing shown in Figure 9.16 from Klein et al. (1987) is frequently used to explain how ion exchange membranes work in electrodialysis and is so used here. It is a representation of a cation exchange membrane. A similar but oppositely charged membrane would serve as an anion exchange membrane.

The cation exchange membrane shown in Figure 9.16 uses sulfonate anions grafted, i.e., chemically attached, to the polymer chains as the fixed cation exchange sites. Positive ions in the solution are attracted by the negatively charged exchange sites and negative ions are repelled, thus being excluded from the membrane structure. Under the influence of an electric field, the positive ions move through the membrane while the negative ions are repelled. Correspondingly, for an anion exchange membrane the fixed ions are cations, negative anions in the feed are attracted, and the cations in the feed solution are repelled. Again, under the influence of the same electric field, the anions move through the anion exchange membrane in the opposite direction from the motion of the cations through the cation exchange membrane. A parallel group of electrodialysis cells can be formed by alternating cation and anion exchange membranes in a stack as illustrated in Figure 9.17.

Figure 9.17 illustrates the desalting of water by electrodialysis. Under the influence of a direct current electric field across the stack in Figure 9.17, the positive sodium ions flow toward the cathode when permitted by the membranes and the negative chloride ions flow toward the anode, again when permitted by the membranes. If saltwater to be deionized is fed in parallel to every other cell, i.e., cells 2, 4, and 6, it will be depleted in ionic content and thus in salt

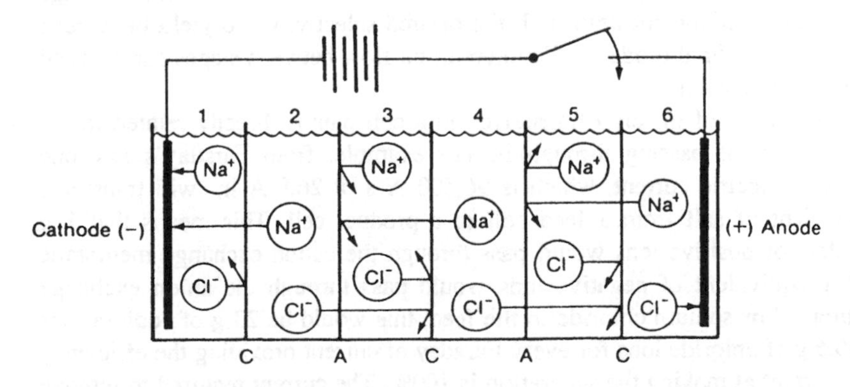

Figure 9.17 Representation of an electrodialysis unit for demineralizing brine. (Courtesy of Ionics, Inc., Watertown, MA.)

content. The adjacent cells 1, 3, and 5 will be enriched in salt content. Manifolding alternate cells permits collection of enriched and stripped product streams.

The basic unit of an electrodialysis stack is the cell pair formed by two adjacent cells, one enriched in electrolyte and one depleted in electrolyte. The current is carried by the positive and negative ions in proportion to their transport numbers which give the fraction of the total current carried by each ion. These transport numbers are based on the mobilities of the ions in solution and in the membranes. For ions of about the same size, positive and negative ions in solution will have transport numbers near 0.5. That is, each ion will carry half the current in solution. However, for acids with a large anion the proton will be much more mobile in solution than the bulky anion and thus carry a much larger share of the current. Correspondingly, the hydroxyl ion will carry most of the current in a solution of a base with a bulky cation. In the membranes the cations will carry all the current through the cation exchange membrane, and the anions will carry all the current through the anion exchange membrane.

The electrical requirements for an electrodialysis stack are directly determined by the number of ions being transferred in a cell pair and by the number of cell pairs in the stack. The DC current requirement is determined by the number of molecules transferred per unit time in the total stack, and the voltage requirement is determined by the total number of cell pairs between the electrodes multiplied by the voltage drop per cell pair. The required power is the product of the stack voltage drop and the current. Because the same current passes through a series of cell pairs in a stack, it appears as if the stack is operating as a multistage process. However, with all the cell pairs connected in parallel, the entire stack represents only a single stage of separation. To get staging in electrodialysis the depleted product from a single stage must be collected and fed to a second single stage unit for further deionization. Correspondingly, the enriched solution from the first stage can be fed to a third stage to get further enrichment until both the desired selectivity and yield have been obtained for the final products. Countercurrent flow between stages can be used to maximize efficiency.

The amount of material transferred in a cell pair is directly related to the electrical current passing through it. For example, from Faraday's law one Faraday of electric current, which is 96,500 A s or 26.8 A h, will transfer 1 g equivalent of salt from a feed cell to a product cell. This means that 1 g equivalent of positive ions would pass through the cation exchange membrane and 1 g equivalent of negative ions would pass through the anion exchange membrane. For sodium chloride in the feed, this would be 23 g of sodium ions and 35.5 g of chloride ions for every Faraday of current providing the efficiency of the current at making the separation is 100%. The current required to remove a given amount of minerals by electrodialysis is provided by Meller (1984) and given by Eq. 7.

$$I = \frac{F^* F_d \Delta N}{e N^*} \tag{7}$$

where:

I	=	direct current (amperes)
F^*	=	Faraday's constant (96,500 ampere seconds/equivalent)
ΔN	=	change in normality of demineralized stream between inlet and outlet of the membrane stack
F_d	=	flow rate of demineralized stream through stack (liters per second)
e	=	current efficiency, usually greater than 85%
N^*	=	number of cell pairs

To get from Ohm's law the voltage required to drive a given system, the current is determined from Eq. 7 and multiplied by the resistance of the total stack. This resistance for the stack is the sum of the component resistances provided by the solutions and membranes in a single cell pair multiplied by the number of cell pairs in the stack. This resistance decreases as temperature is increased and decreases as electrolyte concentration increases.

Concentration Polarization

A problem uniquely severe for electrodialysis, in contrast to the other membrane processes, is concentration polarization. Its magnitude determines most of the difficulties encountered with electrodialysis. Concentration polarization occurs to some extent for all membrane processes where one component of a feed is permeating selectively through a barrier. However, in electrodialysis it is particularly troublesome because it is increased by the very driving force that causes the separation. Klein et al. (1987) have provided an excellent discussion of electrodialysis which details the mechanism of concentration polarization and explains why it is so important. Their explanation will be followed in a somewhat condensed form here.

The problem with concentration polarization comes about as a result of the concentration gradients across the stagnant liquid layers adjacent to the stationary solid ion exchange membrane surfaces past which the fluids are flowing. These gradients are sketched in an idealized fashion in Figure 9.18 for a cell consisting of two electrodes and one anion exchange and one cation exchange membrane.

In Figure 9.18 the central cell bounded by the membranes is the feed cell which is being depleted of electrolyte. Negative ions are moving toward the left positive electrode and positive ions are moving toward the right negative electrode giving a net transport of electrolyte out of the cell. The central portion of the cell is well mixed by the flow velocity of liquid through the cell. In this

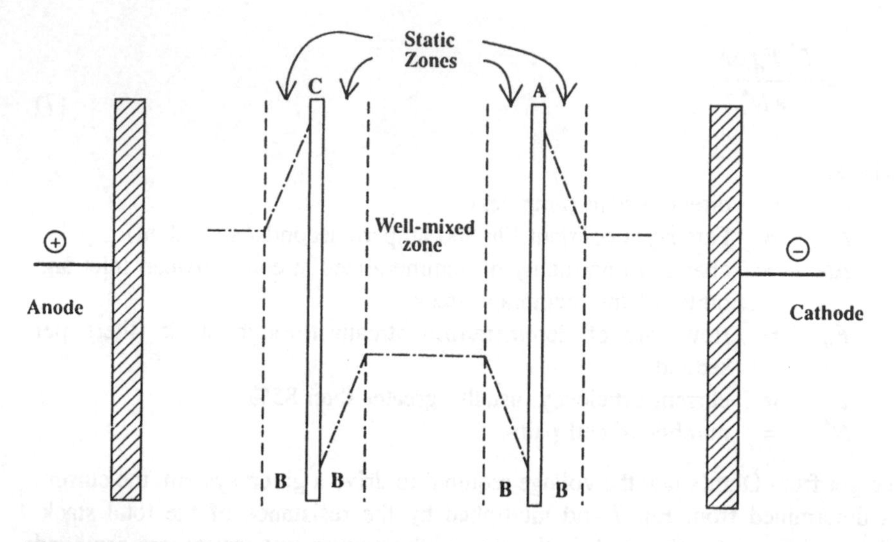

Figure 9.18 Idealized representation of concentration gradients in electrodialysis. (A) Anion exchange membrane, (B) static boundry layers, (C) cation exchange membrane. [From Klein et al. (1987).]

region, ions are transported by physical mixing, diffusion, and the electrical field. However, as the stationary membrane is approached, the velocity of flow decreases, and this region can be ideally treated as a stagnant boundary layer through which ions are transported only by electrical transport and diffusion.

Because the ions have lower transport numbers (approximately 0.5) in the solution than they do in the membranes (approximately 1.0), only half as many ions will be transferred electrically through the solution in the static boundary layers on the inside surfaces of the cell membranes as will be transferred through the membranes. Thus, the solutions on the inside surfaces of the membranes will be depleted of ions. This causes an anion concentration gradient pushing anions to the left and a cation concentration gradient pushing cations to the right. Similarly, the static layers outside the membranes will be enriched in ions, again setting up concentration gradients at a higher absolute level to push the respective ions through the static layer toward the electrodes. The ion concentrations will change until equilibrium concentration gradients are established. These equilibrium gradients are sufficient to drive, by diffusion, the ion flow necessary to carry the extra electrical current not carried by ions driven by the voltage gradient.

As current density is increased, the rate of electrical transfer of ions increases. Therefore, the rate of diffusive ion transfer must also increase to provide the extra ions transferred electrically through the membranes. With any given boundary layer thickness, i.e., fluid velocity, a current density exists, which will totally deplete the transferring ions at the inside surface of the membranes forming the depleting cell. At this limiting current, density any further demand for ions to carry the electrical current will begin to be satisfied by ionization of water to get the necessary protons and hydroxyl ions. This continuous ionization of water is called *water splitting* and causes sharp increases in both fouling and electrical resistance of the electrodialysis stack. In practical operations, current density is usually kept between 50 and 70% of the limiting value. There is an incentive to keep it as high as possible for the sake of productivity, while avoiding the problems associated with approaching the limiting value.

9.5.2 Specific Applications of Electrodialysis

Electrodialysis is basically a process for the demineralization or concentration of aqueous feeds through electrical transport of ions. Its major use by far has been to prepare potable water from brackish water and seawater. It is particularly effective and economical for dilute streams. Therefore, the principal applications in waste minimization occur in the metal finishing and electroplating industries. The recovery of acids and metal salts from dilute rinse water streams is typical of these applications. The low concentration of electrolytes in these waste streams makes electrodialysis especially attractive, and both the recovered water and chemicals can be recycled.

The ability to recover dilute metal salts from aqueous streams has led to the use of electrodialysis to detoxify waste streams to meet environmental requirements before disposal. It is also used to recover valuable materials from waste process streams.

Electrodialysis has also found a variety of applications in the food industry such as (1) adjusting acidity of citrus products and (2) removing minerals from whey so that it can be used in food products rather than being discarded as a waste stream.

9.5.3 Equipment for Electrodialysis

The equipment for electrodialysis resembles very closely a plate and frame filter press with the plates aligned either vertically or horizontally. The stack is made up of alternating spacers and membranes with the membranes, alternately cation and anion exchange membranes. The repeating section of the stack, called a *cell pair*, consists of a cation membrane, a demineralized water spacer, an anion membrane, and a concentrated water spacer. A typical membrane stack may have from 300 to 500 cell pairs. The photo in Figure 9.15 shows what a stack looks like.

Electrodialysis membranes are usually 6 to 24 mils thick and come in either sheets up to 10 ft long by 40 in. wide or continuous rolls up to 44 in. wide.

Figure 9.19 Photograph of an Aquamite XX, a 300,000 gal/day water-desalting plant. (Courtesy of Ionics, Inc., Watertown, MA.)

Table 9.4 Vendors of Electrodialysis Systems

Allied Signal Aquatech Systems 7 Powder Horn Drive Warren, NJ 07059 (908) 563-2800	Andco Environmental Processes 595 Commerce Drive Amherst, NY 14228 (716) 691-2100
HPD Incorporated 305 East Shuman Blvd. Naperville, IL 60563 (708) 357-7330	Ionics, Inc. Water Systems Division 65 Grove Street Watertown, MA 02172 (617) 926-2500
IonPure Technologies Corp. 10 Technology Drive Lowell, MA 01851 (800) 783-PURE	Lancy International, Inc. 181 Thorn Hill Road Warrendale, PA 15086 (412) 772-0044

The spacers usually provide a liquid flow chamber about 40 mils thick. They are die cut to provide a tortuous high velocity path for the liquids across the membranes. Data provided by Ionics, Inc., show that a typical single stage stack for desalination of brackish water may operate with about 14 psi pressure drop and a DC voltage and current of 640 V and 29 A, respectively, to process 167,000 gal/day of water containing 3000 ppm of salt. The stack contains 4500 ft^2 of 18×40 in. membrane sheets and operates at a power consumption of less than 5 kw•h per 1000 gal of purified water. The salt removal for a single stage is 52% per pass, but use of three stages in series can bring it up to 90%. Figure 9.19 shows a photograph of the IONICS Aquamite XX, which is an electrodialysis plant consisting of two parallel lines of three stages each. This unit can process up to 300,000 gal/day of saline water. Units four times as large as this are available. The major suppliers of electrodialysis membranes and equipment are given in Table 9.4.

9.6 COMMON PROBLEMS AND TROUBLESHOOTING OF MEMBRANE PROCESSES

All membrane separation processes have the potential for fouling of the upstream membrane surface due to sedimentation of suspended solids or precipitation of dissolved solids. In addition, hollow fiber membranes are susceptible to plugging if the feed is inside the small diameter fibers. The remedies for fouling are usually one or more of the following:

1. Maintain high flow velocities and use turbulence-enhancing spacers in the low clearance flow paths.
2. Pretreat feed streams to remove particulates and potential precipitates. Ultrafiltration and microfiltration can be used to remove particles from feeds to the other processes. Chemical precipitation before filtration can be used to remove dissolved solid solutes likely to precipitate in the membrane modules.
3. Clean the membrane modules either chemically or mechanically on a regularly scheduled basis. Only tubular modules can be easily cleaned either way, while the rest of the modules, which have a much greater penchant for fouling, cannot be readily cleaned mechanically and are even frequently difficult to clean chemically.
4. Fouling of electrodialysis stacks can be greatly reduced by using current reversal to clear the membrane surfaces. Similarly, filtration-type membranes can be back-flushed to remove deposits with varying degrees of success.

Like fouling, another relatively universal problem encountered, to varying degrees in membrane processing is concentration polarization. This was discussed in some detail for electrodialysis. It is also encountered to some degree wherever a stagnant or sludge layer is formed at the surface of a membrane through which substances are permeating under the influence of a concentration gradient. The concentration gradient across the stagnant layer of liquid next to the membrane surface reduces the concentration gradient available across the membrane itself. This reduces the permeation rate through the membrane. This polarization can be minimized by maintaining high velocities as described in item 1 above.

To get maximum separation out of a given membrane process, the integrity of the membrane against pinholing and of the module against feed bypassing (due to inadequate sealing) must be insured. The onset of these problems will be indicated by a sharp drop in purity of the output of the process due to feed leaking directly to product. By isolating various modules sequentially and testing their product purity, the faulty ones can be located and repaired or replaced.

Membrane resistance to the operating temperature and to various solvents in the feed is essential for long membrane life. If temperatures are too high for the membrane, it may soften, swell, and flow leading to alteration of the membrane properties and thus to its ineffectiveness as a separation medium. Under extreme conditions, it might melt or burst. Certain organic solvents have the same effects as high temperature on organic polymer membranes. Excess temperature coupled with organic solvents in the feed can be disastrous. The remedy is to use membranes which can easily withstand the highest temperatures anticipated and which are chemically resistant to all the solvents anticipated in the feed. If the latter is not possible, the feed must be pretreated to remove the troublesome solvent. This can frequently be done using carbon adsorption to remove the organics from the feed stream.

9.7 ECONOMICS

The cost of any separation process depends on the composition of the feed to be processed, the size of the plant, the local cost of materials, manpower, and utilities, and the cost of money to the owner. Therefore, it is difficult to compare economic evaluations made by different authors using different accounting methods for different sized plants at different locations. However, having made this caveat, I have tried to assemble cost estimates for the membrane processes discussed in this chapter from the literature and equipment supplier sources. These are listed in Table 9.5 for membrane separation of liquids and in Table 9.6 for separation of gases. Separate values are given for capital and operating costs for a given size plant. The capital values are given in dollars per unit of plant capacity, and the operating costs are in dollars per unit of product. The operating costs given in Table 9.5 were estimated including the costs of capital recovery. When a range of values for the costs is given, it is to account for the

Table 9.5 Approximate Costs for Membrane Separation Processes for Liquids

Liquid process application	Plant size, thousand gpd.	Capital cost, $/gpd	Operating cost,* $/1000g.	Reference
Reverse osmosis				
Seawater	> 500	2.50–4.00	2.50–3.50	Pareckh (1991)
Seawater	1000–5000	4.00–10.00	2.50–4.00	Riley (1990)
Brackish water	> 500	0.50–1.20	0.50–1.50	Pareckh (1991)
Brackish water	1000–5000	0.60–1.60	1.00–1.25	Riley (1990)
Pervaporation				
10 ppm trichloroethane from water	63	1.10–2.20	2.00–4.00	Lipski and Cote (1990)
1000 ppm benzene from H_2O	20	9.0	14.1	Wijmans et al. (1990)
Various solvents from wastewater	10	6–12	7–13	MTR (1992)
Ultrafiltration				
Oily waste	< 10	7–10	25	Koch (1992)
Oily waste	> 50	2–4	< 10	Koch (1992)
Microfiltration				
Industrial laundry wastewater	288	2.4	3.4	Estimated from MEMTEK (1992)
Brass wire mill wastewater	72	2.4	3.5	Estimated from Pardus (1987)
Electrodialysis				
Most applications	> 500	0.60–1.0	0.40–1.50	Ionics, Inc. (1992)
Most applications	< 588	0.80–1.90	0.40–1.70	Ionics, Inc. (1992)
Brackish water	1000	1.6–2.0	1.4–1.8	Applegate (1984)

* Includes capital recovery.

Note: Microfiltration examples include chemical pretreating and sludge handling costs.

Table 9.6 Approximate Costs for Membrane Separation Processes to Produce Nitrogen from Air for Various Applications

Nitrogen application	Plant size, thousand CF/H	Capital cost, $/1000 CF/H	Operating cost,* $/thousand CF	Reference
Aluminum extrusion	1	1.35	0.45	Spehn (1992)
Reactive injection molding	0.5	1.60	0.50	Spehn (1992)
Chemical blanketing	10.4	0.57	0.64	Spehn (1992)

* Power only at $0.05/KWH.

variability of conditions mentioned above. Because of the differences in accounting methods used by different sources of the cost information, general comparisons between costs for different suppliers of similar equipment should not be made from Tables 9.5 and 9.6. The numbers given should be used merely as order of magnitude estimates. For detailed cost information, the equipment suppliers should be contacted with respect to the specific application being considered. Frequently the value of the recovered materials can greatly offset the costs of the required separation.

Among the few generalizations that can be made about costs for membrane processes are:

1. Membrane processes are cost competitive with alternative processes.
2. Unit capital costs usually increase with the 0.6 power of the size increase of the plant. Thus, doubling the plant size only increases costs by 52%, i.e., the 0.6 power of 2 is 1.52.
3. Costs go down as membrane flux goes up for a given membrane selectivity.
4. Costs go down as membrane life goes up. Membrane life goes up as the required frequency of cleaning goes down. Membrane life is also a strong function of the composition of the feedstock and its degree of chemical attack on the membrane.

9.8 SUMMARY

Membrane-based separation processes are increasingly finding a home in waste minimization and recycling applications. Reverse osmosis and electrodialysis can be used to remove low concentrations of toxic inorganic salts from wastewaters, making them suitable for disposal or recycling. Furthermore, they can be used to recover valuable dissolved metal salts from effluents from metal finishing operations. These two processes are primarily demineralizing processes for aqueous streams and can be expected to find much wider use in waste management in the future as environmental engineers become more familiar with their potential.

Permeation applied to gas separation has found extensive use in the production of nitrogen. It is finding application in other gas processing applications such as recovering hydrogen in refineries, removing carbon dioxide from natural gas, and recovering carbon dioxide for recycle to enhanced petroleum production. Liquid permeation in the form of pervaporation has found a lot of applications on a small scale. These are mostly for recovery of volatile organic materials such as ethanol from relatively high concentrations in aqueous streams. However, recent research has been directed to removing low concentrations of volatile organics from wastewaters with promising results. We may expect more applications of this type to come to fruition.

Ultrafiltration and microfiltration are already widely used in waste management. Both processes are versatile, reliable, and inexpensive. Ultrafiltration often finds use as a pretreatment process for more sensitive processes such as reverse osmosis. It can be used to break emulsions and recover oil from waste oil and oily wastes, as well as to concentrate particulate streams such as paints and latex. Microfiltration is used effectively in conjunction with chemical precipitation to clarify the final effluent from the precipitation. This technique is very useful for the metal finishing industry. Furthermore, microfiltration can be used to remove particulate bacteria to sterilize wastewater streams.

REFERENCES

Baker, R. W., Cussler, E. L., Eykamp, W., Koros, W. J., Riley, R. L., and Strathman, H. (1990). Membrane Separation Systems: A Research Needs Assessment, DOE Report DE90-011771.

Cooley, T. E. and Dethloff, W. L. (1985). Field Tests Show Membrane Processing Attractive, *Chemical Engineering Progress*, October:45.

Eykamp, W. and Steen, J. (1987). Ultrafiltration and reverse osmosis, *Handbook of Separation Process Technology* (R. W. Rousseau, ed.), Wiley, New York, p. 830.

Fick, A. (1855). Uber diffusion, *Pogg. Ann.*, *94*:59.

Klein, E., Wood, R. A., and Lacey, R. E. (1987). Membrane Processes–Dialysis and Electrodialysis, Chapter 21, *Handbook of Separation Process Technology* (R. W. Rousseau, ed.), Wiley, New York, pp. 954–981.

Lipski, C. and Cote, P. (1990).The Use of Pervaporation for the Removal of Organic Contaminants from Water, *Environmental Progress*, *9(4)*:254.

Loeb, S. and Sourirajan, S. (1962). Sea Water Demineralization by Means of an Osmotic Membrane, *Advances in Chemistry Series*, *38*:117.

MacNeil, J. and McCoy, D. E. (1988). Membrane Separation Technologies, *Standard Handbook of Hazardous Waste Treatment and Disposal* (H. M. Freeman, ed.), McGraw-Hill, New York, p. 6.91.

Meller, F. H., ed. (1984). Electrodialysis-Electrodialysis Reversal Technology, Ionics, Inc., Watertown, MA.

Pardus, M. J. (1987). Material substitution: The effects of waste minimization on effluent metal concentrations, M.S. Thesis, Penn State University (August, 1987).

Parekh, B. S. (1991).Get Your Process Water to Come Clean, *Chemical Engineering*, January:76.

Porter, M. C. (1979). Membrane filtration, *Handbook of Separation Techniques for Chemical Engineers* (P. A. Schweitzer, ed.), McGraw-Hill, New York, p. 2–78.

Riley, R. L. (1990). Reverse osmosis, *Membrane Separation Systems, a Research Needs Assessment*, Vol. II, Final Report, DOE/ER/30133-H1. [Also published as Reverse osmosis, *Membrane Separation Systems, Recent Developments and Future Directions* (R. W. Baker et al., eds.), Noyes Data Corporation, Park Ridge, NJ.]

PPG (1990). Note in *Chemical Engineering*, December:19.

Schott Glaswerk (Mainz) (1990). Note in *Chemical Engineering*, December:19.

Stannett, V. T., Koros, W. J., Paul, D. R., Lonsdale, H. K., and Baker, R. W. (1979). Recent Advances in Membrane Science and Technology, *Advances in Polymer Science*, *32*:69.

U.S. Environmental Protection Agency (1976). Treatment of electroplating wastes by reverse osmosis, EPA 600/2-76-261, U.S. EPA, September 1976.

Wijmans, J. G., Kaschemekat, J., Davidson, J. E., and Baker, R. W. (1990). Treatment of Organic-Contaminated Wastewater Streams by Pervaporation, *Environmental Progress*, *9(4)*:262.

10

Electrostatic Precipitation and Electrochemical Processes

10.1 INTRODUCTION

There are many separation processes either driven by electric fields or dependent on the electrical properties of matter for their success. For example, electrodialysis, discussed in Chapter 9, depends on the ionic dissociation of molecules in solution to remove such molecules from liquids by the influence of direct current electrical fields. Another example is the separation of mixtures of solids by electrostatic separators. This is done by electrostatic charging the solids and then dropping them under freefall between oppositely charged electrodes. The electrostatic field of the electrodes deflects the solids, depending on their charge, into separate receiving bins. This technique was discussed in Chapter 4. A third example is ion exchange, which depends on the ability of molecules to ionize into positively and negatively charged ions. These ions are then removed from waste streams by replacing them with more desirable ions. Ion exchange will be discussed in Chapter 14.

The present chapter deals with electrostatic precipitation for the removal of dusts, fumes, and mists from gas streams as well as with electrochemical processes for the removal and recovery of metals and other materials from liquid waste streams.

10.2 ELECTROSTATIC PRECIPITATION

Electrostatic precipitators are used for the collection of dusts, mists, and fumes. These are defined as follows:

- *Dust*—solid particles from 0.1 to 100 μm in diameter
- *Mist*—liquid droplets suspended in a gas
- *Fume*—solid or liquid particles formed by condensation from a vapor

Electrostatic precipitation is one of the most effective processes for removing such small particles of liquids and solids from gas or vapor streams. It is capable of particulate recoveries of greater than 99% for particles down to 0.01 μm in diameter. It is usually used in industrial operations as a final cleanup process after larger particles have been removed by other methods. This makes it an alternative to scrubbers as a final cleanup technique.

Electrostatic precipitators are used both (1) to recover valuable materials such as the precious metals in gold and silver processing and (2) to remove pollutants such as acid droplets or fly ash from exhaust streams being discharged to the atmosphere. The precipitators vary in size from small units to remove smoke, dust, and fumes from the air in a single room to large industrial units processing gas at over 4 million ft^3/min, (Schneider et al., 1979).

10.2.1 Fundamentals

When particles in a gas are exposed to gaseous ions they become electrically charged. This charge makes them mobile under the influence of an electric field. They then drift toward an oppositely charged or grounded collector plate where they have their charge neutralized. This is illustrated by the sketch in Figure 10.1. Depending on the size of the particle and the amount of charge it collects from the ions, the particle drift velocity may be from 0.1 to over 1.0 ft/sec (Semrau 1984). Once the charge is neutralized at the collector plate, the solid particles can be either shaken or washed off into a bin. Liquid droplets will coalesce on the collector plate and simply run off under the influence of gravity. Electrostatic precipitation uses these principles to remove particles of liquids and solids from gas streams. Thus, the electrostatic precipitator must have:

- A method for ionizing the carrier gas
- Directed flow of the feed gas through the ionizing region
- An electric field to cause particle drift to the collecting electrode
- Adequate gas residence time in the drift field to let particles reach the collecting electrode

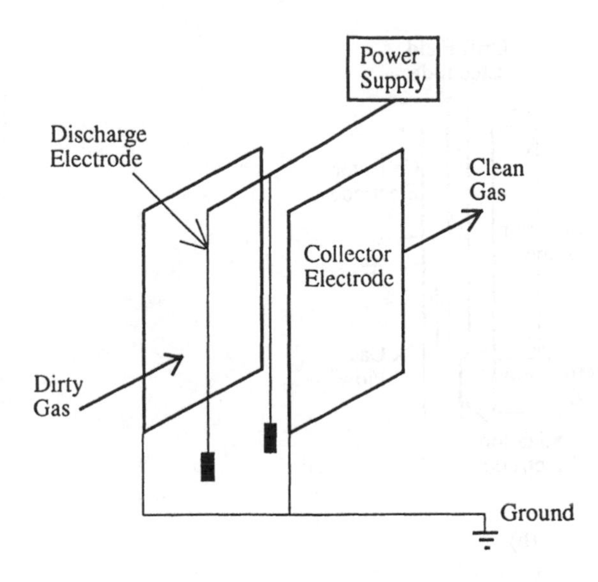

Figure 10.1 Sketch of a simple electrostatic precipitator.

- Means for removing the collected particles from the collecting electrode

The charge to the carrier gas is supplied by a high-voltage direct current corona generated by a non-uniform electrical field. A corona is the local electrical breakdown of a gas into electrically charged gaseous ions due to the high voltage. It is made possible by the non-uniformity of the field, which is very strong in the vicinity of thin wires or sharp points, but weakens with distance from these regions. If the same voltage were applied as a uniform field between flat electrodes, no breakdown would occur. If the voltage were increased until breakdown did occur, it would not be localized. In this case continuous breakdown would occur and would cause sparks between the plates rather than a corona.

The common techniques for generating a non-uniform field are to use either a thin wire or a rod equipped with many very sharp points as the discharge electrode. The thin wire gives a relatively weak corona from the few small sharp imperfections in the wire. On the other hand, the sharp points intentionally placed on the rod literally spray a corona into the gas phase. A number of different electrode shapes have been tested to increase corona discharge from sharp edges and points. However, if the percent of particles escaping precipitation is plotted against corona power input per unit of gas flow, all the data fall on a single correlating line. Thus, corona power input determines collection efficiency for electrostatic precipitators.

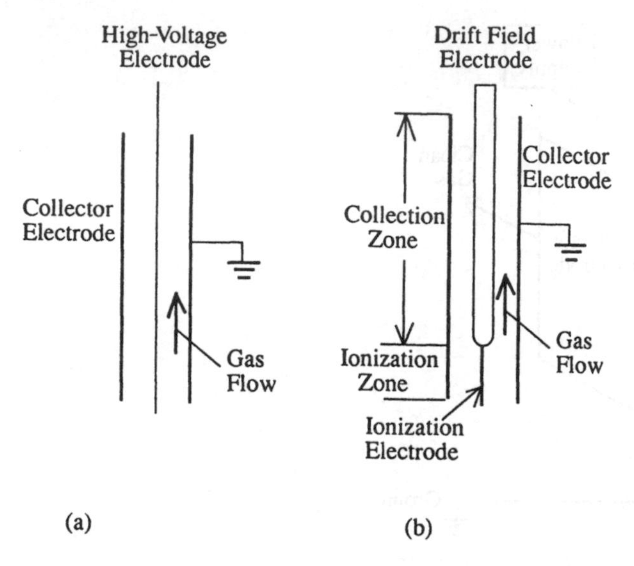

Figure 10.2 (a) Single-stage and (b) two-stage configurations.

Eletrostatic precipitators are divided into two broad classes, depending on whether the ionizing and collecting functions are combined or separated. In the single-stage unit illustrated in Figure 10.2a, the central wire ionizing electrode is also one of the electrodes establishing the field that causes the particle drift. In the two-stage unit shown in Figure 10.2b, the ionizing electrodes and the drift field electrodes are of widely different diameters so that corona discharge and ionization occur only in the vicinity of the small diameter portion of the central electrode. This type of two-stage unit is used industrially with a negative corona and washing of the collector plates is done to recover the particles (Semrau, 1984).

The major market for two-stage precipitators is in the air conditioning and ventilation field. In these units the ionization field and the collection field are controlled independently. A separate pair of electrodes is used to establish a positive corona through which the feed gas is passed. Immediately downstream of the corona the now-charged particles enter the collection zone with alternating positively charged and grounded electrodes which establish the collection field. Since the gas flow velocity in the separator is usually 3 to 10 ft/sec, the charged particles are readily blown out of the ionizing corona field and into the collector field without much loss of particle-charging effectiveness. Since the ionizing field and the collecting field are electrically independent, the field strengths can be optimized separately for each part of the precipitator.

The corona in a precipitator can be generated from either a positively or negatively charged source. However, it has been found that negative coronas generate somewhat faster-moving particles as well as more ozone by-product (Jacobs and Penney, 1987). Therefore, most industrial electrostatic precipitators are single-stage units with the corona electrode negative. Advantage is taken of the faster negative particle drift with industrial units when ozone formation is not a serious problem. On the other hand, all heating, ventilating, and air conditioning precipitators use two-stage positive ionization systems to minimize ozone.

The DC voltage applied must be high enough to establish a corona but not high enough to cause excessive arcing. Arcing causes greatly increased power consumption as well as lower collection efficiency. Commonly used corona voltages are from 30,000 to 100,000 V for large flat plate single-stage units using fine wires as corona electrodes. With two-stage precipitators the corona voltages are more on the order of 13,000 V. The required high voltage is provided by a transformer-rectifier set.

One of the advantages of electrostatic precipitation is that the electric current is carried directly by the particles. Thus, it is quite small and directly proportional to the amount of precipitated material. This makes the corona power input a measure of the effectiveness of the particle precipitation. Typical corona power might range from 300 to 1200 W per thousand cubic feet per minute of feed gas (Beltran, 1987).

The flow of feed gas is directed through the corona by either (1) using a large number of corona electrodes or wires spaced in series between two parallel collector plates or (2) placing a central annular corona electrode inside a tube through which the feed flows parallel to the corona electrode. These configurations are called duct (or plate) and tubular collectors, respectively. They are illustrated in Figure 10.3. The flow through a duct collector can be either horizontal or vertical. The flow through a tubular collector is always vertical and usually

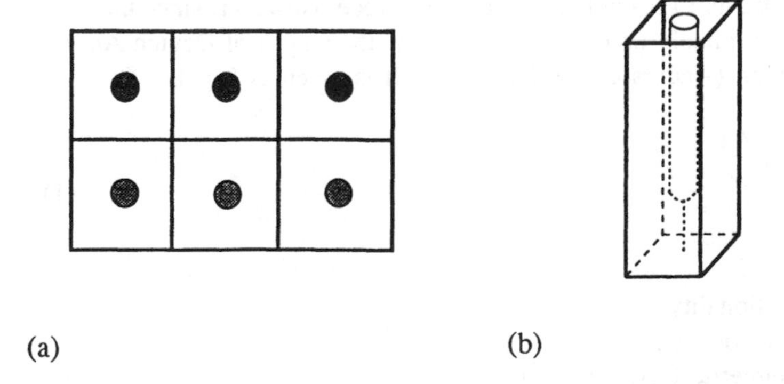

Figure 10.3 Configuration of (a) duct and (b) tubular precipitators.

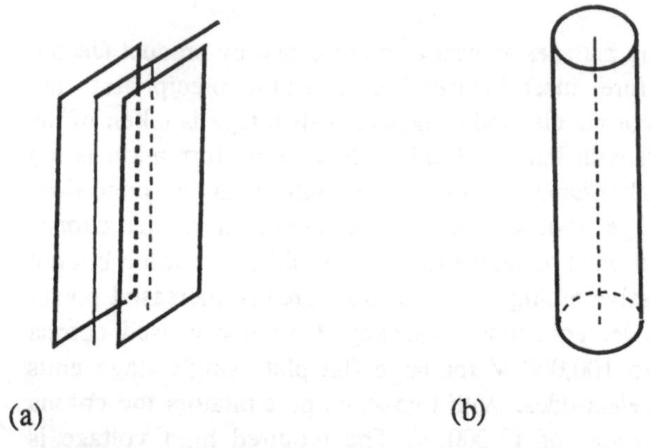

Figure 10.4 Sketch of square tubular configuration. (a) Top view of multitube unit, (b) single tube of two-stage square multitube unit.

upflow. A modification of the tubular unit has been developed which uses a honeycomb of square tubes so that both the inside and outside of the tube surface can be used for collection. This is illustrated in Figure 10.4. The pressure drop for flow through the precipitator is usually less than 1 in. of water column.

As mentioned, in single-stage units the electrostatic field causing the drift of the charged particles is between the corona electrode and the collector plate. It can be independent of the corona electrode in two-stage units. The mobility of the particles in an electric field is relatively low compared to the gas velocity. Therefore, in a two-stage unit the particles blow out of the first stage corona field into the collection stage without appreciable loss of charging efficiency.

The residence time of the gas is controlled by the gas velocity and length of the collector region. It is usually from 1 to 15 s. The residence time needed to remove the particles is determined by the drift velocity and the distance the particles have to travel to reach the collector electrode. However, since the drift velocity cannot be reliably predicted from theory, the empirical Deutch-Anderson design equation (Anderson, 1941), is used. It is given as Eq. 1.

$$\text{Efficiency} = f\frac{Aw}{V} \tag{1}$$

where:

f = functionality
A = collector area
V = volumetric flow rate of gas
w = drift velocity or precipitation rate

The functionality, *f*, collects the other design features that affect efficiency, such as electrode spacing, corona power, and particle properties.

White (1963) has shown by calculating the *A/V* term for different configurations that for a given collector efficiency a tubular precipitator can be operated at twice the gas velocity of a duct precipitator of equal length and electrode spacing. Alternatively, a tubular unit exhibits twice the area of a duct unit for equal collector length.

The square tube or honeycomb design discussed by Beltran (1987) is a distinct alternative to the duct and conventional tubular precipitator. However, it has the advantage of collecting particles on both the inside and outside of each tube. This gives it a twofold area advantage over conventional tubular units and a fourfold area advantage over duct units.

Once the particles have reached the collector electrode they can be removed in a variety of ways. For solid particles, the collected particles build up into a thin layer on the collector electrode. The electrode can be rapped or vibrated to shake off this layer and let it fall into a collecting bin. Alternatively, if the product does not have to be maintained dry, the collector electrode can be washed by sprays to remove the collected solids. Either of these techniques can be continuous or intermittent. Washing is particularly useful for vertical tube collectors.

When collecting solid particles, the resistivity of the particles is very important. If it is too high, the particles do not discharge rapidly as they deposit on the collector electrode and thus form a thick insulating layer. This reduces the flow of current from the discharge electrode to the collector electrode. To compensate for this reduction in current, the voltage is increased by the control equipment until sparking occurs. On the other hand, if the resistivity of the particles is too low, the particles lose their charge as soon as they reach the collector electrode and are immediately re-entrained in the gas stream.

Fortunately, as White (1963) points out, the resistivity of the solids can be modified by (1) the addition of chemicals and/or water to the gas stream, (2) modification of the material producing the dust, (3) changing the temperature, or (4) some combination of all the above. Addition of chemicals or water can have a very profound effect, even when added in trace amounts.

When collecting liquid droplets, the drops merely coalesce on the collector plate, run down the plate, and drip off. Thus, mists are much less of a collection problem than dusts.

10.2.2 Specific Applications

Electrostatic precipitators are probably the most versatile of all types of dust collectors. Very high collection efficiencies can be obtained regardless of the fineness of the dust. Some applications of electrostatic precipitation include:

- Small units for removing smoke, dusts, and pollens from the air of one room
- Local shop units to collect soldering or welding fumes, coolant oil mists, and grinding dusts in industrial shops
- Large central air conditioning and ventilation units
- Very large units for collecting emissions from power plants, waste incinerators, metallurgical refineries, cement plants, smelters, coke ovens, fertilizer plants, chemical manufacturing plants, etc.
- Collectors for valuable products such as precious metals and intentionally produced very fine powders

Wet precipitators are also useful for collecting mists from the effluent streams from scrubbers. This has led to the development of combination scrubber-precipitators, which remove soluble gaseous impurities along with particulates.

Single-stage precipitators, named *Cottrell precipitators* for the inventor of electrostatic precipitation, are most likely to be used with process gases from industrial processes. By maintaining the ionization throughout the collection region, re-entrainment of particles from the collector electrode is eliminated or minimized. The plate, or duct, type single-stage precipitator is usually used to collect dry dusts because of its simplicity, while the tube type is used to collect liquids and volatile fumes. Typical plate type applications are collection of cement dust and recovery of catalyst fines in petroleum refining catalyst regenerators. A typical tube type application is removal of sulfuric acid or oil mists.

Two-stage precipitators have the disadvantage when used for collecting solids that re-entrained particles cannot be recharged for reprecipitation. Consequently, their loss reduces the collection efficiency of the precipitator. Therefore, some other technique must be used to ensure that re-entrainment does not occur. This can be accomplished by continuous or intermittent washing of the collector electrode to remove any discharged particles before they can be re-entrained. Obviously, this would not be a problem with collection of mists of liquid droplets. Typical two-stage tubular applications are treating of scrubber effluent from ammonia-recovery boilers used in paper making and cleanup of stack gases from waste incinerators.

10.2.3 Equipment

Electrostatic precipitators cannot be reliably designed from first principles or theory. Therefore, experience becomes extremely important. This can be obtained either through a vendor who has already built units for the same type of application or through an adequate experimental program on the exact material to be collected. This requires longer lead times to install an electrostatic precipitator compared to other dust collection equipment. However, their low operating costs and ability to collect very small particles at high efficiency more than offset their greater complexity and higher initial cost, especially for large volume units.

Table 10.1 gives a list of vendors for electrostatic precipitators. Single-stage, two-stage, wet, dry, and combination precipitator-scrubber units are all commercially available.

10.2.4 Common Problems and Troubleshooting

Some typical operational problems encountered with electrostatic precipitators for dusts are given by Jacobs and Penney (1987) along with their solutions. These are discussed in abbreviated form below.

- Dust re-entrainment
- Excessive arcing
- Too low dust resistivity
- Too high dust resistivity

Dust Re-entrainment

This can be caused by a variety of factors, such as high gas velocity, bad gas flow distribution, electric arcing, and improper dust resistivity either singly or in combination. It usually occurs when the dust is electrically discharged at the collector electrode or when it is removed by rapping the collector electrode. Good gas flow distribution and proper electrode geometry can eliminate local high velocities and eddies, which contribute to dust re-entrainment.

Excessive Arcing

This condition both lowers collector efficiency and increases power consumption. It is desirable to have the voltage as high as posible without sparking to maximize the corona. Modern electrical control circuits can be obtained to give the optimum voltage to produce occasional sparks. This insures that the voltage is high enough for maximum corona discharge.

Too Low Dust Resistivity

If resistivity is too low, the dust quickly loses its charge when it hits the collector plate. This permits the neutralized particles to be sheared back off into the gas phase. To raise resistivity of the dust requires upstream process changes which are not always possible. However, the precipitator will normally cause agglomeration of the particles so that they can be captured in a downstream cyclone.

Too High Dust Resistivity

Having too high a dust resistivity allows formation of an insulating layer of dust on the collecting electrode and excessive voltage drop across the dust layer. This can cause a corona discharge in the dust layer causing it to explode from the collector surface and be re-entrained. Increasing temperature or adding trace amounts of water or certain chemical additives (such as ammonia, sulfur tri-

Table 10.1 Vendors of Electrostatic Precipitators

ABB Environmental Systems 31 Inverness Center Pkwy Birmingham, AL 35243 (800) 252-2832	Beltran Associates, Inc. 1133 E. 35th St. Brooklyn, NY 11210 (718) 338-3311
Ceilcote Air Pollution Control 140 Sheldon Rd. Berea, OH 44017 (216) 243-0700	Croll Reynolds Co., Inc. Box 668 Westfield, NJ 07091 (908) 232-2146
Ducon Environmental Systems 110 Bi-County Blvd. Farmingdale, NY 11735 (516) 420-4900	Joy Environmental Technologies 10700 N. Freeway Houston, TX 77037 (800) 966-9100
Lurgi Corporation 3700 Koppers St., Suite 101 Baltimore, MD 21227 (410) 644-1307	Powrmatic, Inc., Elton/Airomax 2906 Baltimore Blvd. Finksburg, MD 21046 (800) 966-9100
Research Cottrell Box 1500 Somerville, NJ 08876 (908) 685-4000	Terra Products, Inc. 150 Banjo Drive Crawfordsville, IN 47933 (317) 362-7367
United Air Specialists, Inc. 4440 Creek Road Cincinnatti, OH 45242 (800) 992-4422	United McGill Corporation 1779 Refugee Rd., Box 820 Columbus, OH 43216 (614) 836-998
Wheelabrator Air Pollution Control 441 Smithfield St. Pittsburgh, PA 15222 (412) 562-7300	

oxide, or hydrogen chloride) can greatly reduce resistivity and optimize collector performance.

Other Problems

In addition to the problems discussed by Jacobs and Penney, there can be difficulties in getting sticky solid particles to fall off the collector electrode. This can be overcome by changing the rapping system, by mechanical scraping, or more preferably, by either periodic or continuous washing of the collector electrode.

10.2.5 Economics

The installed cost of an electrostatic precipitator is highly variable because it is so site-specific. This is in spite of the fact that the cost of the unit itself is predictable because it is basically a function of the collector area required. However, the collector area required is getting more difficult to determine reliably because the collection efficiencies now required by the EPA regulations are so high—in the region of 99 to 99.99%, which is outside the historical experience of 90 to 98% efficiency in the precipitator industry. This makes small uncertainties in drift velocity lead to conservative design and frequent overkill. However, in spite of this, the reliability, normally low maintenance, and limited operator attention required make electrostatic precipitators a preferred process for collecting or removing very fine particles from gases.

The major operating cost for electrostatic precipitators is the cost of electricity to operate them. This is primarily due to the corona power requirements which are given by Schneider et al. (1975) as 0.2 to 0.4 and Beltran (1987) as 0.3 to 1.2 $W/ft^3/min$ of gas flow. A general guideline from Schneider is that operating costs are normally about 10% of installed cost per year.

10.3 ELECTROCHEMICAL PROCESSES

Electrochemical processes are widely used in the chemical process industries for the manufacture of materials in an electrolytic cell. Electrochemical processes differ from the other electrically driven processes mentioned in the introduction to this chapter. Electrochemical processes use the electron as a reagent rather than simply using a voltage gradient as a driving force for transport of material. They are starting to find many uses in waste minimization for converting toxic chemicals, recycling process streams, nuclear decontamination, and wastewater treatment (Pletcher and Weinberg, 1992a). These processes use electrochemistry to either generate reagents or carry out conversions at electrodes. They can either make direct recycling possible or make separation by other processes more effective. The main uses of electrochemistry directly as a *separation process* in waste minimization are for the recovery of metals. This can be done by electroplating the metal ions out of an aqueous waste stream. Such waste streams come from the metal finishing, electronics, wire manufacturing, or mining industries. Here, electrochemistry has the advantage that the metal can be recovered in its most valuable form, that is, as the pure metal. This metals recovery application is what will be discussed here.

10.3.1 Fundamentals

The basic apparatus of an electrochemical process is the electrolytic cell. It consists of an anode where oxidation reactions occur, a cathode where reduction reactions occur, an electrolyte which contains in ionic form the material to be converted, and a power supply to provide a direct current voltage between the electrodes. It may also contain one or more membranes between the electrodes which permit the passage of either cations or anions, but not both. The types of cells used for metals recovery (electrowinning) are sketched in Figure 10.5. Figure 10.5a shows an undivided cell, while Figure 10.5b shows a cell with an ion exchange membrane inserted between the electrodes.

Using the circuit shown in Figure 10.6 for the recovery of copper by electroplating from a sulfuric acid solution, the functions of the various parts of an electrolytic cell are described. The negative electrode, or cathode, is the source of the electrons flowing through the cell while the positive electrode, the anode, is the sink from which they leave the cell. The electric current between the electrodes is carried by the positive copper and negative sulfate ions in the electrolyte solution between the electrodes. The negative ions flow

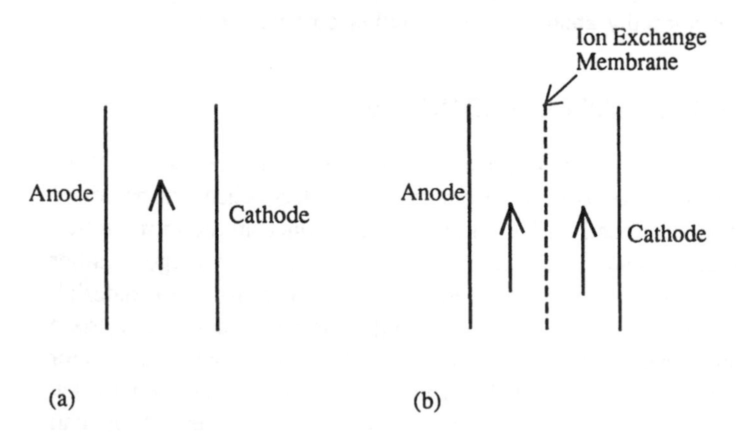

Figure 10.5 Sketch of electrolytic cells for metal recovery. (a) Undivided cell, (b) membrane-separated cell.

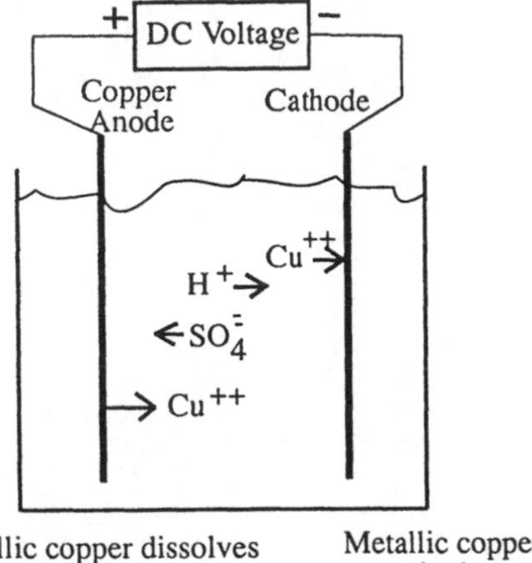

Figure 10.6 Schematic for recovery of copper from a sulfuric acid electrolyte.

toward the anode while the positive ions flow toward the cathode under the influence of the voltage applied between the electrodes.

The reaction at the cathode is the reduction of the cupric ion to metallic copper by the addition of two electrons to the copper ion. This causes the copper to plate on the cathode as copper metal. The corresponding reaction at the anode is an oxidation reaction, the exact nature of which depends on the applied voltage, the material from which the anode is constructed, and what compounds are present in the electrolyte. If the anode is also copper, the electrode reaction will be the dissolving of copper into the electrolyte. This occurs by oxidizing the copper metal to cupric ion with the loss of two electrons. These electrons then flow into the external circuit. The net effect is the transfer of copper metal from the anode to the cathode. This is the common electroplating reaction. It can also be used with chromium, nickel, gold, and silver to provide high quality metal finishes on less attractive or less corrosion-resistant metals. It can also be used in metal refining to recover high purity metals from less pure metal anodes.

However, if the anode is made from a material which will not dissolve in the electrolyte, some other oxidation reaction will occur to release the two electrons into the external circuit. This can be, for example, the formation of sulfuric acid from sulfate ion and water at the anode, with the release of oxygen gas. The reactions at each electrode are called half-cell reactions and the overall reaction of the electrolytic cell is obtained by adding the two half-cell reactions. Similarly, the overall cell voltage is the cathode half-cell voltages minus the anode half-cell voltage. For example, the half cell reactions and standard voltages for the example above with the inert anode are:

Cathode:

$$Cu^{2+} + 2e = Cu^0$$

where Cu^0 is copper metal

Half-cell voltage = 0.3419 V

Anode:

$$H_2O - 2e = 1/2\ O_2 + 2H^+$$

Half-cell voltage = −1.229 V

Overall:

$$Cu^{2+} + H_2O = Cu^0 + 1/2\ O_2 + 2H^+$$

Overall cell voltage = 1.571 V

In the overall cell reaction, the copper metal is plated onto the cathode, the oxygen is released at the anode, and the hydrogen ions maintain the electric neutrality of the sulfuric acid in the solution. However, the overall cell voltage is merely the minimum that will make the reactions go. In actual practice, a somewhat higher voltage is needed to make the reactions proceed at a reasonable rate. This extra voltage is called the overpotential. It is needed to increase the rate of the electron transfer reactions at the electrodes as well as the rate of flow of current carrying ions between the electrodes. A typical cell current versus overpotential curve is illustrated in Figure 10.7. At low overpotentials the current increases exponentially with overpotential. At high overpotentials

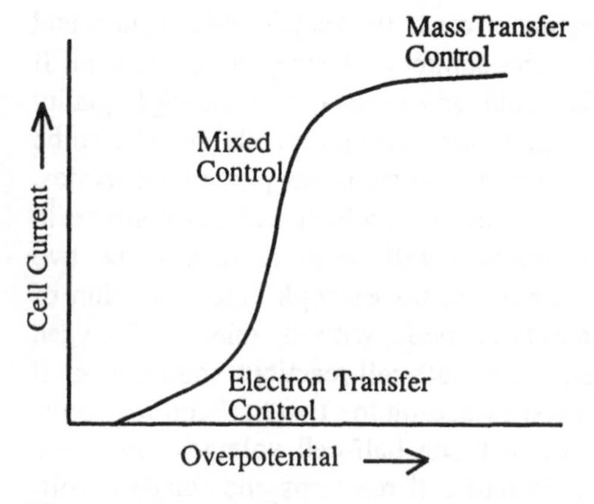

Figure 10.7 Variation of cell current with cell overpotential for a single ion.

the current is limited by the rate of mass transport by the ions. At intermediate overpotentials the current is controlled by a mixture of both effects.

From Figure 10.7 it is clear that the maximum cell current is limited by mass transport. This means that ways of increasing mass transport will be beneficial in increasing the rate of reactions in the cell. Ionic mass transport may occur by (1) diffusion caused by a concentration gradient, (2) convection enhanced by turbulence due to mixing, and (3) migration under the influence of the electric field. Of these three, convection mass transport is by far the most rapid. Therefore, high levels of turbulence in the cell are used to increase the rate of mass transfer by convection. This gives mass transport levels much higher than can be achieved by diffusion or ion migration.

In most cells, attempts are also made to minimize the mass transport problem by minimizing the gap between electrodes. This maximizes the ratio of electrode surface to electrolyte volume. This also minimizes the cell resistance and thus the power consumed by the cell without the electrode reactions.

The mass transport limitation is at its worst when the material to be converted is present in the solution at very low concentrations. This situation can be remedied either by (1) preconcentrating the feed before feeding it to the cell, (2) using the cell to generate an oxidation/reduction (redox) reagent at the electrode, which will react in solution with the low concentration material, or (3) designing high-surface electrodes operating under high turbulence. The preconcentration is frequently done by either ion exchange or solvent extraction. Typical redox reagents which can be elecrolytically generated are metal ions, such as ferric ions generated from an iron electrode. High-surface electrodes are three-dimensional electrodes in which high turbulence is obtained by flowing the electrolyte right through the electrode. Some examples of three-dimensional electrodes discussed by Pletcher and Weinberg (1992b) are sketched in Figure 10.8.

Three-dimensional electrodes can be either static or dynamic. The static electrodes are porous solid materials such as stacked felts, foams, cloths, and screens as shown in Figure 10.8a or fixed beds packed with particles of electrode material as shown in Figure 10.8b. Dynamic electrodes can be beds of small particles in motion such as the fluid beds shown in Figure 10.8c or moving beds. The fluid bed shown in the figure uses inert particles, such as glass spheres. It has as its main benefit the improvement in turbulence in the cell. Such cells can reduce metals level to only the 50- to 150-ppm level. The fluidization or other motion can take place as a result of the upflow of electrolyte through the bed as shown in the figure. It can also be enhanced by the application of mechanical energy, such as vibration, tumbling, or pulsed flow. Rotating drums filled with seed metal particles are used as cathodes in some commercially available metal recovery cells. The three-dimensional beds are important to environmental applications because they are capable of removing metals down to below 1 ppm by providing sufficient bed length.

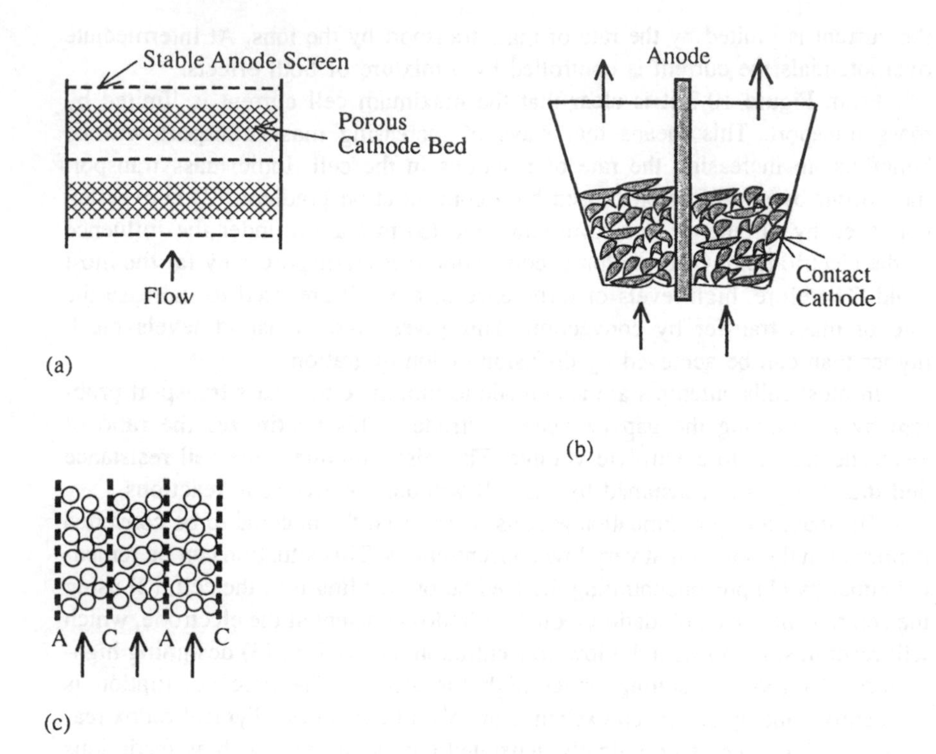

Figure 10.8 Sketches of some special types of electrodes. (a) Porous cathode beds, (b) fixed-bed cathode, (c) fluid bed. A, Anode; C, cathode.

As mentioned above, the anode electrode can be either chemically inert or sacrificial in the cell reaction. Many different electrode materials are available to the cell designer. Therefore, electrode materials are chosen depending on the nature and purpose of the electrolytic cell. Typically, in electroplating as a metal finishing process, the anode is sacrificial to keep the metal ion in the electrolyte at the proper concentration. It is made of the metal being plated. On the other hand, if the plating is being done to remove metal from a waste stream by plating on the cathode, an inert anode is used. This allows depletion of the metal ion in solution and thus its removal. Common electrode materials include metals such as lead, copper, and nickel; alloys such as steels and monels; and coated electrodes, with titanium or carbon being the base material under the coating. Typical coatings are platinum, lead dioxide, or ruthenium dioxide. There is a trend toward using materials that catalyze the electrode reactions as coatings to improve cell performance.

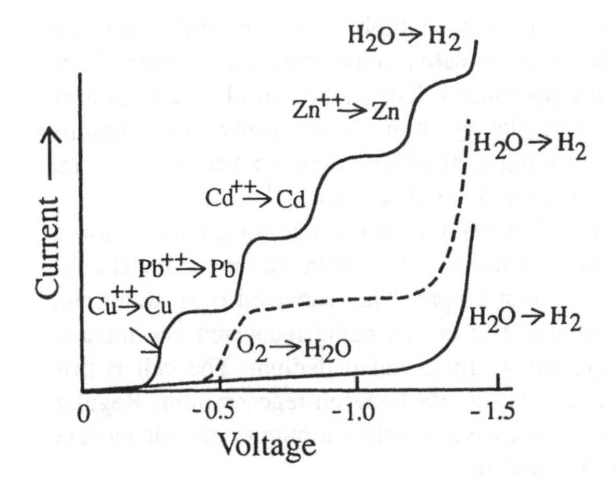

Figure 10.9 Variation of cell current with cell overvoltage for a multicomponent electrolyte.

Electrolytic cells are capable of carrying out a wide variety of reactions depending on the applied voltage and the components present in the electrolyte. For example, Figure 10.9, adapted from Pletcher and Weinberg (1992a), shows an experimental current versus voltage plot for a rotating disk electrode. It is operating in an electrolyte containing divalent ions of copper, lead, cadmium, and zinc in a neutral aqueous chloride solution. A saturated calomel electrode is used as a reference electrode.

As the voltage is increased, the first metal to be deposited is copper. If voltage is further increased the other metals start to plate as their required voltage is reached. Also shown in the figure are the curves for oxygen dissolved in the electrolyte and an oxygen- and metal-free electrolyte as a blank. These curves show that by controlling the applied voltage, either copper could be removed selectively or all the metals could be removed at once without interference by electrolysis of water to form hydrogen. However, the presence of air dissolved in the electrolyte would interfere with the recovery of all the metals except copper. This illustrates how voltage and electrolyte composition can determine the results of electrolytic reactions.

10.3.2 Specific Applications

Because of the wide variety of reactions which can be driven by appropriately designed electrolytic cells, the number of applications possible is limited only by the ingenuity of the cell designer and the economics of alternative processes. For metals recovery, the environmental applications can be to either process wastewaters or recycle streams. Either low or high metals concentrations can be involved. Low metal concentration streams come from rinse waters from plating, pickling, or photographic operations. The higher metal concentrations are in waste process solutions from plating or in waste regeneration solutions from ion exchange. A new and growing application is the recovery of zinc, tin, and steel from galvanized steel scrap and scrap tin cans.

A typical metal recovery operation from a rinse water stream would use a packed bed of carbon granules as the cathode. This could reduce the metal content from about 10–50 ppm to less than 1 ppm. The recovered rinse water can often be recycled to the rinse circuit. The metals being recovered can include lead, copper, cadmium, mercury, silver, gold, and palladium. The cell is primarily used for dilute streams to extend its life between regenerations. Regular in situ regeneration of the bed is done by either acid washing or electric current reversal to remove metals from the carbon.

The more concentrated metals, up to 20% by weight, in process solutions are more likely to be recovered using a simpler cell. The required cell may be simply parallel plates in a simple tank with air sparging to provide some turbulence to increase mass transfer. The metals are electroplated onto the cathode, and the cathode is then removed for further processing when an adequate amount of metal has built up. The anode used is probably chemically inert in the electrolyte.

According to Pletcher and Weinberg (1992a), a new, but likely to grow, application is the large-scale electrolytic recovery of metal coatings from a steel base metal. This process has been applied in two new plants to tin removal from scrap tinplate and zinc removal from scrap galvanized steel at levels of 40,00 and 20,000 tons per year, respectively. In these applications, the electrolytic cell is used to strip the coating from the steel at the anode and plate it out onto the cathode. The coated steel scrap feed is chopped up into small pieces and loaded into rotating plastic barrels where the coating is anodically removed from the steel and plated out onto cathode sheets. The bare steel and the recovered sheets of coating metal are all recycled.

10.3.3 Equipment

The equipment used for electrochemical recovery of metals varies greatly in size. Small off-the-shelf units are used for recovering precious metals in the jewelry and photographic businesses. Large custom-designed units for electrowinning of metals are used in the mining industry.

Some of the vendors of electrochemical metal recovery equipment are listed in Table 10.2. It is advisable to contact vendors early in the planning stage when considering electrochemical processing of wastewaters for metals removal.

Table 10.2 Vendors of Electrochemical Recovery Equipment

ABB Raymond 650 Warrenville Road Lisle, IL 60532 (708) 971-2500	Andco Environmental Processes 595 Commerce Drive Amherst, NY 14228 (716) 691-2100
Ionics, Inc. Separations Technology 65 Grove St. Watertown, MA 02172 (800) 338-9238	Lancy International, Inc. 181 Thorn Hill Road Warrendale, PA 15086 (412) 772-0044
Met-Pro Corp, Systems Div. 160 Cassell Rd. Harleysville, PA 19438 (215) 723-6751	Mobile Process Technology 2070 Airwys Blvd. Memphis, TN 38114 (800) 238-3028
Osmonics, Inc. 5951 Clearwater Dr. Minnetonka, MN 55343 (800) 848-1750	Remedial Systems, Inc. 56 Leonard St. Foxboro, MA 02035 (508) 543-1512
Serfilco, Ltd. 1777 Shermer Rd. Northbrook, IL 60062 (800) 323-5431	Trionetics, Inc. 2021 Midway Drive Twinsburg, OH 44087 (216) 425-2846

10.3.4 Common Problems and Troubleshooting

When operating an electrolytic cell for metal recovery, the primary indications of problems are (1) excessive metals content in the effluent or (2) excessive power consumption. Either of these symptoms can be attributed to many causes, either resulting from cell design or from unexpected feedstock characteristics. The causes and remedies are now discussed.

Excessive Metals Content of the Effluent

Excessive metals content of the effluent can have many causes. Those related to cell design are inadequate electrode area, gas accumulation in the cell, inadequate residence time, too low an applied cell voltage, and poor mass transfer in the cell electrolyte.

Inadequate electrode area can be due to error in the original estimate of the area required. It can also be due to accumulation of gas in the cell. This lowers the level of electrolyte in contact with the electrodes, thereby reducing active electrode area. Proper venting of the cell and installation of a membrane cell divider near the gas-producing electrode can insure that gas accumulation in the electrolyte does not occur. Furthermore, reduction in flow rate through the cell, i.e., increased residence time, can improve metal removal when electrode area is marginal. Poor mass transfer in the cell electrolyte can be fixed by increasing turbulence in the electrolyte. This can be done by increased mixing, by installing three-dimensional electrodes, or both. Three-dimensional electrodes also increase electrode area, giving an added benefit.

The cell performance failures related to feedstock composition are usually the result of an upset in the operations upstream from the cell. This could be in the form of a concentration surge in the feed metals content due to dumping of a high metals content bath. It could also be due to a surge in concentration of impurities which compete electrochemically with the reactions for which the cell was designed. Mixing of incompatible feeds to the cell could cause this. Such feedstock upsets can be minimized by using a holding tank for feed equalization before it is fed to the cell. The cell may need to be operated at reduced feed rate or increased voltage until the altered feed has been either processed by the cell or removed from the system. Inadequate overvoltage can also lead to low metal deposition rates and thus to incomplete removal of metals.

Excessive Power Consumption

Excessive power consumption is usually the result of either (1) unexpectedly high cell resistance, (2) competing reactions of impurities in the electrolyte, or (3) polarization at the electrodes. The high cell resistance can be caused by large distances between the electrodes, low ion mobility, or poor mixing. These can be improved by increasing the turbulence in the cell, which also minimizes polarization. Competing reactions can be stopped by lowering the voltage to below the level needed to make the undesirable reactions occur. However, this may not give enough overvoltage to make the desired reactions occur. If this is the case, the interfering reagents will need to be removed from the feed if that is possible. For example, dissolved air can be removed from the feed quite readily, but water reactions in aqueous electrolytes are a natural consequence of high-voltage operation.

10.3.5 Economics

In evaluating the economic attractiveness of electrochemical recovery of metals, the important variables are the cost of the installation and operation of the process, the value of the recovered metals, and the cost of alternative treatment and disposal. The annual amortization cost of the installation is obtained by totalling the cost of the cells with all their auxiliaries, including the cost of predictable replacements. This total is then divided by the expected life of the installation. In this capital cost area, electrochemical processes are not particularly superior to alternative chemical processes. However, they often have a clear advantage in operating cost. The operating cost is the cost of electric power, operating manpower, and maintenance. In these areas electrolytic processing is relatively inexpensive. Furthermore, the value of the recovered materials helps offset the costs of operation. The value of recovered materials can be very high when recovering precious metals or relatively small for non-precious metals. However, in the latter case, the cost savings from reduced disposal costs through recycling of the treated stream can frequently give short payout times for the process.

10.4 SUMMARY

Electrostatic precipitation and electrolytic metal recovery differ greatly even though they are both electrically driven processes. The electrostatic process is a high-voltage gas-phase process in which particles are electrically charged by gaseous ions generated from the carrier gas by a high-voltage corona. The charged particles then drift under the influence of an electrostatic field to a collector plate where they are discharged and recovered. Either solid or liquid particles can be collected down to very low effluent concentrations. The environmental uses are primarily to remove particulates from waste gases being discharged to the atmosphere. However, electrostatic precipitation is also widely used to recover valuable materials in industrial operations. It is one of the most effective ways to remove small particles from gas streams.

In contrast to the above, electrolytic metal recovery is a low-voltage, electrode process. Chemical reactions are caused to occur at the electrodes by a direct current voltage applied to the electrodes. The electrodes are immersed in a well-stirred electrolyte. The cathode reaction is the plating out of the metal ions in the electrolyte onto the cathode electrode. This is done by reducing the ion to the metal with electrons from the cathode. The reaction at the anode is the oxidation of an equivalent amount of material to give up electrons to the anode. This reaction can be electrolysis of water to release oxygen, formation of chlorine gas from chloride ions, oxidation of metals to ions, and others. Effective removal of metals to very low concentrations is achieved by using three-dimensional high surface area cathodes and good mixing of the cell electrolyte. Electrolytic cells can be used for many other reactions beside electroplating. They find use in (1) modifying the composition of electrolytes to permit recycling in industrial processes, (2) the production of specific reagents to treat industrial wastewaters, and (3) the manufacture of many chemical products.

REFERENCES

Anderson, E. (1941).Separation of Dusts and Mists, *Chemical Engineers' Handbook*, 2nd ed. (R. H. Perry, ed.), McGraw-Hill, New York.

Beltran, M. (1987). "Wet and Dry Tubular Electrostatic Precipitators for Control of Incinerator Flue Gas Emissions," presented at the Air Pollution Control Association Specialty Conference on Thermal Treatment of Municipal, Industrial, and Hospital Wastes, Nov. 6, 1987, Pittsburgh, PA.

Jacobs, L. J. and Penney, W. R. (1987). Electromotive Devices, *Handbook of Separation Process Technology* (R. W. Rousseau, ed.), Wiley, New York, p. 144.

Pletcher, D. and Weinberg, N. L. (1992a). The green potential of electrochemistry. Part 1, The fundamentals, *Chemical Engineering*, August: 98–103.

Pletcher, D. and Weinberg, N L. (1992b). The green potential of electrochemistry. Part 2, The applications, *Chemical Engineering*, August: 132–141.

Schneider, G. G., Horzella, T. I., Cooper, J., et al. (1975). Selecting and specifying electrostatic precipitators, *Chemical Engineering*, May 26: 94–108.

Schneider, G. G., Horzella, T. I., and Striegl, P. J. (1979). Electrostatic precipitators, *Handbook of Separation Techniques for Chemical Engineers* (P. A. Schweitzer, ed.), McGraw-Hill, New York, pp. 6-31 to 6-55.

Semrau, K. T. (1984). Gas-solids separations, *Perry's Chemical Engineers' Handbook*, 6th ed. (R. H. Perry and D. W. Green, eds.), McGraw-Hill, New York, pp. 20-75 to 20-121.

White, H. J. (1963). *Industrial Electrostatic Precipitation*, Addison-Wesley Publishing Co., Reading, MA.

11

Evaporation and Crystallization

11.1 INTRODUCTION

Evaporation and distillation are similar processes in that a liquid is vaporized, and the vapor is usually condensed to recover a liquid of different composition. The distinguishing feature between them is that in evaporation the material left behind is non-volatile. In distillation, all components of the feed may be at least slightly volatile and require further separation. Thus, evaporation is used to remove volatile solvents, usually water, from non-volatile impurities. Such non-volatile impurities may be inorganic salts, heavy tars and sludges, or radioactive wastes. Evaporation leaves behind either a thickened slurry of solids, a viscous non-volatile liquid, or a dry solid product.

Historically, evaporation has been used as a waste disposal technique for dilute aqueous solutions, slurries, and sludges. The waste was pumped into a holding pond where it was allowed to evaporate under the influence of solar radiation. When the pond was full of residue, it was covered with soil and usually planted with grass. Recent concerns about air and groundwater pollution have increased the cost of constructing holding ponds (Delaney and Turner, 1989). This is because of the need for leak-proof linings and monitoring as well as the rising price of land. The recent trend has been away from ponds and toward more intensive, totally enclosed evaporation equipment.

Crystallization and evaporation are also related in that many industrial crystallizers use evaporation of the solvent to precipitate crystals from an aqueous solution. Thus, the large-scale production of many organic and inorganic salts depends on evaporator-crystallizers for control of purity, particle size distribution, and appearance. Typical products recovered in this way are sugar and table salt.

Evaporation is widely used in waste management either to reduce the volume of a liquid or slurry waste or to recover liquid solvent for recycle. This often brings the concentration of dissolved solids in the liquid to the point where they begin to precipitate as crystals. Thus, crystallization can occur in the evaporation processes used in waste minimization. However, crystallization can also be used intentionally to recover solids from waste or recycle streams. Crystallization by freezing can be used to remove solvents from solutions. This is called *freeze crystallization* and is starting to find use as an alternative to evaporation for removing water from wastewaters.

This chapter will discuss both evaporation and crystallization primarily as they apply to waste management.

11.2 EVAPORATION

Evaporation has been widely used for many years to recover a desirable product as either the overhead stream, the bottoms stream, or both. The evaporation of seawater to recover potable water is a typical example of a desirable overhead stream. It is widely used both on ships and ashore in arid countries. In evaporative salt crystallizers, the bottoms stream will contain the valuable product, i.e., the salt crystals. However, in waste management, the reduction in volume of the bottoms stream to reduce the cost of disposal is the main economic driving force. The recovery of the overhead water or organic solvent may also have a secondary value for recycling to the process generating the waste.

11.2.1 Fundamentals

Evaporation is the vaporization of a liquid by the addition of heat. The heat can be supplied to a pool of liquid in a variety of ways. Perhaps the simplest is the holding pond, which lets the heat from the sun and wind evaporate water from the surface of the pond into the air. Salt is frequently recovered from seawater in this way. In contrast, industrial evaporation is basically an intensive boiling process, and as such its major concerns are (1) getting the heat into the liquid, (2) forming bubbles of vapor, (3) separating the vapor from the liquid, and (4) condensing the vapor. However, the high heat fluxes required by industrial evaporators require special heat transfer surfaces. These are usually in the form of tubes, but may be coils, jackets, double walls, or flat plates. With tubular surfaces, the heating medium may be either inside or outside the tubes, with the evaporating feed on the other side. Coils and jackets normally contain the heating medium inside the coil or jacket.

The rate of heat transfer is given by Eq. 1 (Bennett, 1979).

$$Q = UA\,\Delta T \tag{1}$$

where:

Q = quantity of heat transferred, Btu/h
U = heat transfer coefficient, Btu/h/ft^2/°F
A = heat transfer area, ft^2
ΔT = temperature difference, °F

Thus, the rate of heat transfer increases with the heat transfer area, with the temperature driving force, and with increasing heat transfer coefficient. Unfortunately, things are not quite this simple because the heat transfer coefficient is a strong function of temperature difference. Furthermore, the temperature difference itself requires some operational corrections to the apparent overall temperature difference (Sandiford, 1984). The apparent temperature difference is the difference between the condensing temperature of steam at the pressure on the steam side of the heater and the boiling point of water at the surface of the boiling liquid. These temperatures can be obtained from the pressures in the steam chest and the evaporator body, respectively, using the steam tables. Reductions that must be applied to the apparent temperature difference are (1) the increase in boiling point of water due to the dissolved solute in the water and (2) the increase in boiling temperature due to the increased pressure caused by hydrostatic head at the submerged boiling surface. These two values additively decrease the value of the apparent temperature difference to the value of the actual operating temperature difference, which may be much lower.

The boiling point of the evaporating liquid can be controlled by controlling the pressure at the vapor-liquid disengagement surface. This can be reduced by maintaining a subatmospheric pressure in this space. Conversely, if it is desired to maintain the vaporization at a higher temperature, the evaporator can be operated at higher back pressures. Many industrial evaporations are normally run under vacuum to minimize thermal damage to the residual product.

Two common types of vaporization used in evaporators are pool boiling and flash vaporization. These are illustrated in the sketches in Figure 11.1. Pool boiling occurs when the heat transfer surface of the evaporator is immersed in a large pool of liquid which is mixed only by natural convection and the flow of rising vapor bubbles. Flash vaporization occurs when the liquid is pumped through a heater under sufficient pressure to prevent vaporization in the heater

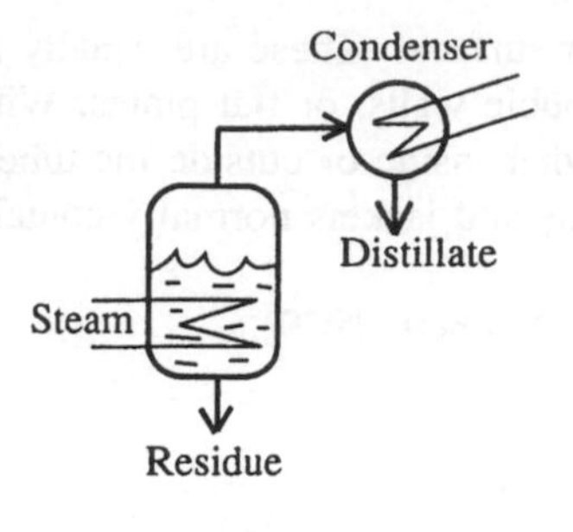

(a)

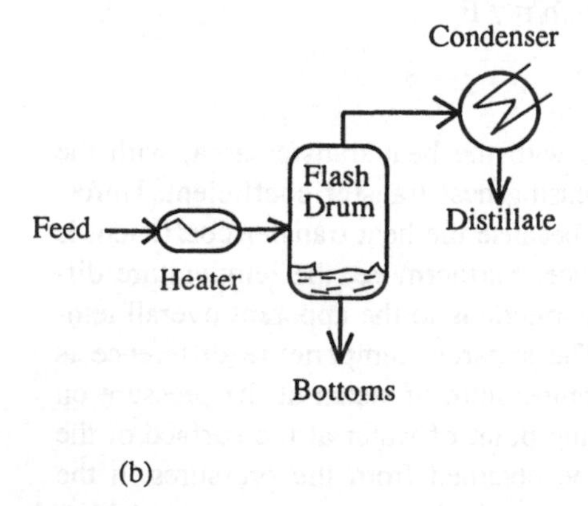

(b)

Figure 11.1 Sketches of (a) pool boiling and (b) flash vaporization.

followed by release of the pressure downstream. The pressure can be maintained by hydrostatic head using deep submergence of the heating surfaces. It can also be provided by installing an orifice plate or back pressure regulator downstream from the heater. As the pressure is released, all the sensible heat carried by the liquid is used to vaporize liquid. The vapor then flashes out of the liquid in a liquid-vapor separator.

In pool boiling the nature of the boiling changes as temperature difference is increased. These different types of boiling have different heat transfer coefficients. The major types of boiling involved are (Bell, 1984):

- Natural convection boiling
- Nucleate boiling
- Film boiling

There are also transition regions from one type to another.

Natural convection boiling occurs when the liquid is heated as it flows past a submerged hot surface. The flow is driven by thermal currents caused by heating the liquid. The liquid rises to the pool surface and vaporizes there. With this type of boiling, the heat transfer coefficient increases with temperature difference, but is relatively low.

In *nucleate boiling* the bubbles of vapor form at the liquid-solid interface of the heat transfer surface. This type of boiling occurs in kettles or thermosiphon reboilers used in the process industries. Very high heat transfer coefficients are obtained with this type of boiling.

Film boiling occurs when the temperature difference is so high that a continuous vapor film is formed on the boiling liquid side of the heat transfer surface. The heat transfer resistance of this vapor film causes the overall heat transfer coefficient to decrease during the transition from nucleate boiling to stable film boiling, that is, while the amount of heat transfer surface covered by the film is increasing. Once total coverage has been achieved, the boiling is called *stable film boiling*. In this region the heat transfer coefficient again increases as temperature difference is increased. However, the absolute values are lower than for nucleate boiling.

The formation of vapor bubbles in a heated liquid can occur either on the heated surface or in the bulk of the liquid. When formed on the surface, the bubble is initiated, or nucleated, by imperfections in the smoothness of the surface. If bubbles do not form at the surface, they can be nucleated in the bulk liquid phase by trace solids such as dust or small crystals, or by intentionally introduced rough surfaces, such as boiling chips. In the absence of nucleation of bubbles, the liquid will superheat until nucleation is forced by the superheat level. At this point a violent explosive vaporization will occur converting all the superheat into heat of vaporization. This is the cause of "bumping" which may occur repetitively and can damage equipment. It is avoided by adequate nucleation of boiling.

The circulation of liquid past the heating surface may be caused either by the boiling itself or by mechanically induced flow. If the boiling occurs inside tubes heated on the outside, the lift of the vapor bubbles causes liquid flow up through the tubes. This is the thermosiphon effect which is illustrated in Figure 11.2. The vapor and liquid are separating in a disengaging space above the exit from the tubes. The high flow velocities in the tubes provide high heat transfer coefficients for the evaporation.

Forced circulation also gives the high heat transfer coefficients afforded by high flow velocities. However, the flow is now caused by pumps, propellers, or hydraulic head. This is illustrated for pumped circulation in Figure 11.3.

Liquid-vapor separation can be relatively simple, such as by settling the entrained liquid droplets from the vapor phase in the vapor space of the evapor-

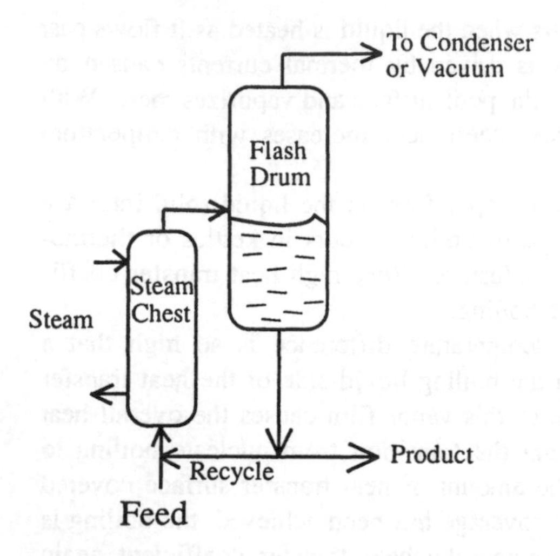

Figure 11.2 Sketch of a thermosiphon boiler.

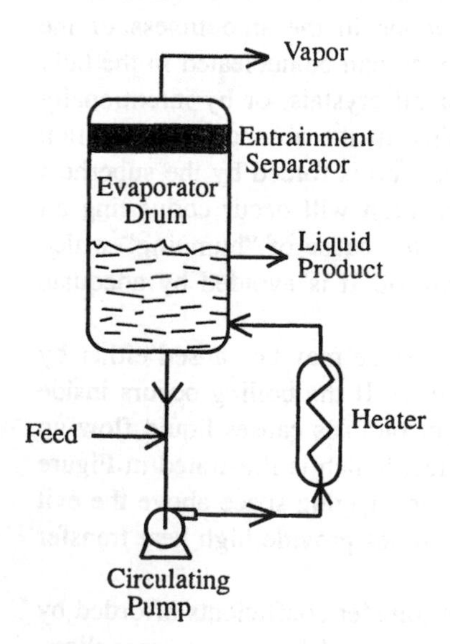

Figure 11.3 Sketch of a forced circulation boiler.

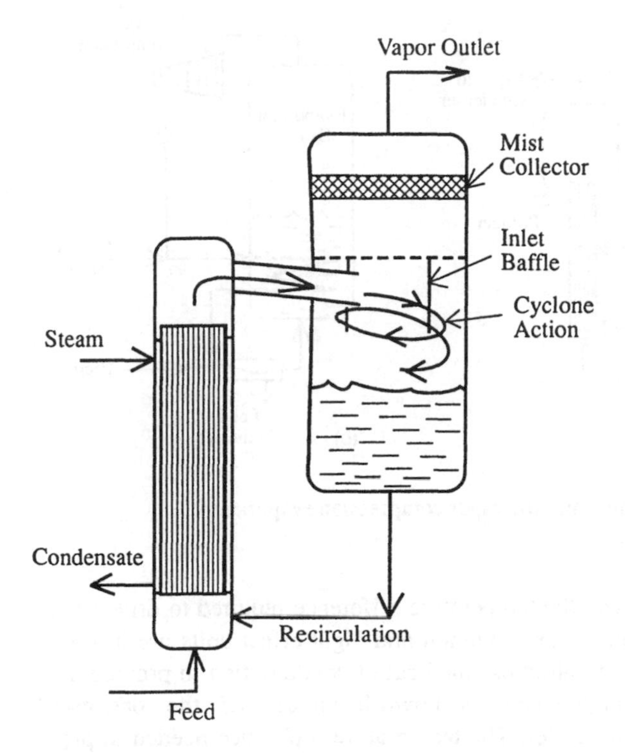

Figure 11.4 Vapor-liquid separation techniques.

ator. However, with intense boiling or flash vaporization, additional vapor-liquid separation may be required. This can utilize cyclonic separators, mesh mist collectors, or direction changing baffles, as shown in Figure 11.4 for all three. Vapor-liquid separation is desirable to prevent the loss of valuable product. However, it may also be required to prevent fouling or corrosion of condenser surfaces due to deposition of entrained solids.

The cost of heat for evaporation is a major part of its operating cost. Because of this, ways of using this heat of vaporization repeatedly have been developed using two different techniques. These are multiple-effect evaporation and vapor compression evaporation. They are illustrated by the sketches in Figure 11.5. Multiple-effect evaporation uses the latent heat recovered by condensation in a first evaporator stage, or "effect," as the heat of vaporization in a second effect. This is done by making the condenser for the first effect the boiler for the second effect, and so on. The number of effects in series is limited only by the initial temperature and pressure available, the final temperature and

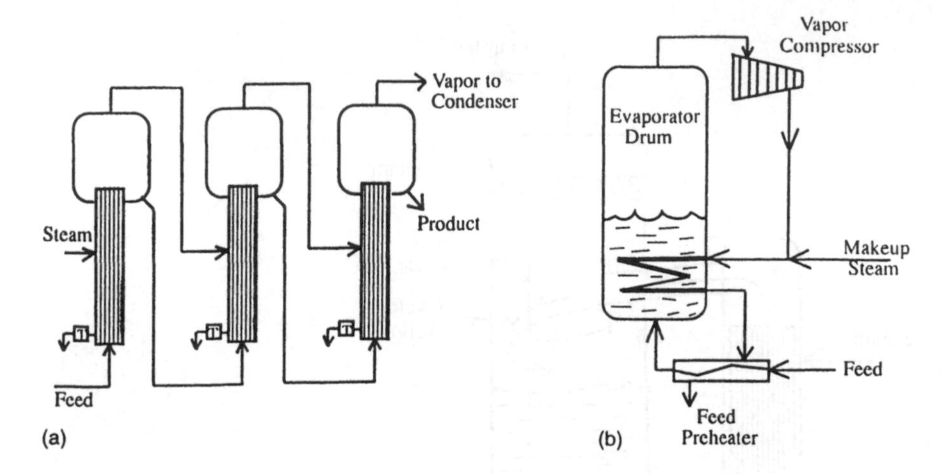

Figure 11.5 (a) Multi-effect and (b) vapor compression evaporation.

pressure for condensation, and the temperature difference required to drive each effect. Triple-effect evaporators are common and eight-effect units are in use.

Vapor compression evaporators use the heat of condensation to provide the heat of vaporization in a single stage. As shown in Figure 11.5, the condenser for the evaporator is also the boiler. The temperature difference needed to provide the heat transfer is obtained by compressing the vapor before it is fed to the condenser-boiler. This gives it a higher condensation temperature than its lower-pressure boiling point in the evaporator and makes the whole process possible.

11.2.2 Specific Applications

Evaporation is widely used in the process industries to concentrate solutions containing dissolved solids or liquids of very low vapor pressure. As the solvent is removed, the dissolved solids concentration can reach the saturation concentration in the solution and begin to precipitate out as a separate solid phase. If precipitation is caused, the process is called *evaporative crystallization*. Evaporative crystallization is one of the major applications of evaporation and is widely used in the manufacture of crystalline solid materials.

Among the applications of evaporation to waste minimization are:

- Removal of dissolved gasoline and solvents from waste lubricants
- Solvent recovery from spent wash solutions
- Concentration of sludges going to landfills

- Recovery of solids from solutions

The first two of these applications are basically single-stage batch distillations although they fufill the definition of evaporation. This is because the residue is essentially non-volatile at evaporation temperature. Gasoline and solvents usually boil below 400°F. while the initial boiling point of waste lubricating oils is about 700°F. The spent wash solutions obtained from cleaning painting equipment, rinsing pipe, and flushing machinery in repair shops can be evaporated to recover the wash solvents for recycling to the washing operation. The residual paint or oils are much reduced in volume and can be disposed of much less expensively.

The removal of water from sludges going to landfills is done to reduce the volume, reducing the cost of disposal. The sludges as produced may contain only a few percent solids. Thus, the volume can be greatly reduced by removing most of the water in a pure enough form to either recycle or discard to the sewer.

In some cases it may be necessary to recover the solids for further processing before disposal. If the solids are highly toxic or radioactive, it may not be possible to dispose of them without further processing to render them safe, for example, by encapsulation. Evaporation can concentrate them so that much less material needs to be further processed.

11.2.3 Equipment

The equipment used for evaporation comes in many sizes, from small laboratory units for recovering solvents to large multi-effect units for recovering potable water from seawater. Furthermore, the variety of arrangements of the heat transfer surface and the vapor disengagement volume is also very large. However, most modern evaporators are either long tube vertical (LTV) or forced circulation (FC) types. Examples of these units are illustrated in Figure 11.6 for LTV units and Figure 11.3 for forced circulation units. In each case a bundle of vertical tubes is placed inside a steam chest much in the manner of a shell and tube heat exchanger. The LTV type is used for low-viscosity systems, while the FC unit is used either for viscous liquids or where scaling or solids precipitation are expected. In the rising film LTV unit shown in Figure 11.6a, the boiling takes place inside the tube and, as vapor is formed, it causes a rapid flow of liquid-vapor mixture up the tube to the vapor disengaging region. The high velocity gives good heat transfer rates and the flow suction pulls fresh feed into the bottom of the tube.

The falling film evaporator illustrated in Figure 11.6b is used for feedstocks which can be exposed to high temperatures only for a short time. In these units, the feed is allowed to run down the inside of the heated tube as a film of liquid

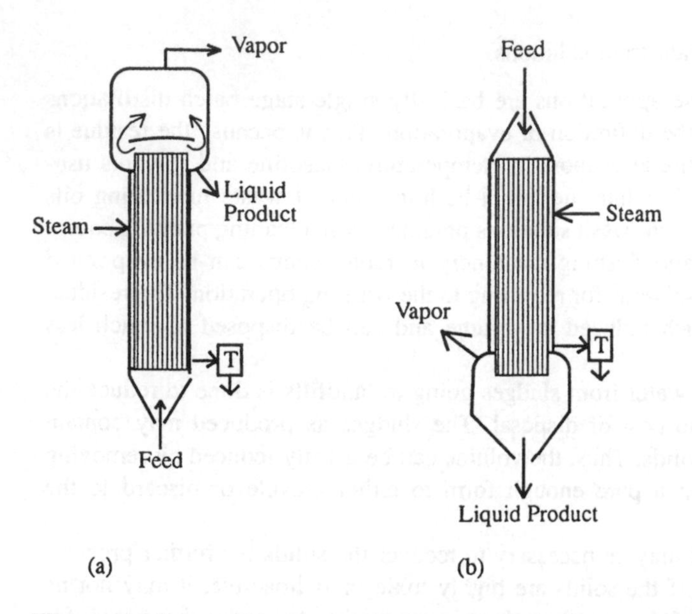

Figure 11.6 Long tube vertical (LTV) evaporators. (a) Rising film, (b) falling film.

under the influence of gravity. The residence time in the hot tube is short and is determined by the length of the heater tube.

Many special designs exist to take care of specific problems unique to particular feeds. For example, the wiped film evaporator illustrated in Figure 11.7 is primarily for very viscous feeds which cannot be properly circulated through tubular heating surfaces. In this case, the feed is spread as a thin film

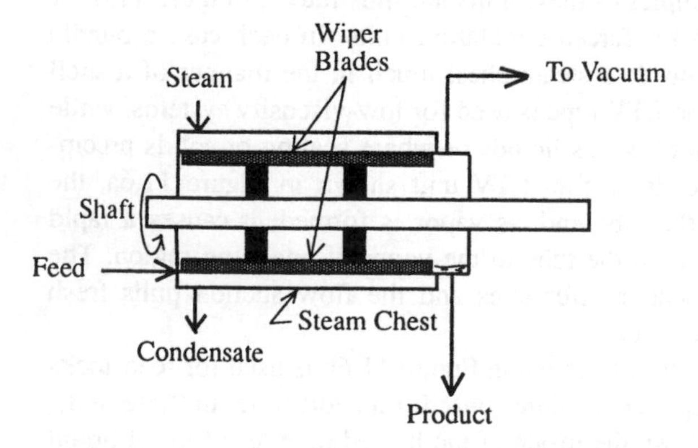

Figure 11.7 Sketch of a wiped film evaporator.

on the heating surface by a mechanical wiper blade. The volatile material then evaporates from the thin film.

There are very many vendors of evaporation equipment of all types. Only a few of them are listed in Table 11.1. The *Thomas Register* lists many more and should be consulted, if necessary.

11.2.4 Common Problems and Troubleshooting

The common problems encountered in evaporation are (1) loss of capacity, (2) loss of condensate quality, and (3) degradation of solids quality. The first of these can be caused by fouling of the heat exchange surface, which causes a reduction in heat transfer coefficient, or by reduction in temperature difference across the heating surface. Fouling can be either salting or scaling of the heat transfer surface. Salting is the buildup of solids on a surface due to excessive supersaturation of the liquid because of either local vapor formation or change in liquid temperature. The solids must be periodically removed to regain normal capacity.

Loss of condensate quality is caused by carryover of liquid into the vapor line either by liquid entrainment or by foaming. Entrainment can be reduced by installing mesh between the liquid surface and the vapor exit. The mesh coalesces the small liquid droplets in the vapor and lets them drip back into the bulk liquid. Thus they are removed from the vapor phase before condensation of the vapor. Carryover of liquid can also occur when a stable foam is formed from vapor bubbles encapsulated in liquid. The foam builds up as evaporation proceeds until it exits into the vapor condenser. Antifoam agents which reduce the surface tension of the liquid can be effective in preventing foam formation as well as in breaking foams already formed. They are needed in only trace concentrations to be effective. When the feedstock regularly gives very severe foaming, special equipment using thermal or mechanical foam breakers are available or can be designed into existing equipment. These consist of either heated surfaces or slowly moving mechanical mixers in the vapor space of the evaporator shell.

Degradation in quality of the solid product can be caused by overheating, excessive residence time, or inadequate control of supersaturation in the liquid. This degradation can take the form of scorching, improper crystal shape or size, and wrong crystal size distribution. These will be discussed in Section 11.3 on Crystallization.

Bumping, which was mentioned earlier, generally occurs only with relatively quiescent boiling under vacuum and is not encountered with high-intensity evaporators. However, in situations where it does occur, it is easily overcome by adding bubble nucleation surfaces to the boiler vessel. These can take the

Table 11.1 Vendors for Evaporation Equipment

APV Crepaco Inc.
395 Fillmore Ave.
Tonowanda, NY 14150
(716) 692-3000

Aqua-Chem Inc.
Water Technology Division
Box 421
Milwaukee, WI 53201
(414) 577-2942

Arkay Environmental Protection
228 S. First Street
Milwaukee, WI 53204
(414) 276-9196

Artisan Industries, Inc.
73 Pond Street
Waltham, MA 02254
(617) 893-6800

Emtec, Inc.
22622 Lambert #309
El Toro, CA 92630
(714) 583-0512

Filcorp Industries
Box 2304
Concord, NH 03302
(603) 225-6638

Giant Industries,
Distillation Division
900 N. Westwood Ave.
Toledo, OH 43607
(419) 531-4600

Gooch Thermal Systems, Inc.
P.O. Box 5064
44 Old Hwy. 22
Clinton, NJ 08809
(908) 735-9350

HPD Inc.
HPD Place Box 3032
Naperville, IL 60566
(708) 357-7330

LICON Inc.
2442 Executive Plaza
P.O. Box 10717
Pensacola, FL 32524
(904) 477-0334

Pope Scientific, Inc.
Box 495
Menomonee Falls, WI 53052
(414) 251-9300

Samsco Inc.
18 Cote Avenue
Gaffstown, NH 03045
(603) 668-7111

Swenson Process Equipment, Inc.
15700 S. Lathrop Ave.
Harvey, IL 60426
(708) 331-5500

Technotreat Corp.
6212 South Lewis
Tulsa, OK 74136
(918) 742-5052

Universal Process Equipment
Box 338
Roosevelt, NJ 08555
(609) 443-4545

Wallace & Tiernan, Inc.
25 Main Street
Belleville, NJ 07109
(201) 759-8000

Wastemizer Corp.
55 Southern Boulevard
Nesconset, NY 11767
(516) 979-8338

form of broken pieces of china, strips of wire screening, or any other surface with many sharp points for bubble nucleation.

11.2.5 Economics

The capital cost of evaporation equipment is largely determined by the amount of heat transfer surface needed and the type of heating used. The heat transfer surface is usually in the form of a tube bundle heated by steam. The operating cost is dependent on the amount of heat needed, its price, and the efficiency of its use. The efficiency of an evaporator is usually given in terms of steam economy, that is, the amount of solvent evaporated per pound of steam used for heating. The higher this number, the higher the steam economy.

Heat is used to raise the temperature of the feed to its boiling point, to overcome any chemical energy binding the solvent in the solution, and to provide the latent heat of vaporization of the solvent. Normally, the heat of vaporization of the solvent is large compared to the other two and is the major heat requirement. Thus, while good heat integration of heating and cooling around the evaporator can help keep energy costs down, by far the major saving comes from reusing the heat of vaporization carried by the solvent vapors. This is achieved by using multiple-effect evaporators or by vapor compression. Both of these practices increase capital costs for equipment. However, the overall savings in steam costs for large installations more than offset the increased capital costs.

11.3 CRYSTALLIZATION

Crystallization is the process of generating a solid crystalline material from either a vapor, a melt, or a solution. In the first two cases, the crystallization is caused by cooling the medium to below the melting point of the crystals, as shown in the phase diagrams in Chapter 2. They then separate from the vapor or melt as solid crystals rather than as liquids and can be removed by settling, filtration, centrifuging, cycloning, and the like. When crystallizing from a saturated solution, the amount of solids that can remain dissolved in the liquid is reduced either by cooling or by removal of liquid by evaporation. Cooling is used where the solubility of the solids is a strong function of temperature. Evaporation of solvent is used when solubility is relatively insensitive to temperature.

The major advantage of crystallization over other separation methods is its ability to precipitate essentially pure solids from a solution in a single stage. However, it does not completely remove all of the solid material from the solution. Thus, further treatment of the residual saturated mother liquor is usually required.

While crystallization is very widely used in the manufacture of solid materials, its occurrence in waste minimization is usually incidental to evaporation processes. The major exceptions are where the solvent is being recovered by freeze crystallization or where a valuable solid is being recovered from a recycle stream. These applications are just starting to be used, but are expected to grow in popularity.

This chapter briefly discusses crystallization from solution as well as freeze crystallization.

11.3.1 Fundamentals

The three requirements for solids crystallization from a solution are (1) the achievement of supersaturation, (2) the nucleation of crystals, and (3) crystal growth. Figure 11.8 shows a crystallization equilibrium graph devised by Miers (1927) to describe formation and growth of crystals from solution. This graph plots the concentrations versus temperature and shows the regions in which crystal nucleation and crystal growth occur. Below and to the right of the saturation curve, no formation or growth of crystals occurs. If seed crystals are added to the solution, they merely dissolve. Between the two curves is the region of moderate supersaturation. That is, the concentration of dissolved solids is slightly above the equilibrium solubility value. In this region crystal growth can occur on already formed crystals, but no new crystals are formed by nucleation. Above the nucleation curve, both nucleation and crystal growth occur at the same time.

The saturation curve is a true equilibrium solubility curve. However, the nucleation curve is really a band which depends on such operating conditions

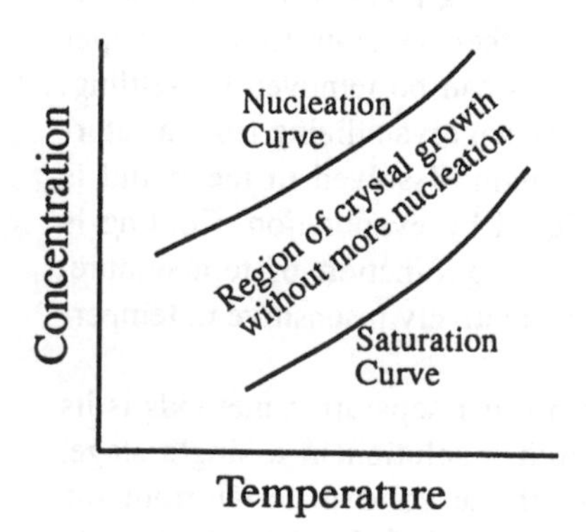

Figure 11.8 Miers plot of supersaturation.

as rate of cooling, mechanical shock, and crystal surface area present. According to Singh (1979), nucleation occurs by the following three mechanisms:

1. Homogeneous nucleation
2. Heterogeneous nucleation
3. Secondary nucleation

Homogeneous nucleation occurs in the bulk fluid phase without any influence of a solid-fluid interface. Heterogeneous nucleation is nucleation in the presence of surfaces other than the crystals themselves. Such surfaces might be the walls of the containment vessel, the mixer, or foreign particles such as dirt or dust. Secondary nucleation is nucleation caused by the presence of the surface of already formed crystals, that is, by seeding the solution with small crystals of the same type being grown.

Homogeneous nucleation occurs when the molecules of the solute in solution gather into clusters of sufficient size to form particles. This process is enhanced as supersaturation is increased. Once the nucleus is formed, other molecules in the solution can then diffuse to these nuclei and attach to their surfaces, thus building new surface layers. As additional layers are added, the crystal is said to grow. Foreign solid particles can also serve as nuclei by providing an ordered surface to start the growth of the crystals.

The shape, or *habit*, of the crystals depends on the identities and amounts of impurities present in the mother liquor during the growth period. Impurities can compete with the crystallizing molecules for adsorption sites on the faces of the growing crystals. If they are more strongly adsorbed on certain faces than the crystallizing molecules, those faces will be blocked from growth and all growth will be on unblocked faces in unblocked directions. In the absence of impurities, the crystal shape will be determined by the energy of attraction between the various crystal faces and the depositing molecules, that is, the relative rates at which growth occurs on the individual faces. If very rapid relative growth occurs on a particular face, the crystals will be needles. On the other hand, if growth is relatively equal on all faces, the resulting crystals will be more like cubes, bricks, or spheres. The exact shape depends on the directions in space of the molecular interactions forming the clusters leading to the nuclei.

The size distribution of the crystals formed is a function of how nucleation and growth are balanced in the crystallizer. If very high nucleation is induced by excessive supersaturation at the start of a batch crystallization, the solute will be depleted in forming so many nuclei and the resulting crystals will all be small. Conversely, if only a few nuclei are formed or a few seed crystals are added, the crystals will be very large. Large single crystals are grown using one seed crystal to start the growth.

In continuous crystallizers the nucleation and growth are usually occurring simultaneously in different regions of the crystallizer. The balance between them can be controlled by controlling the level of supersaturation in the crystallizer. Furthermore, excessive fines can be removed by heating or adding solvent to the slurry.

Supersaturation is achieved and maintained by any of the following methods:

- Removal of solvent from the slurry by evaporation
- Cooling of the slurry by either heat transfer or autorefrigeration due to evaporating solvent
- Addition of an anti-solvent
- Forming the crystallizing material by chemical reaction

As the crystals grow in the slurry, the supersaturation tends to disappear as the solute concentration is reduced due to deposition on the growing crystals. Thus, the method for sustaining supersaturation must be continuous throughout the crystallization. For most crystallizations from solution, either the removal of solvent or temperature reduction causes the precipitation of the solid solute. However, for freeze crystallization the solvent is the precipitating solid and the supersaturation is achieved by cooling or refrigeration. Crystallization by chemical reaction occurs when the reagents giving the reaction are soluble in the solvent but their product is not. The supersaturation can be controlled by the rate of addition of one of the reagents to the other one dissolved in the solvent. The rate of formation of the crystallizing material versus its rate of crystal growth determines the supersaturation.

In all cases the precipitated crystals are collected and washed free of the solution from which they were recovered. Finally, they are dried if the product is a solid, or remelted if the product is normally liquid. Thus, the recovery of water by freeze crystallization would include a crystal melting step.

11.3.2 Specific Applications

The two applications discussed here are examples of evaporative crystallization from solution and freeze crystallization of wastewater. In the first example, the material to be crystallized is usually the minor component in the feed. In contrast, freeze crystallization is usually used to crystallize the major component, that is, the solvent, usually water, from relatively dilute solutions such as seawater.

Figure 11.9 shows a process flow sheet for an evaporative crystallization process to recover ammonium sulfate from a gas scrubber solution. The sulfuric acid in the spent scrubber solution is converted to ammonium sulfate by the addition of ammonia. The solution then goes to an evaporator-crystallizer where water is removed and ammonium sulfate crystalls are formed. The crystals are recovered from the slurry by centrifuging and then dried to give the final product. The mother liquor is filtered and returned to the crystallizer. The water recovered from the evaporation can be recycled to scrubber solution make-up.

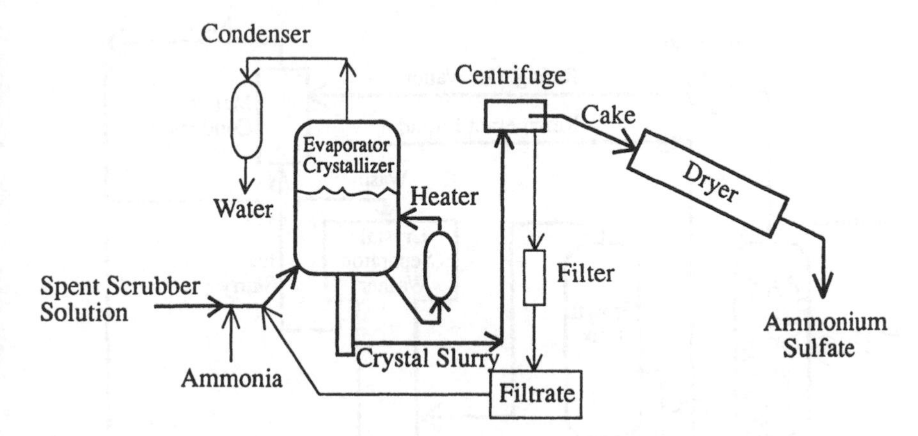

Figure 11.9 Simplified flow plan of ammonium sulfate recovery process.

Freeze crystallization separates water from waste solutions by freezing it out as ice. Its advantages are that ice usually freezes out as pure water and its heat of melting (or freezing) is only about one-seventh of the heat of vaporization. Thus, freeze crystallization can be used to concentrate waste brines and at the same time recover pure water for recycle. Since it rejects all the impurities without fractionation, the final brine contains all the impurities in the same proportions they had in the feed, but at higher total concentration. The process consists of (1) a crystallizer in which the feed is cooled either by direct contact with a refrigerant or through indirect heat exchange, (2) a separator for removing the solid crystals from the residual liquid, (3) a melter for the product crystals, and (4) heat integration of the chiller and melter. Direct contact heat exchange with refrigerants is more efficient than indirect heat exchange. Very effective containment of the refrigerant is essential because such refrigerants are either chlorofluorocarbons or hydrocarbons.

A schematic of a simplified freeze crystallization process is shown in Figure 11.10.

11.3.3 Equipment

The types of equipment used for crystallization are increasingly being classified according to the method used to suspend the growing crystals (Singh, 1979). On this basis, the four types of crystallizers are as follows:

1. Circulating magma
2. Circulating liquor
3. Scraped surface

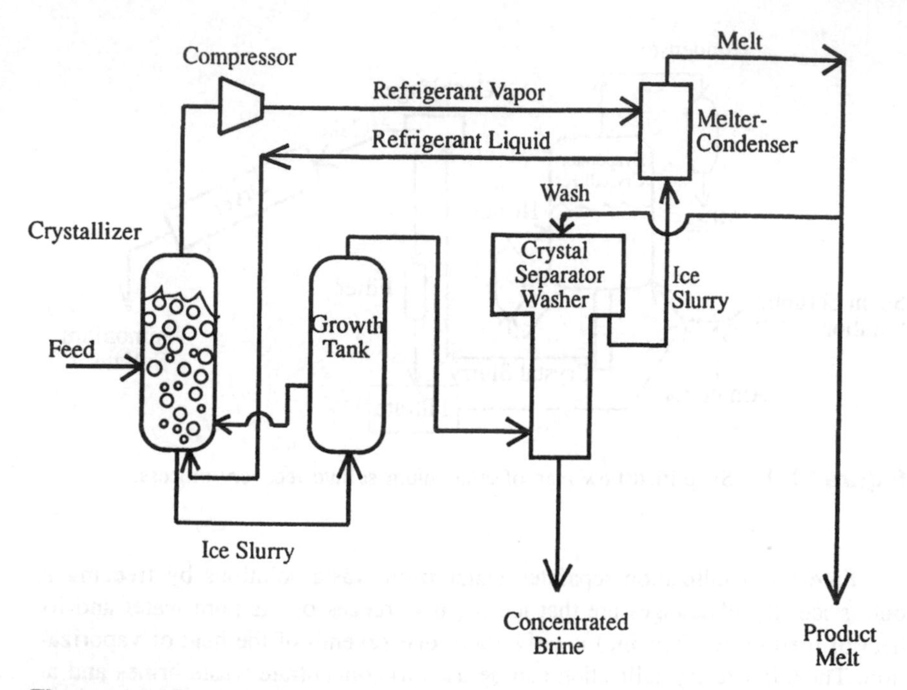

Figure 11.10 Sketch of freeze crystallization.

4. Tank crystallizers

In the circulating magma crystallizer, all of the growing crystals are circulated through the region where supersaturation is generated. With the circulating liquor unit, most of the crystals are maintained in the growth part of the crystallizer, while only the mother liquor is circulated through the part generating the supersaturation. It is then returned to the growth region where it gives up the supersaturation through crystal growth. Either of these units can be equipped for size classification and/or fines destruction.

Scraped surface crystallizers use direct heat exchange through a scraped cooling surface to form crystals which are continuously scraped off the heat transfer surface to minimize fouling and solids deposition. They are often concentric double pipe units with the scraper rotating inside the inner pipe, which contains the crystallizing suspension. The outer pipe contains the cooling medium. Another variation uses a jacketed round-bottomed trough as the heat transfer surface. The scraper consists of a ribbon mixer which rotates inside the trough and both scrapes the crystals from the cooled surface and moves them down the length of the trough through the mother liquor.

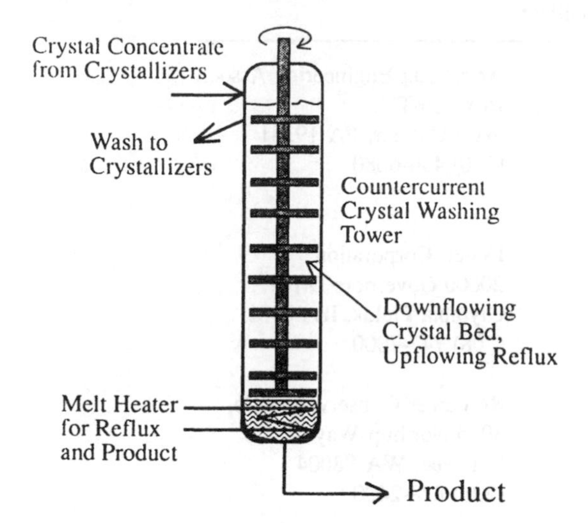

Figure 11.11 Countercurrent crystal washing tower.

The tank crystallizers are usually used for smaller capacities. They get their supersaturation by cooling with submerged coils or jackets and are usually stirred to improve control of heat transfer rate, eliminate localized concentration gradients, and reduce fouling. The level of agitation must be controlled to prevent excessive secondary nucleation and the consequent production of undersized crystals.

A relatively new entrant into the types of crystallizers is the continuous countercurrent crystallizer. In this type of unit crystals are formed in scraped surface chillers and then fed to a countercurrent washing tower, as illustrated in Figure 11.11. In this tower the crystal bed is washed with a reflux derived from melting the crystals from the high purity end of the wash tower. Table 11.2 gives a list of some of the vendors of crystallization equipment.

11.3.4 Common Problems and Troubleshooting

When crystallizing from solution, the problems encountered with evaporator-crystallizers are generally the same as those encountered with evaporators, that is, loss of capacity due to fouling of heat transfer surface, loss of product by entrainment, and thermal degradation of the solids. However, with crystallizers, loss of valuable solids due to poor separation from the mother liquor also becomes a consideration. Furthermore, the quality of the precipitating solids is now a major concern because it is the valuable product. Poor quality can take the form of unacceptability of crystal size distribution, crystal shape, or crystal purity.

Table 11.2 Vendors of Crystallizers

Aqua-Chem Inc. Water Technology Division Box 421 Milwaukee, WI 53201 (414) 577-2942	Armstrong Engineering Assoc. Box 566-T West Chester, PA 19381 (610) 436-6080
C. W. Nofsinger Co. 4600 East 36th Street Kansas City, MO 64141 (816) 361-7999	Dedert Corporation 20000 Governors Drive Olympia Fields, IL 60461 (708) 747-7000
HPD Inc. HPD Place Box 3032 Naperville, IL 60566 (708) 331-5500	Resources Conservation Co. 3006 Northup Way Bellevue, WA 98004 (206) 828-2400
Swenson Process Equipment, Inc. 15700 South Lathrop Ave. Harvey, IL 60426 (708) 331-5500	Unitech Division of Graver Co. 2720 U.S. Highway 22 Union, NJ 07083 (908) 964-2600
Wheelabrator Engineered Systems P.O. Box 64118 St. Paul, MN 55164 (612) 636-3900	

As discussed by Bennett (1984), crystal growth is a relatively slow process. It takes 2 to 6 h compared to nucleation, which can occur in a fraction of a second. It is estimated that it takes from four to six retention periods for an upset to dampen out in a crystallizer. Thus, recovery from a burst of excessive nucleation may take from 8 to 36 h. This makes control of nucleation and seed crystal introduction, i.e., crystallizer stability, the single most important variable when crystal size distribution is important. When large crystals are wanted, all feed and recycle streams must be carefully controlled to prevent unwanted small seed crystals from entering the crystallizer. Furthermore, operation at a high magma density tends to make larger crystals both through longer retention time for growth and reduced nucleation rate through reduced supersaturation. When small crystals are desired, control of nucleation rate at low levels is not as critically required.

Crystal shape, or habit, is largely controlled by the amount and type of impurities in the mother liquor as well as the strength of their adsorption on the various crystal faces. Crystal purity depends on avoiding inclusions in the crystal through eutectic formation (see Chapter 2) or excessive supersaturation.

With freeze crystallization, the major problems are freezing point depression, fouling of heat transfer surface when using indirect cooling, emulsification with direct cooling, and formation of eutectics or additional liquid phases. Freezing point depression is the temperature lowering of the ice freezing point due to the presence of solutes in the solution. It increases with concentration and ionization of solute in the wastewater and can amount to 36°F in strong brines.

11.3.5 Economics

The cost of crystallization equipment varies widely depending on size, materials of construction, heat transfer method, and special auxiliaries. The equipment is usually custom designed. However, Singh (1979) has given an approximate order of capital costs for different types of basic evaporative units. Starting from the most expensive, they are circulating liquor, draft tube baffle, forced circulation, and tank units. The addition of heat-saving techniques such as using multiple effects and vapor compression to evaporative crystallizars can more than double the cost of the basic equipment. However, this extra capital cost can often be offset by the reduced operating heat requirement afforded by these methods. It is important to consult with a vendor to see what tradeoffs are desirable for the particular application being considered.

With freeze crystallizers there is a basic advantage in operating costs due to the low heat of fusion, which is only 15% of the heat of vaporization for water. Furthermore, there is a great cost advantage for units using direct contact heat exchange over indirect contact chillers. According to Heist (1989), capital costs for a direct contact process are less than half those for an indirect heat exchange process. The primary operating costs are for electricity and labor. The electricity is to produce the refrigeration, and the manpower is required for operation and maintenance. Again, electric costs are about twice as high for the indirect freeze process. The cost of manpower is highly dependent on degree of automation and sharing manpower with other operations. Costs per unit of throughput decrease as capacity goes up.

11.4 SUMMARY

Evaporation and freeze crystallization are used to concentrate impurities in wastestreams to reduce the cost of either disposal or further treatment. They can be used to recover valuable solvents for either recycle or use as a raw material for other processes. One example of such an application for either process is the recovery of pure water from wastewaters. Both processes have a history in recovery of drinking water from seawater. However, with the more recent emphasis on zero effluent industrial plants, they are both being increasingly applied to wastewaters.

Crystallization by solvent evaporation or by cooling can be used to recover valuable solids from waste or recycle streams. It can also be used to adjust the composition of recycle streams by removing some of the solute present at excessive concentration.

REFERENCES

Bell, K. J. (1984). Boiling of Liquids, *Perry's Chemical Engineers Handbook*, 6th ed., (R. H. Perry and D. W. Green, eds.), McGraw-Hill, New York, pp. 10-21 to 10-24.

Bennett, R. C. (1979). Evaporation, *Handbook of Separation Techniques for Chemical Engineers* (P. A. Schweitzer, ed.), McGraw-Hill, New York, pp. 2-131 to 2-149.

Bennett, R. C. (1984). Crystallization from Solution, *Perry's Chemical Engineers Handbook*, 6th ed. (R. H. Perry and D. W. Green, eds.), McGraw-Hill, New York, pp. 19-25 to 19-40.

Delaney, B. T., and Turner, R. J. (1989). Evaporation, *Standard Handbook of Hazardous Waste Treatment and Disposal* (H. M. Freeman, ed.), McGraw-Hill, New York, pp. 7.77 to 7.84.

Heist, J. A. (1989). Freeze Crystallization, *Standard Handbook of Hazardous Waste Treatment and Disposal* (H. M. Freeman, ed.), McGraw-Hill, New York, pp. 6.133 to 6.143.

Miers, H. A. (1927). Growth of Crystals in Supersaturated Liquids, *British Journal of the Institute of Metals*, *37*:331.

Moyers, C. G., Jr., and Rousseau, R. W. (1987). Crystallization Operations, *Handbook of Separation Process Technology* (R. W. Rousseau, ed.), Wiley, New York, pp. 578–643.

Sandiford, F. C. (1984). Evaporators, *Perry's Chemical Engineers Handbook*, 6th ed. (R. H. Perry and D. W. Green, eds.), McGraw-Hill, New York, pp. 11-31 to 11-43.

Singh, G. (1979). Crystallization from Solutions, *Handbook of Separation Techniques for Chemical Engineers* (P. A. Schweitzer, ed.), McGraw-Hill, New York, pp. 2-151 to 2-182.

12
Distillation

12.1 INTRODUCTION

Distillation is the most important process that is available today for separating liquid mixtures. It is used to purify liquid products in chemical manufacturing, wine and liquor production, petroleum refining, and in the pharmaceutical industries, among many others. Its particular use in waste minimization is for recovery of used oils and solvents for recycling. It is also widely used for removal of undesirable toxic or hazardous materials from waste streams before their disposal.

Distillation consists of heating a liquid to its boiling point, separating the vapors from the liquid, and then condensing the vapors to form a new liquid phase. The condensate contains a higher concentration of the more volatile component than did the original feed. If the difference in boiling point between the components of the feed is large, it may be possible to achieve the desired purification in a very simple apparatus. On the other hand, if the difference is only a few degrees, a satisfactory separation may require a very complex apparatus and a lot of energy. Fortunately, many of the distillations encountered in waste management involve large differences in boiling point between the components in the feed. Thus, the components in the feed can usually be separated in a single-stage batch still.

The range of liquids that can be distilled runs from liquefied gases at low temperatures and/or high pressure up to high boiling liquids under vacuum just below their thermal decomposition temperature. An example of the former is purification of ethylene under pressure at cryogenic temperatures, while an example of the latter is vacuum distillation of lubricating oil fractions. The boiling point of the liquids can be adjusted by changing the pressure at which the distillation is run. By raising the pressure, the boiling point of condensed gases can be raised to permit convenient condenser temperatures. The boiling points of high boiling liquids can be reduced to prevent thermal decomposition by running under vacuum.

Distillation processes were discussed broadly as examples in Chapter 2 on the basic principles of separation science. Chapter 2 includes discussions of enriching, stripping, and compound columns as well as extractive and azeotropic distillation. The present chapter will cover only the types of distillations encountered in the waste minimization area. These tend to be of the simpler types found in on-site solvent recovery and waste treating operations. They include batch distillation and the simpler forms of continuous distillation, such as simple equilibrium flash and continuous stripping.

12.2 BATCH DISTILLATION

A batch distillation consists of placing a quantity of feedstock in a boiling vessel, heating it to generate vapors, and then condensing the vapors in a separate vessel or condenser by cooling them to below their condensing temperature. The condensed vapors are the distillate, or overhead, and the residue left in the boiler when the distillation is finished is called the still bottoms or residue. In its simplest form, a batch still is the combination of a boiler and condenser. However, it can also have large numbers of vapor-liquid contacting stages inserted between the boiler and the condenser in the form of a distillation column or tower. This greatly increases the composition difference between the distillate and the contents of the boiler and makes the separation much sharper. However, when this type of still is started, it must be run at total reflux for several hours before starting to collect product. This allows time for the concentration gradient to be established in the column. Many small-scale laboratory distillations use such an "enriching" column, while most solvent recovery columns for waste management are of the simple boiler and condenser type. Product can be taken immediately after startup for the simple batch still because only one theoretical stage is involved and no time is required to establish a concentration gradient.

12.2.1 Fundamentals of Batch Distillation

Every pure material has a vapor pressure which is a constant at constant temperature and increases as temperature is increased. This value is equal to atmospheric pressure at the normal boiling point of the material. In general, a plot of the logarithm of the vapor pressure against the reciprocal of the absolute temperature is nearly a straight line, with its slope defined by the heat of vaporization. This relationship can be used to approximate vapor pressures over a wide temperature range by plotting the boiling points at atmospheric pressure and one other pressure. Alternatively, the heat of vaporization can be used with the normal boiling point to estimate vapor pressure at other temperatures. This is illustrated for toluene in Figure 12.1.

Materials having a high vapor pressure at room temperature are volatile, while materials having a low vapor pressure at room temperature are non-volatile. The larger the difference in the vapor pressures of two materials at the boiling point of the mixture, the greater the relative volatility and the degree of separation achieved in a simple distillation, that is, the distillate is richer in the more volatile component as the difference in vapor pressures increases. If one component of the mixture has a very high boiling point, and is, thus, essentially non-volatile, while the other boils a little above room temperature, the distillate is essentially the pure component of low boiling point. An example of this is the recovery of solvents boiling below 200°F from mixtures with

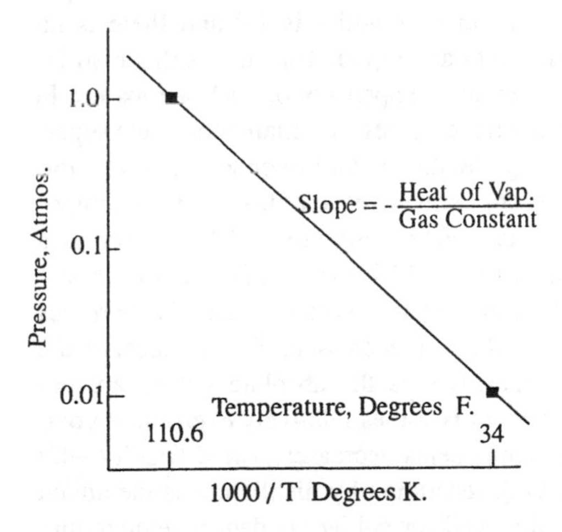

Figure 12.1 Variation of vapor pressure of toluene with reciprocal of absolute temperature.

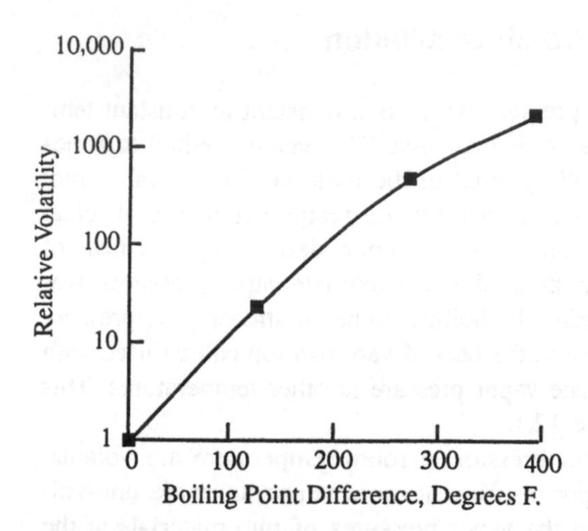

Figure 12.2 Variation of separation factor with boiling point difference for paraffinic hydrocarbons.

lubricating oils boiling above 700°F. At atmospheric pressure, the oil is essentially non-volatile and the distillate is pure solvent.

Chapter 2 discusses the way that vapor-liquid equilibrium, as defined by relative volatility, can be calculated from vapor pressures for ideal systems using Raoult's law. When both components of a binary mixture have the same vapor pressure at the same temperature, the relative volatility is 1.0 and there is no difference in composition between the vapor and liquid. This means that distillation will not give any separation between the components of such a mixture. In contrast, the oil-solvent mixture just mentioned gives essentially complete separation in a single stage and has a very high relative volatility or separation factor. This suggests that the ease of separation, or relative volatility, of two components is related to the difference between their boiling points. This is true as a first approximation and is illustrated in Figure 12.2 for paraffin hydrocarbons.

Another generality in distillation is that relative volatility usually increases as distillation pressure, and thus temperature, is decreased. This is because the ratio of the vapor pressures usually increases as the absolute values are decreased. Thus, the more volatile component becomes relatively even more volatile as the absolute volatility of both components decreases. The difficulty with using this as a general way of improving separation by distillation is the added cost of operating under vacuum and the need for colder condenser temperature and possibly refrigeration. Furthermore, there are, although rare, some exceptions to this generality.

Simple batch, or Rayleigh, distillation is the most common type used in solvent recovery operations for waste management. The feed is placed in a boiler and the distillate is condensed and collected in a receiver. As the distillation proceeds, the boiling temperature in the still gradually rises as the more volatile components in the feed are removed. At some predetermined temperature, corresponding to nearly complete removal of the volatile materials from the feed, the heat is shut off and the boiler is allowed to cool. The residue is then drained from the boiler and the distillate is drained from its receiver. Either the distillate, the residue, or both can be the desired products. If desired, the distillate can be taken as fractions when there is sufficient change in composition from one fraction to another.

When the separation factor between components to be separated is not sufficiently large to give a good separation in a single stage, an enriching column can be inserted between the boiler and the condenser. This is illustrated in Figure 12.3. However, this adds the requirement that a downflowing liquid must be present in the column to contact the upflowing vapors from the boiler. This liquid is called reflux and is obtained by feeding part of the condensate from

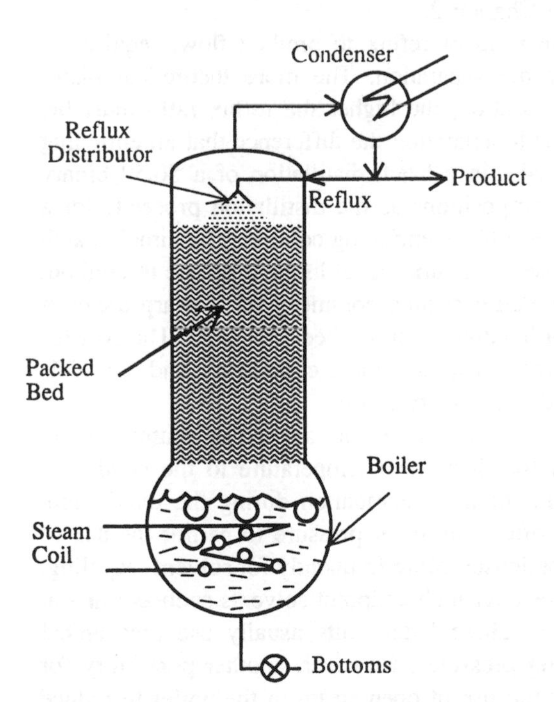

Figure 12.3 Sketch of a batch distillation apparatus using an enriching column.

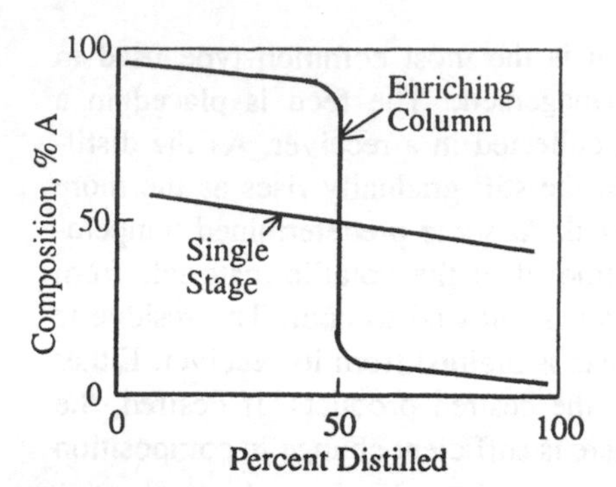

Figure 12.4 The effect of multi-stage enriching on product purity.

the overhead condenser back to the top of the enriching column. This requires a device to split the condensate into reflux and product streams as well as a liquid distributor at the top of the column to make sure the vapor and liquid are properly contacted. The column may be a packed column or any of a variety of plate columns as discussed in Chapter 2.

The reflux ratio, that is, the ratio of reflux to product flow, required is determined by the difficulty of the separation. The more theoretical plates required to make the desired separation, the higher the reflux ratio must be. Figure 12.4 illustrates for a difficult separation the difference that an enriching column properly operated can make in a batch distillation of a 50:50 binary feed. It compares the distillate compositions as the distillation proceeds for a single-stage batch unit with that of a batch enriching column. The simple batch still gives a steady gradual increase in impurity level in the distillate throughout the distillation. In contrast, the batch enriching column gives a sharp break in composition when the more volatile component has been removed. The cost for this improved separation is the increased equipment complexity and the additional heat required to generate the necessary reflux.

Batch stills can be run under pressure or vacuum as well as at atmospheric pressure. This is done to adjust the distillation temperature to the condenser temperature available or to avoid thermal degradation during the distillation. Low boiling point materials are often run under pressure to permit the use of available water for cooling the condenser. More frequently for solvent recycling, the still is run under vacuum to recover high endpoint solvents such as mineral spirits from waste oils or paints. These latter units usually use mechanical vacuum to get their low operating pressures. However, another possibility for simulating vacuum distillation is the use of open steam in the boiler to reduce the partial pressure of the other components in the boiler vapor. This is called steam distillation.

Steam has often been called the poor man's vacuum in the distillation of high boiling point materials. The more steam that is blown through the boiler, the lower the temperature at which a high boiling point material will distill. This is because the steam makes up most of the pressure in the boiler, and only a small partial pressure of high boiling point material is needed for it to come over with the steam. The steam and its content of the high boiling point material condense together and if the liquids are immiscible, they form two layers and can be decanted to recover separately. The price paid for this method of simulating vacuum distillation is the cost of the large amounts of steam that must be generated.

12.2.2 Specific Applications of Batch Distillation

The major applications for batch distillation in waste minimization are for on-site recovery of used oils and solvents. Just a partial list of these applications includes:

- Recovery of paint thinners and equipment-cleaning solvents from painting operations
- Recovery of parts cleaning solvents from oils and dirt in machine shops and automotive shops
- Recovery of wash solvents from their impurities
- Removal of solvents, water, and other low boiling point contaminants from waste oils

Paint thinners can be recovered from waste paint by segregating the waste paint by the type of solvent used in manufacturing the paint or in thinning it for use. For example, paints containing hydrocarbons, such as varsol, mineral spirits, or toluene and xylenes, can be distilled together to recover these solvents. the recovered mixed solvent can then be used to clean the painting equipment used to apply these paints. Segregation by type is necessary because not all paints use the same types of solvents for compounding or thinning. Some paints use hydrocarbon solvents, while others use organic solvents like ketones, esters, ethers, and alcohols. These are not always compatible with the hydrocarbon solvents and often form either azeotropes in distillation or two phases in the product receiver on condensation. While the distillate product can be recycled as a cleaning solvent, the boiler residue must normally be disposed as a hazardous waste.

Parts-cleaning solvents are used to remove oils and coolant residues from parts manufactured in machine shops. They are also used to wash oils and greases from parts handled in automotive repair shops. In the past, these solvents have been primarily high-flash mineral spirits, chlorinated hydrocarbons, or fluorocarbons. However, due to the regulation of ozone-depleting chemicals, the trend now is toward either high-flash mineral spirits, other hydrocarbons, oxygenated solvents, or aqueous detergents. Distillation is used to recover the non-aqueous solvents which can then be recycled to the cleaning operation.

Wash solvents can vary from the high purity types used for cleaning printed circuit boards to those of less stringent requirements for cleaning pipes and equipment prior to repair or servicing. The preferred solvents for these uses were usually fluorocarbons or chlorinated hydrocarbons because of their high purity or their safety from fire hazards. However, as with the parts-cleaning solvents, the trend is toward aqueous solvents or high-flash hydrocarbon solvents. Again, the hydrocarbon solvents can be recovered by distillation for recycling.

Finally, low boiling point contaminants such as solvents and water can be removed from waste hydraulic or process oils by batch distillation. This can improve the flash point of the oil sufficiently to be used either directly as a fuel or as a blending component in the blending of fuels from waste oils. In the case of especially clean waste oils, they may even be recycled to the original use.

12.2.3 Equipment for Batch Distillation

There are many suppliers of batch distillation equipment for recovery of used solvents and oils. These stills usually operate at atmospheric pressure but may also be equipped for operation under moderate vacuum. A photo of a typical batch still is shown in Figure 12.5. To obtain a still for operation under pressure, it is usually necessary to have it custom built. Table 12.1 lists some of the suppliers of batch distillation equipment. The suppliers generally manufacture their stills in a variety of standard sizes, varying from a few gallons per 8-h shift to over 25 barrels per shift. Special features such as wiped heating surfaces, explosion-proof electrical circuitry, automatic filling and shutdown, and open or closed steam heating can also be obtained.

12.2.4 Common Problems and Troubleshooting

Simple Batch Stills

The most common problems with simple one-stage batch stills are entrainment, foaming, and bumping. Entrainment is the carryover of liquid droplets from the boiler along with the vapor going to the condenser. It can be caused by either the design of the still or the nature of the material being distilled. In its simplest form, entrainment is caused by driving the still too hard, that is, boiling the liquid so fast that the vapor velocity is too high for the liquid droplets generated by the intensity of the boiling to settle out of the vapor before it goes to the condenser. Clearly, diluting the distillate with boiler liquid contaminates the distillate and usually makes it unsuitable for recycling. This type of entrainment

Figure 12.5 Photo of a typical batch still. (Courtesy of Finish-Thompson Engineering, Inc.).

can usually be eliminated by reducing the heat input to the boiler. This reduces the intensity of boiling and thus the number of droplets formed. It also reduces vapor velocity and allows more time for letting the liquid droplets settle out of the vapor before it is condensed. Wire mesh mist eliminators can also be installed in the vapor space of the still to collect and remove liquid droplets from the vapor.

Foaming is generally caused by the nature of the feedstock. For example, when the feed contains two phases, such as liquid water in a hydrocarbon solvent, the liquid in the boiler may foam as it boils. This foam is a bed of vapor bubbles in a liquid network around the bubbles. As the bubbles break, the vapor is released and passes out of the bed of foam. If the foam is quite stable and breaks very slowly, the bed of foam may actually pass out of the boiler to the condenser. This again contaminates the distillate with boiler liquid, thus ruining the effectiveness of the distillation. This foaming can continue until the water phase is finally removed from the boiler by the distillation. Then the foaming stops and the distillation proceeds normally. Foaming can also be caused by solid particles in the feedstock and is frequently encountered when recovering solvents from paint.

Table 12.1 Vendors of Distillation Equipment

Batch Stills Only

Bowers Process Equipment, Inc.
P.O. Box 322
Marysville, MI 48040
(313) 982-0584

Disti Environmental Systems
525 Boulevard
Kenilworth, NJ 07033
(201) 272-7600

Finish Thompson, Inc.
921 Greengarden Road
Erie, PA 16501-1591
(814) 455-4478

Giant Industries, Inc.
900 N. Westwood Ave.
Toledo, OH 43607
(419) 531-4600

PBR Industries
400 Farmingdale Road
West Babylon, NY 11704
(516) 422-6004

Progressive Recovery, Inc.
1976 Congressional Drive
St. Louis, MO 63146
(314) 567-7963

Recyclene Products, Inc.
8038 Fernham Lane
Forestville, MD 20747
(301) 420-8500

Renzmann, Inc.
310 Oser Ave.
Hauppage, NY 11788
(516) 231-3030

Solvent Kleene, Inc.
131-1/2 Lynfield St.
Peabody, MA 01960
(508) 531-2279

Continuous Stills Only

Artisan Industries, Inc.
73 Pond Street
Waltham, MA 02254-9193
(617) 893-6800

Pall Industrial Hydraulics Corp.
2200 Northern Blvd.
East Hills, NY 11548
(516) 484-5400

Pope Scientific, Inc.
N 90 W 14337 Commerce Drive
Menomonee Falls, WI 53051-9990
(414) 251-9300

UIC, Inc.
P.O. Box 863
Joliet, IL 60434-0863
(800) DIAL-UIC

Batch or Continuous Stills

Alternative Resource Technologies, Inc.
1221 East Houston
Broken Arrow (Tulsa) OK 74012
(918) 251-0880

Westport Environmental Systems
251 Forge Road
Westport, MA 02790-0217
(800) 343-9411

The simplest way to avoid contamination of the distillate due to foaming caused by water is to reduce the distillation rate until the water has distilled over and foaming stops. This allows enough time for the foam to break in the vapor space of the still so that the foam does not get into the condenser. However, there are also a variety of ways to either prevent foaming or to break foams more rapidly. The simplest is to remove any solid or liquid second phase by filtering or decanting the feedstock before putting it in the still. It is also possible to break many foams by putting small amounts of additives that reduce surface tension into the feed to the still. Typical of such defoaming additives are low molecular weight silicones and alcohols. These additives lower the strength of the foam network and let the foam break as fast as it is formed.

When the foam cannot be controlled by additives, it is often possible to break foams using heat or mechanical agitation. Putting a heater in the vapor space of the still to heat the foam above the boiling point at which it was generated weakens the foam and lets it break faster. Furthermore, a mild drag on the surface of the bubbles in the foam distorts them and causes them to break faster. This drag can be supplied by installing a wire mesh layer in the vapor space of the still or installing mild mechanical agitation in the vapor space. The mechanical agitation can be generated by slowly rotating paddles in the vapor space of the still. These methods work best when installed some distance above the boiling liquid surface. This gives the foam some time to drain and weaken before the mechanical or thermal energy is applied.

Another problem that is often encountered in distillation, especially when the still is pushed to high distillation rates, is "bumping." This occurs when the liquid in the still does not form vapor bubbles easily and thus the boiling is not smooth. Under these conditions the heat input to the liquid is not dissipated smoothly by boiling but collects to superheat the liquid. Then when the superheat is high enough to force vapor formation, the extra heat stored in the liquid is released almost explosively, making a large amount of vapor at once. This lifts the liquid layer in the boiler, throwing it into the vapor space with a loud thump and strongly jolts the still. It may even entrain the boiler liquid into the condenser. To avoid bumping promote smooth, steady vapor formation as heat is supplied to the still. This can be done by putting rough surfaces or boiling chips into the liquid in the boiler. These rough or pointed surfaces supply sites for easy vapor bubble formation and prevent superheating of the boiler liquid. Typical promoters of smooth boiling are pieces of broken ceramics, cut up wire screens, and the like.

Batch Enriching Columns

With batch enriching columns, the most common problems are foaming, flooding, bumping, and poor liquid distribution. Foaming and bumping, if present, occur in the reboiler. However, if not corrected, foaming can fill the fractionating column with foam and nullify the separation. Foaming and bumping in enriching columns are corrected in the same way as simple batch stills. Flooding occurs when the still is driven so hard that the vapor velocity in the column is too high for the downflowing liquid to pass it in the column. Thus, the liquid collects in a layer held up by the vapor and eventually is pushed out the top of the column. The efficiency of the separation is reduced to essentially one stage, or theoretical plate, and the quality of the overhead product is ruined. Flooding is relieved by lowering the heat input to the still. This lowers both the vapor velocity and the liquid flow rate, thus permitting the two phases to pass each other in the enriching column. When the flooding has been relieved, the column must be put back on total reflux so that the concentration gradient across the column can be re-established before product flow is started. Flooding occurs most commonly in packed columns, but can also occur in plate columns if vapor velocities are high enough.

12.2.5 Economics for Batch Distillation

The economic incentives for batch distillation in waste minimization are dictated largely by the cost of disposing the untreated waste. A secondary factor is the value of the recovered purified material, if it can be recycled. In many cases it costs more to dispose of waste materials than the original purchase price of those same materials. For example, it may cost as much as $20/gal to dispose of waste solvents that cost only $1/gal to buy in the first place.

The economic value of recycling by distillation can be calculated as either a time required to pay for the purchase of the distillation equipment or as an annual saving. In the first case, the operating costs—such as extra manpower and utilities needed for the distillation, solvent make-up to account for losses to residue, residue disposal, and so on—are added together to get a total operating cost for the distillation. Similarly, the operating costs of present practice, such as cost of waste disposal and cost of fresh solvent, are also totaled. The total operating costs for distillation are then subtracted from those of the present practice to get an annual operating cost differential. This net annual value is then divided into the capital cost of the installed equipment to get the payout time for the capital equipment. This may be from a few months to a few years.

To get the net annual savings for converting from present practice to recycling by distillation, the capital cost of the distillation equipment must be charged against the distillation case. This can be done by using the expected life of the equipment and the company's cost of money to get an annualized expense required to retire the capital cost. This is usually on the order of 10 to 20% of the capital cost. The total operating cost of the present practice minus the total cost for the distillation case is the net annual savings. The net annual savings usually increases with the volume of solvent recycled and can be very large. However, if there are no savings and present practice is legal, the distillation option will clearly not be installed.

The cost of energy is usually not a critical factor in simple batch distillations. However, when fractionating columns must be operated at high reflux ratios, energy costs can become important. For example, distillation in a single-stage batch still only requires that the distillate be vaporized one time. Thus, the heat required is only the heat of vaporization of the distillate. On the other hand, an enriching column may operate at a reflux ratio (R) of twenty to one. This would require a heat input of (R + 1) times the heat of vaporization, or 21 times as much heat as for the simple batch distillation. Fortunately, most waste minimization distillations are not that difficult.

In addition to the increased cost of energy for complex distillations, the cost of the equipment needed to perform the separation also increases with the difficulty of the distillation. For example, multistage towers cost extra money; the vacuum and pressure requirements demand extra equipment; and the use of vapor compression to minimize energy costs requires a large compressor. These factors all need to be evaluated on a case by case basis to establish the overall cost of a distillation plant for on-site recycling.

12.3 CONTINUOUS DISTILLATION

Continuous distillation is so named because the feed stream is continuously fed to the distillation apparatus. Either gravity flow or a suitable pump can be used along with flow rate control to provide the continuous feed. The continuous distillation can vary in complexity from a simple one-stage equilibrium flash separation to a compound column having many enriching and stripping stages in a single apparatus. The equilibrium flash is generally used where the separation is easy, that is, where there is a large difference in volatility between the components to be separated. It is also very useful when a small amount of volatile material must be removed from a large amount of high boiling point material as, for example, in removing solvents from used oil.

Multistage distillation is usually used to separate and recover materials which have relatively close boiling points, that is, less than 20°F apart. Typical continuous multistage separations can use enriching columns, stripping columns, or compound columns. Enriching columns are used to recover high purity lower boiling point products (such as pure solvents) from higher boiling point mixtures (such as fuels, lubricants, and hydraulic oils). In contrast, stripping columns are used to remove lighter materials from heavier products that are desired in high purity. A typical example of a stripping separation would be removal of small amounts of solvents or gasoline from recyclable hydraulic or lubricating oils.

When only two materials are in the feed and both can be recycled if they can be separated in adequate purity, a compound distillation can be used. This is in essence combining an enriching column and stripping column in the same distillation unit. By doing so, both the overhead and bottoms products can be obtained in good purity as well as good yield. However, only two high purity products can be obtained from one compound column.

12.3.1 Fundamentals of Continuous Distillation

One of the most important variables in continuous distillation is the material balance around the distillation unit, that is, the cut point or percent of feed material taken overhead. This is illustrated in Figure 12.6 for an equilibrium

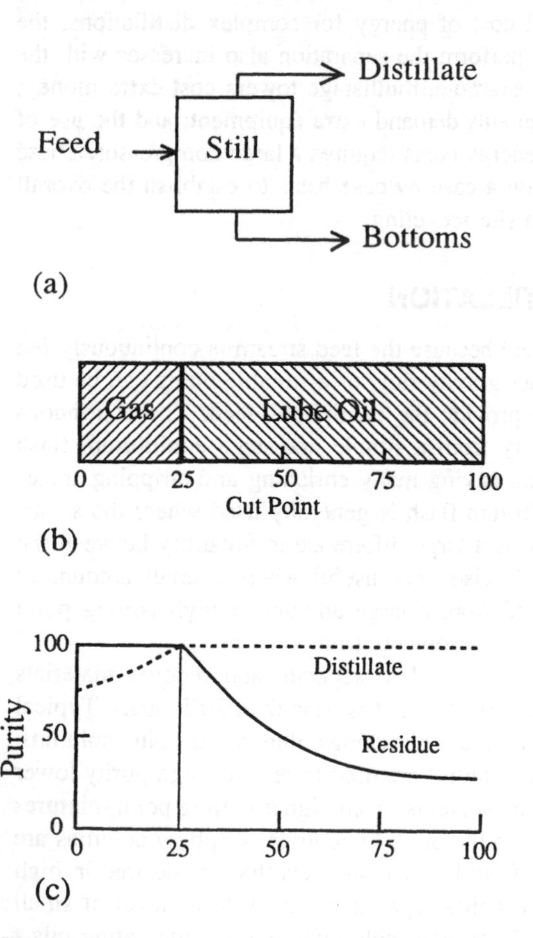

Figure 12.6 Control of cut point in equilibrium flash of 25% gasoline/75% lube oil mixture. (a) Equilibrium flash distillation, (b) feed composition, (c) product composition versus cut point.

flash distillation of a 25% gasoline 75% lube oil fraction. Each continuous still has one input stream, the feed, and two output streams—the overhead, or distillate, and the bottoms. This is illustrated in Figure 12.6a. Since it is usually difficult, if not impossible, to exactly balance the flow rates of all three of these streams, one of them is allowed to "float" by difference. This is done by using either temperature or level control to adjust the difference. In equilibrium flash distillation, the amount taken overhead, or cut point, is controlled by the flash temperature. This is the temperature to which the feed is preheated before the vapor and liquid in equilibrium are separated in the flash drum. The higher the temperature, the higher the percentage of the feed taken as distillate.

Figure 12.6b is a sketch of the feed composition as related to the cut point, showing the 25% gasoline and 75% lubricating oil in the feed. Figure 12.6c shows how the composition of the distillate and bottoms from a perfect distillation is controlled by the cut point. If less than 25% is taken overhead, the distillate will be pure, but the bottoms will contain some material that would have gone overhead if the proper cut point had been used. Conversely, if more than 25% is taken overhead, the bottoms will be pure, but the distillate would be contaminated with material that should have gone to the bottoms. Only at the 25% cut point are both products at their optimum purity.

In a typical separation of gasoline from lube oil, the final boiling point (400°F) of the gasoline may be 300°F lower than the initial boiling point (700°F) of the lubricating oil. In such a case, the separation is easy and the temperature of the flash controls the cut point. With a flash temperature below 400°F, not all the gasoline will be taken overhead as distillate. Thus, the bottoms will still contain some gasoline. On the other hand, if the flash temperature is maintained above 700°F, the cut point is moved to the right and some of the lube oil is taken overhead with the gasoline distillate. Either of these conditions may be desirable, depending on which product is to be recovered in the more highly purified state. However, to get an optimum level of separation, a combination of flash temperature and vacuum is desirable equivalent to a corrected atmospheric boiling point of 550°. This would minimize both gasoline in the bottoms and oil in the distillate.

Temperature can also be used to control cut point and product purity in a more complex distillation, such as that illustrated in the compound distillation illustrated in Figure 12.7a. In this case part of the distillate is returned to the top of the column as reflux. A reboiler is also provided at the bottom of the column to generate the extra vapor needed to operate at a reflux ratio above zero. Such columns are often controlled by instrumental analysis of the composition of the distillate stream. Reflux ratio is then adjusted through product flow rate to maintain product purity. However, product purity can also be controlled less expensively using temperature in the enriching section of the tower to control distillate rate and, thus, reflux ratio. The temperature sensor is

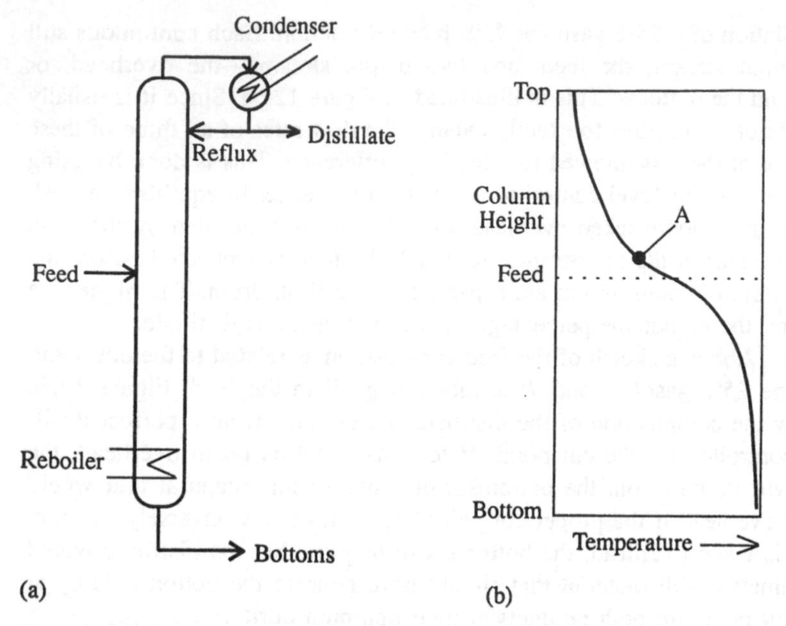

Figure 12.7 (a) Compound distillation column showing (b) variation of temperature with height.

placed at an intermediate point in the enriching column where the temperature varies rapidly with height in the column. If the temperature rises, the distillate rate is decreased, thus raising the reflux ratio but primarily adjusting the cut point by forcing lower boiling material down the column. If the temperature falls, more distillate is taken off, re-establishing the temperature at the control point. Figure 12.7b illustrates the temperature profile through the column. The control point position is shown as A, where the temperature changes rapidly with column height. This makes it sensitive and effective for controlling cut point. The column can then use all of its separating power effectively.

The reflux ratio is controlled by the amount of heat supplied to the reboiler. With the cut point controlled by temperature in the enriching column, increasing reflux ratio can be used to improve the quality of the distillate and bottoms products. The effects of relative volatility, reflux ratio, number of theoretical plates, and various special forms of distillation are all discussed in Chapter 2.

12.3.2 Specific Applications of Continuous Distillation

Most of the small-scale on-site recycling operations are run in simple batch stills. These units range in capacity from a few gallons per shift to a few drums

Figure 12.8 Photo of a major used oil re-refinery. (Courtesy of Safety-Kleen Corporation).

per shift. However, larger solvent users such as military bases, large machine shops, commercial recyclers, and used oil re-refiners are of sufficient size to use continuous distillation. Typically, as feed rates reach or exceed 50 gal/h, there is a trend to switch to continuous operation. A large used oil re-refiner may have a feed rate of 20,000 gallons per day. Figure 12.8 shows a large solvent recycling plant using many distillation towers.

12.3.3 Equipment for Continuous Distillation

Continuous distillation equipment is almost always custom designed. This is because of the larger sizes and the additional complexity of separation-specific pumps and controllers. In addition, flow rates, liquid levels, heating rates, reflux ratios, integrated heat transfer, and temperatures need to be controlled rather than merely measured as with batch stills. Cut points must also be held at the desired levels to maximize the separation.

The simplest continuous still, the flash drum, is illustrated in Figure 12.9. It consists of a feed pump, a controlled heater to maintain the feed at a constant exit temperature, a flash drum to separate the vapor and liquid at equilibrium with each other, a condenser for the distillate product, and a cooler for the bottoms product. At the other end of the scale of complexity is the compound

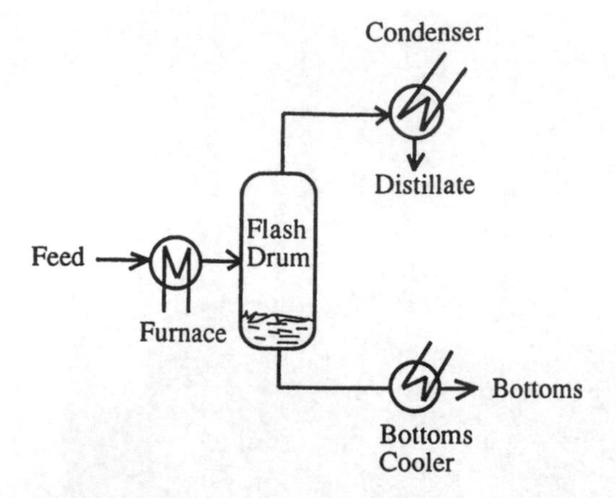

Figure 12.9 Sketch of a continuous equilibrium flash drum.

column with combined enriching and stripping sections. This requires controls for the reflux ratio, boilup rate, cut point, reboiler liquid level, and so on. These units are almost always custom designed around the exact separation contemplated. The required number of plates, the reflux ratio, and the boilup rate all vary with both the desired cut point and the difficulty of the separation desired.

In between these two extremes lie continuous enriching or stripping columns which are sometimes available off the shelf in smaller sizes, but must be custom designed in larger sizes.

12.3.4 Common Problems and Troubleshooting

The problems encountered with continuous distillation units are similar to those common to batch stills, that is, foaming, entrainment, flooding, and control of the cut point. However, bumping is usually not a problem because the steady-state continuous flow tends to prevent accumulation of superheat which leads to explosive vaporization. The remedies for these problems are also the same as for batch stills—lower heating rates to relieve flooding, foaming, and entrainment and additives to decrease or prevent foaming.

12.3.5 Economics of Continuous Distillation

The economics of continuous distillation are much like those of batch distillation. The exception is that control devices become a larger part of the equipment cost and equipment capital cost becomes a larger portion of the total operating cost. Furthermore, since the plant is essentially custom designed rather than an off-the-shelf unit, there will be a design and engineering cost associated with it. This can be on the order of 10% or more of the capital cost. The incentives for installing a continuous distillation plant versus contract disposal or off-site recycling are calculated in the same way as discussed for batch stills earlier in this chapter. However, the amount of feedstock to be processed usually needs to be larger for a continuous still to break even.

12.4 SUMMARY

Distillation is the oldest and still most widely used separation process for liquid mixtures. Capacities for a single still range from a few ounces per day up to ten million gallons per day. The sizes encountered in waste recycling operations vary from a few gallons per day up to about 20,000 gallons per day. Usually batch stills are used at the lower end of the size scale, while continuous stills are used at the upper end. Distillations can be run at atmospheric pressure, at elevated pressure to adjust the boiling point to a convenient level for ease of condensation, or under vacuum to avoid thermal decomposition or chemical reaction at higher boiling points. The separations are easy if the feed components have widely different boiling points, say, 100°F apart. However, they may require many theoretical plates and a high reflux ratio if the boiling points are only 15 or 20°F apart.

The product purities achieved by distillation are sufficient that the products can usually be recycled, either to the original use or to a less demanding replacement use. For example, solvents recovered from waste paints can usually be used for cleaning painting equipment, although they are not recommended for blending or thinning new paint. Wash solvents can usually be returned to their original use.

SUGGESTED READING

Bonilla, J. A., (1993). Don't neglect liquid distributors, *Chemical Engineering Progress*, March:47.

Chen, G. K. and Chuang, K. T. (1989). Recent developments in distillation, *Hyldrocarbon Processing*, February: 37–45.

Harrison, M. E. and France, J. J. (1989). Trouble-shooting distillation columns, *Chemical Engineering*: Part 1, Technique and tools, March:116–123; Part 2, Packed columns, April:121–128; Part 3, Trayed columns, May:126–133; Part 4, Auxiliary equipment, June: 130–137.

Hasbrouck, J. F., Kunesh, J. G., and Smith, V. C., (1993). Successfully troubleshoot distillation towers, *Chemical Engineering Progress* March:63.

Shah, G. C., (1991). Effectively troubleshoot structured-packing distillation systems, *Chemical Engineering Progress* April:49.

13

Absorption, Stripping, and Scrubbing

13.1 INTRODUCTION

Absorption, stripping, and scrubbing are all processes involving the contacting of a liquid and a gas followed by separation of the two phases. Absorption is a unit operation in which a liquid solvent is contacted with a gas and the soluble components in the gas dissolve in the liquid. The liquid is called the absorption solvent. The reverse of this, i.e., the transfer of volatile components from a liquid to a gas, is called stripping. The gas is called the stripping gas. Stripping is accomplished by contacting the liquid with an inert gas such as nitrogen or steam. With aqueous waste streams the stripping gas is usually air. Absorption and stripping are frequently used together. Absorption is used to remove the soluble component(s) from the gas stream, and stripping is used to regenerate the solvent by removing the dissolved component(s) from the spent solvent. Figure 13.1 shows such a plant. Conversely, absorption in a solvent can be used to recover the gaseous product removed in a stripping operation when it cannot be condensed. It is an alternative to adsorption for the recovery of gaseous components. However, it should be remembered that *ad*sorption uses solid surfaces

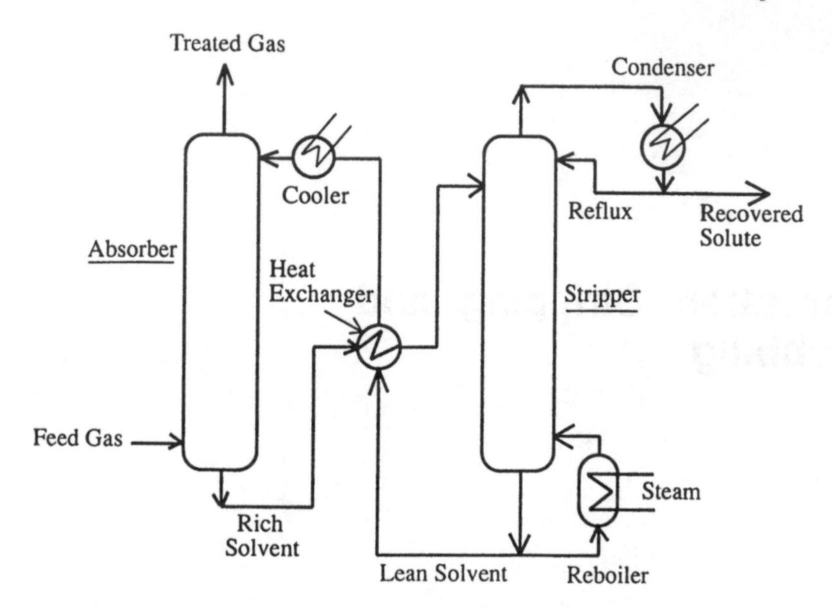

Figure 13.1 Sketch of a typical absorption-stripping plant flow diagram.

on which to capture the desired components, while *ab*sorption uses liquid solvents to dissolve the soluble components.

Scrubbing is primarily concerned with the removal of particulate liquids or solids from gas streams by contacting the gas with a liquid scrubbing medium. That is, the particles are scrubbed out of the gas. However, if the gas also contains some soluble gaseous components, they can dissolve in the scrubber solution. Thus, the scrubber may also be an absorber, just as an absorber may also be a scrubber if the feed contains particulate matter as well as soluble components.

Absorption and scrubbing are commonly used separation processes for preventing air pollution from industrial exhaust or stack gases. Stripping is used for both prevention of water pollution by treatment of wastewaters to remove soluble toxic gases and remediation of contaminated water at already existing hazardous waste sites.

13.2 ABSORPTION AND STRIPPING

Absorption and stripping both require the transfer of mass between gas and liquid phases. Absorption pulls mass into the liquid phase, and stripping takes it out. The absoption solvent can depend on either physical or chemical interactions with the solute for its solubility properties. In general, solvents using chemical interactions are more selective and have higher capacities in the absorption step. Their downside is that they are more difficult to regenerate by stripping. This is because the chemical interactions are stronger than the physical interactions and require more energy to be broken when the solvent is being recovered.

Solvent recovery is a normal part of most absorption processes. It is almost always required unless (1) the solvent is already on its way to waste before it is used in the absorption step and (2) the absorbed solute is also going to waste.

13.2.1 Fundamentals

One of the fundamental theories of mass transfer used in absorption and stripping is based on the concept of a thin, relatively stagnant film on each side of the gas-liquid interface. Mass is transported through these films by diffusion. The driving force for the mass transfer in either film is the concentration difference between the bulk phases and the interface between the phases. Any consistent set of concentration units, such as partial pressures, concentrations, or mole fractions, can be used. These concentration gradients are illustrated for both absorption and stripping in Figure 13.2. At the interface the gas and liquid concentrations are at equilibrium with each other. For absorption, shown in Figure 13.2a, the bulk concentration in the gas phase drives the solute to the interface, where it dissolves. The concentration of solute at the liquid interface then drives it into the bulk liquid phase. With stripping, shown in Figure 13.2b, the reverse sequence operates. The bulk liquid concentration drives the solute to the gas interface where it vaporizes through the film into the stripping gas. There are other theoretical treatments based on mobility of portions of the film, but we will discuss only the simple film theory.

The two films each have their own resistance to mass transport, and either one can be large compared to the other. When the gas phase resistance is large, the absorption rate is called *gas film controlled* and the concentration gradient is essentially all across the gas film. Correspondingly, if the liquid film resistance is much larger, the absorption is called *liquid film controlled* and the concentration gradient is across the liquid film. In general, the overall resistance is the sum of the two individual film resistances.

The amount of a component transferred must be the same in both phases to satisfy material balance. Therefore, the rate of mass transfer through the two films must also be the same. This can be written for a single component as Eq. 1 using partial pressures in the gas phase and concentrations in the liquid phase.

$$N_A = k'_G (p - p_i) = K'_L (c_i - c) \quad (1)$$

where:

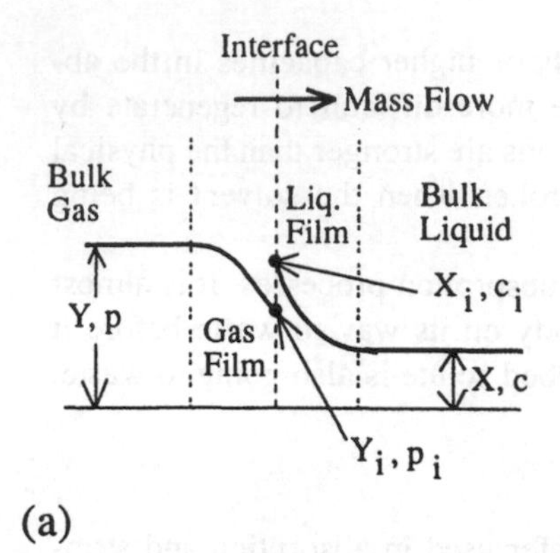

(a)

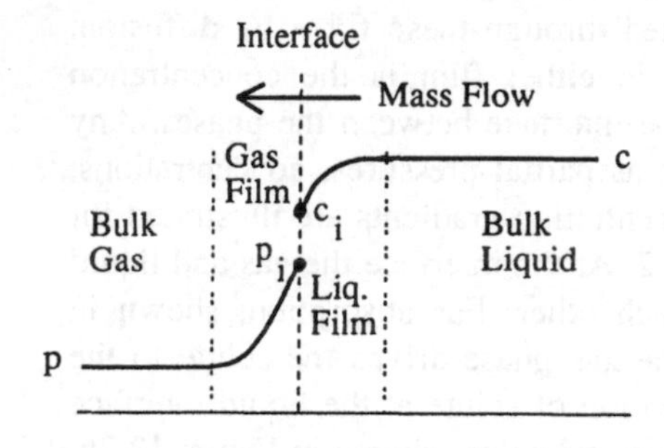

(b)

Figure 13.2 Sketch of concentration gradients across interfacial films for (a) absorption and (b) stripping.

N_A = mass transfer rate of component A
k'_G = gas phase mass transfer coefficient
k'_L = liquid phase mass transfer coefficient
p = solute partial pressure in bulk gas
p_i = solute partial pressure at interface
c = solute concentration in bulk liquid
c_i = solute concentration in liquid at interface

Alternatively, the equation can be written using mole fractions, Y and X, in the gas and liquid phases, respectively, to give Eq. 2.

$$N_A = k_G (Y - Y_i) = k_L (X_i - X) \tag{2}$$

The equilibrium relationship between gas and liquid compositions at the interface for dilute solutions at constant temperature is given by Henry's law. For the preceding equations, the Henry's law relationships are given by Eqs. 3 to 5.

$$p_A = HX_A \tag{3}$$

$$p_A = H'c_A \tag{4}$$

$$Y_A = H''X_A \tag{5}$$

where:

p = partial pressure
H = Henry's law constant
X = mole fraction in the liquid
Y = mole fraction in the gas
c = concentration in the liquid
A = subscript for component A

The different values for the Henry's law constants are merely to show that the numerical value changes as the measuring units change. Henry's law also applies at low pressures of 1 atmosphere (atm) or less, as well as to dilute solutions. Thus, it is fairly accurate for absorptions carried out at a total pressure of 1 atm.

It is usually desirable to use a single overall mass transfer coefficient based on the composition change in only one of the phases if it can be determined that one film is controlling. The film mass transfer coefficients have built into them the film thickness and the diffusivity of the material being transported. These values are not readily measurable or predictable separately. The overall value of the transfer coefficient also includes the film surface area. This overall coefficient can be derived from the individual film coefficients and the slope of the equilibrium line. It is also obtained by dividing the mass transported per unit time by the driving force, i.e., the partial pressure, mole fraction, or concentration difference. Edwards (1984) gives some sources of key gas absorption data.

Calculation of a packed tower absorber performance using overall transfer coefficients involves a rather complicated mathematical integration over the tower height of a differential vertical element of packing.

The ability to visualize what goes on in absorbers and strippers has been greatly simplified by using a graphical method called a design diagram (Edwards, 1984). This graphical method uses molar compositions of the gas and liquid phases based on the amounts of gas and solvent entering the absorber or stripper. That is, the gas composition, Y, is in moles of solute per mole of gas entering

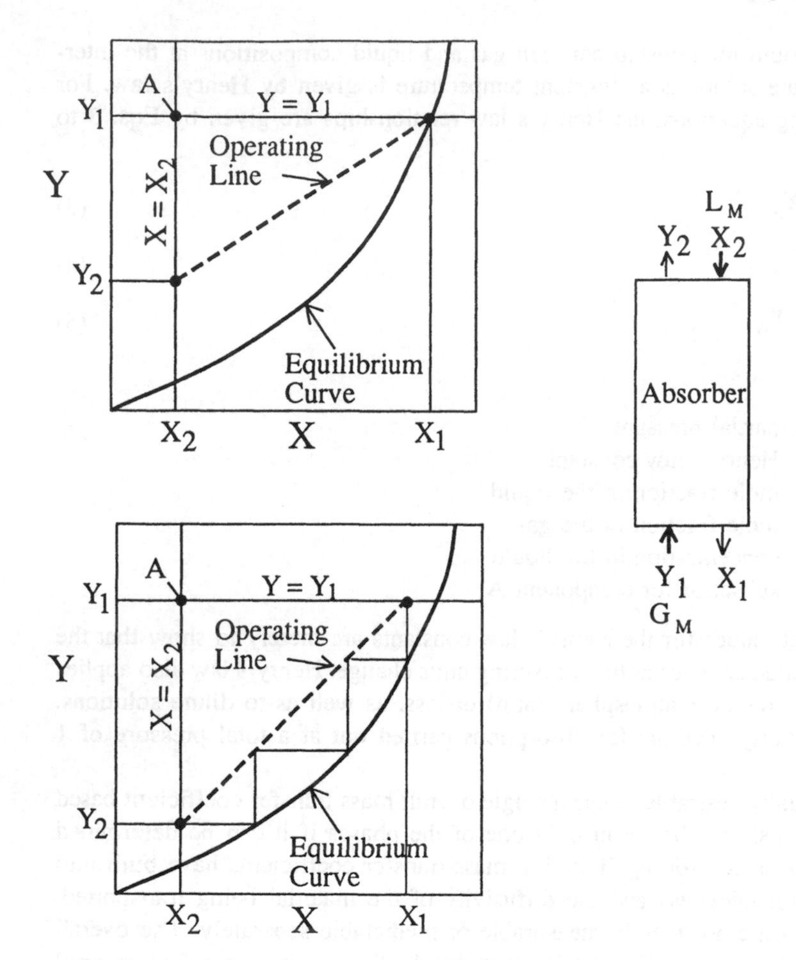

Figure 13.3 Graphical representation of absorber operation.

the absorber. Correspondingly, the liquid composition is in moles of solute in the liquid phase per mole of solvent entering the absorber. The use of these composition units gives straight operating lines because the flow of gas and solvent remains essentially constant through the absorber or stripper if the solute is ignored, that is, on a solute-free basis. This type of diagram is illustrated for absorption in Figure 13.3 and in Figure 13.4 for stripping.

In Figure 13.3 the right side of the figure is a sketch of the absorber. It shows the flows of gas and solvent to the absorber along with the compositions of the gas and liquid streams entering and leaving the absorber. The left side shows a graphical representation of the operation of the absorber based on an equilibrium curve and an operating line. The equilibrium curve is the locus of compositions of vapor and liquid that are in thermodynamic equilibrium with each other. The operating line is the locus of compositions of the gas and liquid streams passing each other at various levels in the absorption column. Thus, one end of the operating line has the coordinates X_1,Y_1 for the compositions of the passing streams of the entering gas phase and the exiting solvent phase. The other end of the operating line has the coordinates X_2,Y_2 for the compositions of the entering solvent and exiting gas at the other end of the column. Intermediate compositions of passing streams inside the column are points on the operating line. The operating line is essentially linear while the equilibrium line may be linear, curved upward, or curved downward. In Figure 13.3 it is shown curved upwards. More details of graphical analysis of separation processes are given in Chapter 2.

The compositions of the entering gas and liquid streams define the coordinates of point A in Figure 13.3. The space between these coordinates and the equilibrium curve define the composition region accessible to the absorpion separation. The ends of the operating line must fall on these two coordinates. Where they will fall depends on the number of theoretical plates in the absorber as well as the molar solvent to gas flow rate ratio in the absorber, i.e., L_M/G_M. Raising the solvent to gas ratio increases the slope of the operating line. Increasing the number of theoretical plates moves the operating line closer to the equilibrium curve. It takes an infinite number of plates for the operating line to touch the equilibrium curve in a pinch point. A pinch point can occur at either end of the operating line or at a tangent point when the equilibrium curve is concave downwards. A pinch point defines the minimum operable liquid to gas ratio by the slope of the corresponding operating line. To operate an absorber containing a finite number of plates requires a higher liquid to gas ratio than the minimum. Practical values are usually 1.2 to 1.5 times the minimum.

The upper graph in Figure 13.3 shows operation at the minimum liquid to gas ratio. The infinite number of plates is in the pinch caused by the operating line falling on the equilibrium curve at its right end. The lower graph shows a more practical absorption with a liquid to gas ratio of 1.5 times the minimum. These operating conditions require only three theoretical plates. Furthermore, the exit gas from the absorber has a lower value of Y_2, yielding better cleanup at the expense of the higher solvent circulation rate.

To determine the height of packing needed to give a specific number of theoretical plates in a packed tower, the height equivalent to a theoretical plate (HETP) must be multiplied by the number of theoretical plates needed. For plate towers, the number of theoretical plates needed must be divided by the fractional efficiency to determine the number of actual plates needed. For example, if five theoretical plates are needed and the efficiency of the actual plates used is 50%, the number of actual plates needed is 5/0.5 = 10 actual plates.

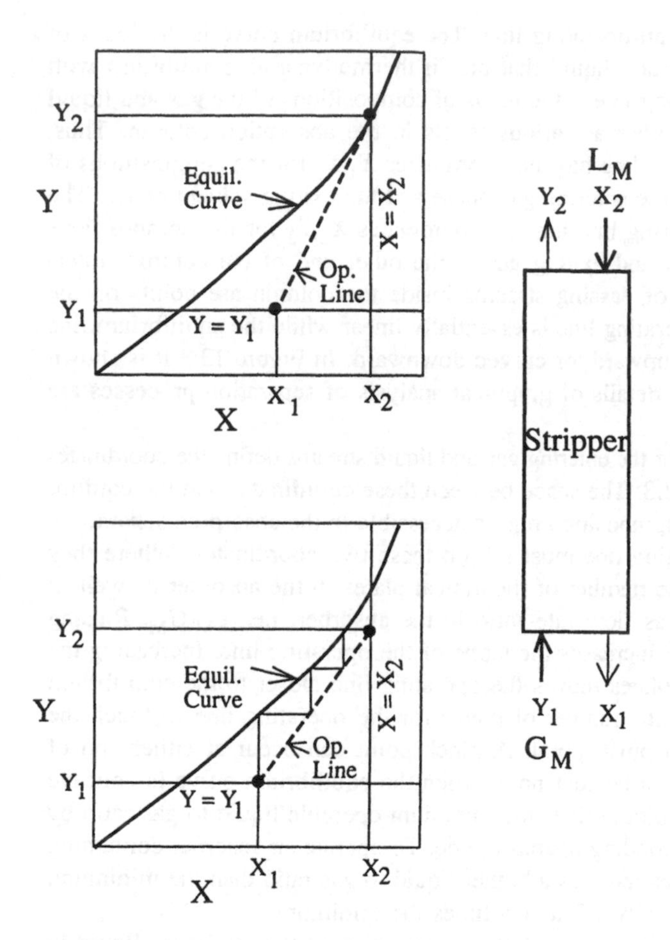

Figure 13.4 Graphical representation of stripper operation.

The graphs shown in Figure 13.4 are similar but represent stripping rather than absorption. Point A again represents the coordinates of the compositions of the streams entering the stripper, and the ends of the operating line must again fall on the coordinates defined by the entering streams. The slope of the operating line is again the liquid to gas ratio, but stripping operators normally think in terms of gas to liquid ratios. Thus, the slope of the operating line is the reciprocal of the gas to liquid ratio.

The upper graph in Figure 13.4 shows operation with an infinite number of theoretical plates at the minimum gas to liquid ratio, i.e., the maximum liquid to gas ratio. Again the plates are used up in the pinch where the operating line intersects the equilibrium curve. Practical gas rates for strippers are, again, about 1.2 to 1.5 times the minimum. This gives a lower liquid to gas ratio and moves the operating line away from the equilibrium curve. The lower graph in Figure 13.4 shows a stripping operation at 1.5 times the minimum gas to liquid ratio and two theoretical plates. Again the quality of the exit stream is improved at the expense of more stripping gas.

Absorption may involve the transfer of only one or two soluble components from an inert gas. In this case the soluble materials may be lumped together and the treatment just described can be used. However, absorption may also involve the transport of a very large number of solutes from the gas phase to the solvent. As pointed out by Diab and Maddox (1982), absorption of natural gas to remove higher boiling liquid hydrocarbons involves a feed in which all of the components of the feed are at least slightly soluble in the hydrocarbon fraction used as a solvent. This type of operation is treated more like a distillation and usually uses a reboiler to generate the stripping vapors in the solvent recovery section. It can be treated graphically on a solvent-free basis, while allowing for the effects of the solvent on the relative volatility of the other components in the feed. This treatment is discussed in Chapter 2.

The ideal solvent for absorption is:

- Highly soluble for the absorbed solute
- Relatively non-volatile
- Inexpensive
- Non-corrosive
- Stable
- Of low viscosity
- Non-foaming
- Non-flammable

The high solubility helps minimize solvent circulation rate, and the low volatility reduces solvent losses with the exit gas from the absorber. Low cost is important both for cost of solvent inventory and for cost of solvent losses. Stable, non-corrosive solvents minimize solvent losses due to degradation and keep maintenance costs down. Low viscosity improves contacting efficiency in the absorber. Foaming causes excessive liquid carryover with the gas flowing between stages and causes great loss of absorber efficiency. Finally, non-flammable solvents are desirable from a safety standpoint, although flammable solvents are used safely in large volume in petroleum and natural gas processing.

No perfect solvent for any process has yet been discovered. Therefore, a compromise must be made to get the best collection of solvent proprties for a given desired separation. As a rough categorization, water is often used for gases soluble in water, higher boiling hydrocarbon fractions are used for recovery of liquids from natural gas or petroleum refining gas streams, and reactive

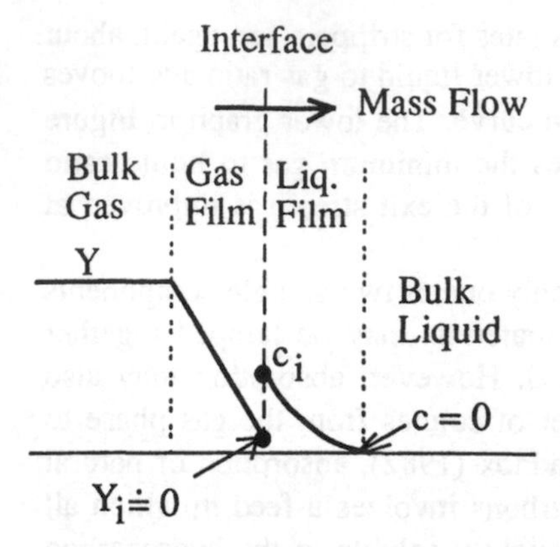

Figure 13.5 Sketch of concentration gradients across interfacial films for absorption with chemical reaction.

solvents are used for acid gases such as hydrogen sulfide, sulfur dioxide, and carbon dioxide. The reactive chemicals may be dissolved in aqueous solution, dissolved in other solvents, or may actually be the solvent when certain organic chemicals are used. When reactive solvents are used, the solute concentration gradients over the liquid and gaseous films are altered by the chemistry of the reaction between the gaseous solute and the liquid solvent. Figure 13.5 shows the solute concentration gradients with a reactive solvent in absorption, and Figure 13.6 shows the gradients during stripping. In Figure 13.5 the gradients are shown for a rapid irreversible reaction of solute and solvent. The gas phase concentration of the solute decreases linearly across the gas film to almost zero

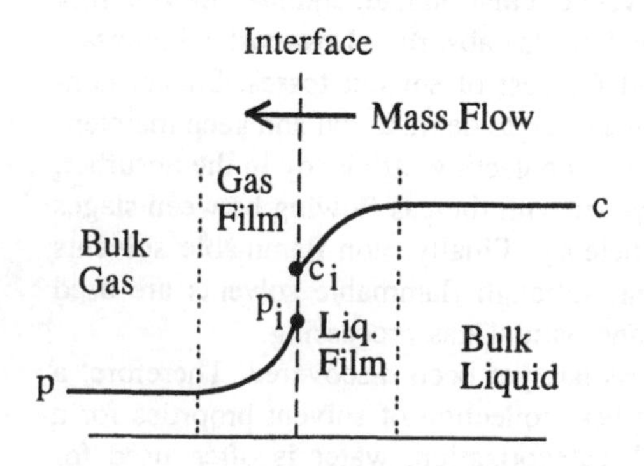

Figure 13.6 Sketch of concentration gradients across interfacial films for stripping with chemical reaction.

at the interface. The interfacial concentration is somewhat higher because the solvent selectively dissolves the solute. This concentration then decreases across the liquid film to zero at the bulk liquid phase boundary of the liquid film due to chemical reaction with the solvent. This is an example of absorption with gas phase controlling.

In Figure 13.6 the temperature has been raised to shift the equilibrium of the chemical reaction and release the solute into the bulk liquid phase. From there it is removed by the stripping gas. As the stripping gas removes solute, more solute is released by the chemical reaction until no more solute remains tied up in the liquid phase. The equilibrium shift can be done by raising the temperature, reducing the pressure, or both. The use of a stripping gas is often called the poor man's vacuum because it reduces the partial pressure of solute needed to vaporize it.

13.2.2 Specific Applications

Absorbers are commonly used for compliance with the regulations of the Clean Air Act. They remove toxic or environmentally undesirable components from otherwise innocuous gases before they are vented to the atmosphere. They can also be used to recover valuable gaseous components from waste gas streams. Strippers are used for removing or recovering dissolved toxic volatile components from wastewaters to comply with the Clean Water Act. They can also be used to recover valuable volatile materials for recycling. At present, use of these processes may be an additional expense in the cost of doing business. However, as uses and markets for the recovered materials develop, the costs of these operations can be largely offset or even show a net return. In addition to the applications driven by environmental regulations, there are many industrial applications driven by economics and technology, such as treatment of natural gas to improve both the quality of the gas and, at the same time, recover valuable hydrocarbon feedstocks for petrochemical manufacture.

Some specific applications of absorbers are:

- Industrial manufacture of strong acid or base solutions by absorbing the gaseous acid anhydrides or bases in water or weak solutions of the same gaseous solute. This technique is used to prepare hydrochloric, sulfuric, and nitric acids as well as ammonium hydroxide.
- Removal of sulfur dioxide from stack gases being vented to the atmosphere. Typical solvents used for these operations are water solutions of sodium hydroxide or sodium carbonate.
- Recovery of acid gases such as hydrogen sulfide, mercaptans, and carbon dioxide from recycle gases in petroleum refining or natural gas production. Typical solvents used here are water solutions of amines such as ethanolamines.

- Removal of water vapor from natural gas to improve heating value, prevent solid hydrate formation, and minimize corrosion. The solvent here is di- or triethylene glycol.
- Removal of odors from vent gases from sewage plants, chemical plants, refrigeration plants, fragrance manufacturers, or meat packing plants. Solvents for these operations are usually water or water solutions of acids, bases, or oxidizing agents.
- Removal of VOCs from vent gases. Solvents used for this application are water or water solutions of chemical reagents.
- Removal of fumes of hydrogen chloride, chlorine, HCN, HF, or ammonia from industrial exhaust gases by absorption with water.
- Removal of oxides of nitrogen as well as other toxic gases from industrial exhaust gases by scrubbing with solutions of oxidizing agents, followed by further treatment, if necessary.

Liquid particulates are usually removed or recovered by scrubbing rather than absorption. However, absorptions carried out for other purposes will also incidentally remove any liquid particulates such as entrainment, mists, or fogs.

In its applications, stripping is usually in competition with evaporation or distillation as alternate options. Factors which favor stripping are low boiling point of solute, low solubility of solute in the solvent, and low heat of vaporization of solute. High values of these properties as well as strong chemical interactions of solute and solvent tend to favor the other options.

Typical environmental applications of stripping are:

- Regeneration of absorber solvents for recycling.
- Soil venting using air as a stripping gas to remove VOCs such as solvents, gasoline, and diesel fuel from contaminated soil or groundwater.
- Air stripping of wastewaters to remove VOCs, odors, and ammonia. Typical VOCs are MTBE, acetone, halogenated solvents, and light aromatic and aliphatic hydrocarbons. To remove ammonia the solution must have a basic pH.
- Steam stripping to recover organic chemicals from solvents and remove volatiles from recycle oils.
- Inert gas stripping of reactive materials from water and solvents using nitrogen, carbon dioxide or steam.
- Removal of radon from water.

In addition, there are many industrial applications of stripping in manufacturing processes.

13.2.3 Equipment

Absorbers and strippers are gas-liquid contactors which use continuous flow of gas and either batch or continuous flow of liquid. Perhaps the simplest unit is a laboratory gas bubbler. When used as an absorber, a batch of liquid solvent is placed in the bubbler and the feed gas is bubbled through the liquid using a submerged gas disperser. The gaseous solutes dissolve in the solvent and are removed from the feed gas before it exits the bubbler. When used as a stripper, a batch of solvent containing the solute is placed in the bubbler and the solute-free stripping gas is passed through the bubbler. The volatile solute in the bubbler solvent vaporizes into the stripping gas and is removed with the stripping gas. Depending on the effectiveness of the bubbler in dispersing the gas, it can approach one theoretical equilibrium stage of vapor-liquid contacting. Bubblers are frequently used in laboratory apparatus for both absorption and stripping.

When continuous flow of both gas and liquid is employed, the gas-liquid contacting devices used for both absorption and stripping are essentially the same as those used for distillation. That is, they are either packed towers or plate columns using sieve plates or bubble cap plates. Single-stage or multistage contacting can be used along with flow patterns of countercurrent, cocurrent, or crossflow contacting of the liquid and gas. Figures 13.7 to 13.9 show sketches of various types of equipment using these flow patterns. Figure 13.7 shows sketches of absorbers and strippers using countercurrent flows. Figure 13.8

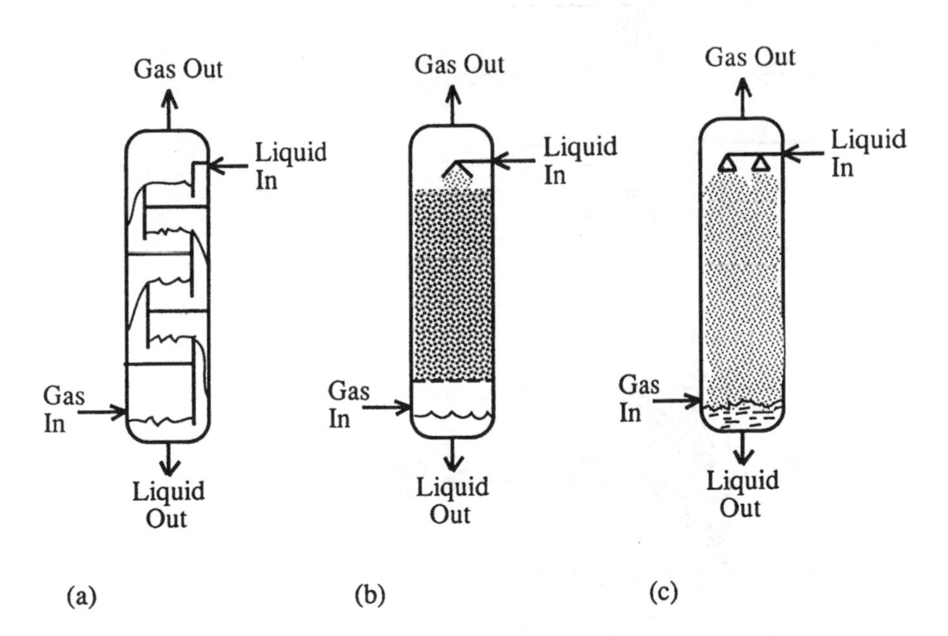

Figure 13.7 Illustrations of countercurrent flow. (a) Tray tower, (b) packed tower, (c) spray tower.

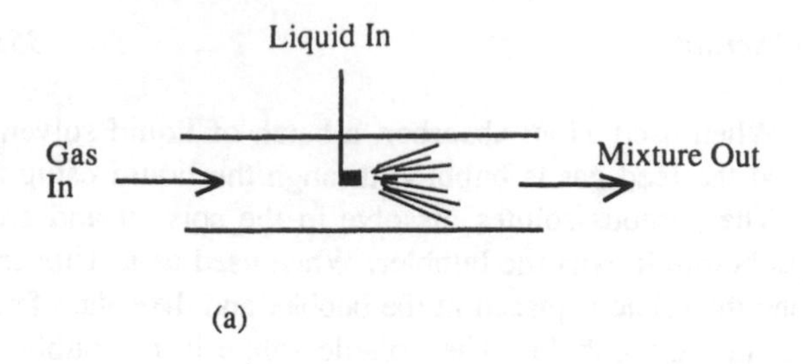

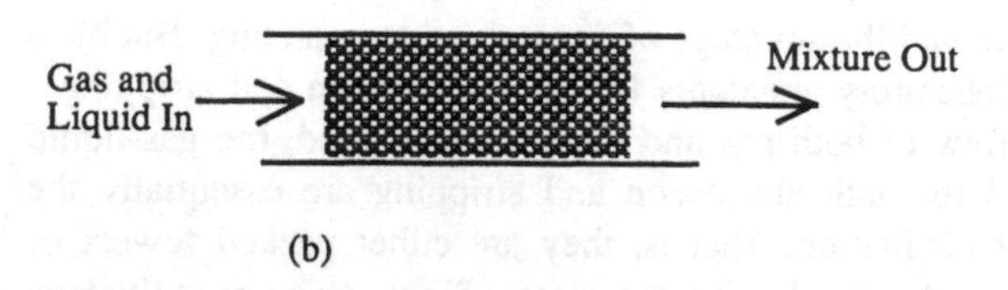

Figure 13.8 Illustrations of cocurrent flow. (a) Cocurrent spray, (b) in-line mixer.

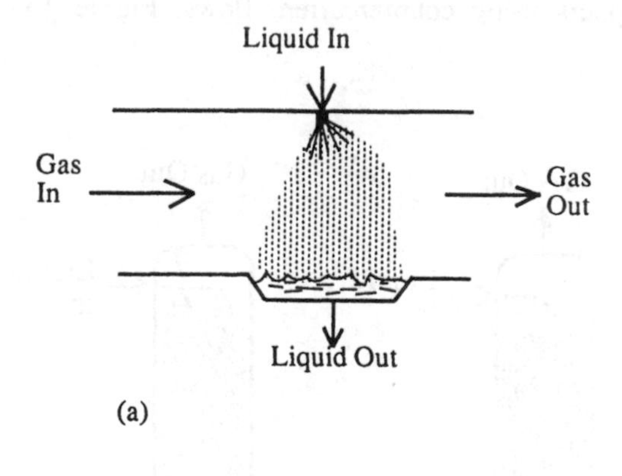

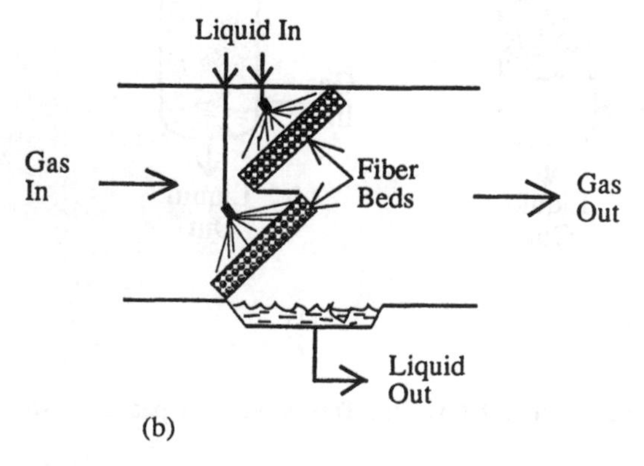

Figure 13.9 Illustrations of crossflow. (a) Crossflow spray, (b) fiber bed design.

shows absorbers and strippers using cocurrent flow, and Figure 13.9 shows a variety of crossflow absorbers and strippers.

In addition to packed and plate towers, both absorbers and strippers can also use sparged or stirred tanks, horizontal liquid crossflow spargers, or falling film gas-liquid contactors. Spray towers and horizontal spray chambers are also sometimes used for absorption. The size range for these units can vary from the small laboratory bubbler to large contactors air stripping over 5000 gal/min of liquid with over 20,000 ft^3/min of air. Absorbers also often process over 7,000 ft^3/min of contaminated air.

Furthermore, reboilers can be used in stripping columns to generate the stripping gas in much the same way stripping columns are operated in distillation. This technique is especially useful when the boiling point difference between the solvent and solute is relatively small and the final stripping gas at the tower bottom is vaporized solvent.

Vendors for absorption, stripping, and scrubbing equipment are listed in Tables 13.1 to 13.4. Table 13.1 lists vendors of all three types of equipment, Table 13.2 lists vendors of absorbers and either strippers or scrubbers, Table 13.3 lists vendors of air strippers, and Table 13.4 lists vendors of strippers and scrubbers or scrubbers only. These tables are not exhaustive.

13.2.4 Common Problems and Troubleshooting

Because the gas-liquid contacting equipment used for absorption and stripping is so much like that used for distillation, as would be expected, the operating problems are also much the same. For example, they include channeling, entrainment, foaming, and flooding as discussed, along with their solutions, in Chapter 12 on distillation. In addition to these mechanical problems, there can be product quality problems such as incomplete recovery of solute from the gases in absorption, or incomplete removal of gaseous solute from the liquid in stripping. Poor exhaust gas quality may be caused in absorption by:

1. A pinch between the operating line and equilibrium curve which makes most of the vapor-liquid contacting useless. It can be alleviated by changing the liquid to gas ratio so that the operating line is more parallel to the equilibrium curve. This is usually done by increasing the liquid rate.
2. Incomplete regeneration of the absorption solvent. This will bring solute into the absorber along with the regenerated solvent. The partially regenerated solvent makes the final contact with the gas leaving the absorber. The exit gas will then be in equilibrium with this impure entering solvent and thus carry out some solute with it. More complete removal of solute from the solvent in regeneration will improve the quality of the exit gas.
3. Too few theoretical plates in the column. In this case the separating power of the absorber is insufficient to produce the desired cleanup of the gas.

Table 13.1 Vendors of Absorbers, Strippers, and Scrubbers

Anderson 2000 Inc. 306 Dividend Drive Peachtree City, GA 30269 (404) 997-2000	The Ceilcote Company, Inc. 9A South Gold Drive Trenton, NJ 08691 (609) 890-2700
Clean Gas Systems, Inc. 707 Broadhollow Rd. Farmingdale, NY 11735 (516) 756-2474	Ducon Environmental Systems 110 Bi-County Boulevard Farmingdale, NY 11735 (516) 420-4900
Met Pro Duall Division 700 S. McMillin St. Owosso, MI 48867 (517) 725-8188	

Table 13.2 Vendors of Absorbers and Strippers or Absorbers and Scrubbers

Absorbers and Strippers

James W. Roderick & Assoc. 4500 S. Garnett #906 Tulsa, OK 74146 (918) 622-7175	Tigg Corp. Box 11661 Pittsburgh, PA 15228 (412) 563-4300
VIC Environmental Systems, Inc. 1620 Central Ave. NE Minneapolis, MN 55413 (612) 781-6601	

Absorbers and Scrubbers

Airotech Inc. Boyle Center 120 9th Ave. Homestead, PA 15120 (412) 462-4404	AMCEC Corp. 2625 Butterfield Rd. Oak Brook, IL 60521 (708) 954-1515
ARI Technologies 600 N. First Bank Dr. Palatine, IL 60067 (708) 359-7810	

Table 13.3 Vendors of Air Strippers

Advanced Air Technology, Inc. 3800 N. Wilke Rd. Arlington Heights, IL 60004 (708) 394-9553	Crane Company 800 Third Ave. P.O. Box 191 King of Prussia, PA 19406 (215) 265-5050
Delta Cooling Towers, Inc. P.O. Box 952 Fairfield, NJ 07004 (201) 227-0300	Geopure Continental Env. Serv. 2300 NW 71 Place Gainesville, FL 32606 (800) 342-1103
Hydro Group Inc. 97 Chimney Rock Rd. Bridgewater, NJ 08807 (201) 563-1400	National Environmental Systems 36 Maple Ave. Seekonk, MA 02771 (508) 761-6611
North East Environmental Prod. 17 Technology Drive West Lebanon, NH 03784 (603) 298-7061	ORS Environmental Equipment 4 Mill Street Greenville, NH 03048 (603) 878-3559
QED Groundwater Specialists Box 3726 Ann Arbor, MI 48106 (800) 624-2026	Remedial Systems Inc. 56 Leonard St. Foxboro, MA 02035 (508) 543-1512

Table 13.4 Vendors of Strippers and Scrubbers or Only Scrubbers

Strippers and Scrubbers

Advanced Air Technology, Inc.
3800 N. Wilke Rd.
Arlington Heights, IL 60004
(708) 394-9553

Air Chem Systems, Inc.
15222 Connector Lane
Huntington Beach, CA 92649
(714) 897-1017

Air Plastics Inc.
1224 Castle Drive
Mason, OH 45040
(513) 398-8081

Branch Environmental Corp.
Box 5265
Somerville, NJ 08876
(908) 526-1114

Heat Systems, Inc.
1938 New Highway
Farmingdale, NY 11735
(516) 694-9555

KCH Services Inc.
Box 1287
Forest City, NC 28043
(704) 245-9836

Pollution Equipment Co.
220-B Old Dairy Rd.
Wilmington, NC 28405
(919) 452-5663

Vanaire
10151 Bunsen Way
Louisville, KY 40299
(502) 491-3553

Scrubbers Only

AMBI INC.
Box Z
Lincoln, RI 02865
(401) 724-6330

Ambient Engineering Inc.
180 Highway 34
Matawan, NJ 07747
(908) 566-6866

American Environmental Intl.
111 Pfingsten Rd.
Deerfield, IL 60015
(708) 272-8635

Beardsley & Piper
5501 West Grand Ave.
Chicago, IL 60639
(800) 726-3700

Beco Engineering Co.
Box 443713
Oakmont, PA 15139
(412) 828-6080

Chem-Pro
P.O. Box 3100
Parsippany, NJ 07054
(800) 524-1543

CMI-Schneible Co.
Box 100
Holly, MI 48442
(800) 627-6508

Compliance Systems Int.
1029 Conshohocken Rd.
Conshohocken, PA 19428
(800) 220-CSI1

Table 13.4 (Continued)

Croll-Reynolds Co., Inc. 751 Central Ave. P.O. Box 668 Westfield, NJ 07091 (908) 232-4200	Davis Water & Waste Industries, Inc. 2650 Tallevast Rd. PO Box 29 Tallevast, FL 34270 (800) 345-3982
Scrubbers Only	
Fisher-Klosterman, Inc. P.O. Box 11190 Louisville, KY 40251 (502) 776-1505	Heil Process Equipment Xerxes 34250 Mills Rd. Avon, OH 44011 (216) 327-6051
Jet Airtechnologies 401 Miles Drive Adrian, MI 49221 (517) 263-0113	Ketema Inc Schuette & Koerting 2233 State Road Bensalem, PA 19020 (215) 639-0900
Komline Sanderson Holland Ave. Peapack, NJ 07977 (908) 234-1000	Monroe Environmental Corp. 11 Port Ave. P.O. Box 806 Monroe, MI 48161 (800) 992-7707
NuTech Environmental Corp 5350 N. Washington St. Denver, CO 80216 (800) 321-8824	Procom Environmental, Inc. 111 N. Second St. Suite. 303 Coeur D'Alene, ID 83814 (800) 368-7937
Sly Manufacturing Co. 2195 Drake Rd. PO Box 5939 Strongsville, OH 44136 (800) 334-2957	Tri Mer Corporation 1400 Monroe St. P.O. Box 730 Owosso, MI 48867 (517) 723-7838

The remedy is to add more packing height or more contacting plates to the absorption column.

4. Inadequate equilibrium between the solvet and gas. This can occur if the solvent has too low an affinity for the solute. The best remedy here is to change either the solvent or the operating temperature or pressure to improve the equilibrium. Economics dictate which is the better solution.

Incomplete removal of solute from the product liquid in stripping operations has the same types of causes as given for absorption. However, the remedies are slightly different. The problems are:

1. For the pinched line described in item 1 for absorption, the solution is the same except that the gas rate is usually raised rather than the liquid rate.
2. Inadequate purity of the stripping gas brings impurities in the stripping gas into final contact with the liquid being stripped, keeping the final liquid from achieving greater purity. The remedy is using a cleaner stripping gas.
3. Too few theoretical plates in the column is solved as explained in item 3 for absorption.
4. Inadequate equilibrium between the solvent and gas can occur if the solvent has too high an affinity for the solute. The best remedy here is to change the operating temperature or pressure or both to improve the equilibrium. Generally an increase in temperature and a reduction in pressure will improve stripping equilibrium.

In addition, with reactive solvents, care must be taken in the stripping operations to be sure that the solution pH has the appropriate value. For example, a basic pH is required to strip ammonia out of water solution, while removal of acid gases by stripping requires an acidic pH to release the acid solute.

13.2.5 Economics

The major economic factors in solvent processes are usually determined by the solvent circulation rate and solvent losses. This is because solvents are often relatively expensive and the most costly part of the plant is usually that associated with solvent regeneration. Regeneration normally requires the input of energy to break any chemical association required to free the solvent as well as to provide heat of vaporization for the dissolved solute. Furthermore, solvent regeneration is normally carried out at higher temperature and lower pressure than the absorption, thus requiring even more energy input. Clearly the choice of solvent plays a major role in the process economics. If a waste stream on its way to treatment can be used as a solvent or stripping gas without upsetting either product quality or waste treatment operations, it can give significant cost savings.

An additional factor in determining costs is the pressure drop through the absorber or stripper along with the absolute pressure at which the equipment operates. This determines the cost of fans or compressors needed to operate the equipment.

13.3 SCRUBBING

Strictly speaking, wet scrubbers are devices which use a liquid stream to recover small particles from a gas stream. However, common usage has come to define a scrubber as any device which uses a liquid to clean up a gas. This causes frequent overlap in common usage terms between absorbers and scrubbers. This is because absorbers can scrub and scrubbers can absorb. Furthermore, some types of dry particle collection devices are called dry scrubbers. What we will discuss in this section is wet scrubbers, which use liquids as the scrubbing medium.

Almost all wet scrubbers involve two distinct zones of operation. In the first, the particles are captured and solutes are dissolved from the gas phase into the liquid. In the second, the liquid containing the particles and dissolved solute is removed from the gas. Gilbert (1977) has divided scrubbers into three basic types:

1. Packed beds
2. Low energy scrubbers
3. High energy scrubbers

Packed beds include countercurrent packed towers and crossflow beds containing packings of fiber, rings, saddles, or other commercial packing materials for gas-liquid contacting. They are useful for collecting mists and soluble solids as well as absorbing gases.

The low energy scrubbers include spray scrubbers, baffle- and centrifugal-type units, self-induced spray devices, and some mechanically aided scrubbers. They typically use 0.75 to 3 hp/1000 ft^3/min of gas. Low energy scrubbers can clean feeds containing particles down to about 3 μm in average particle size. They frequently operate with a cocurrent flow pattern and thus require an extra vessel to separate the liquid from the gas after contacting.

High energy scrubbers generally require 3 to 6 hp/1000 ft^3 of gas treated. They include some plate towers and many gas-atomized spray units. These latter units are venturi units or cocurrent mixers such as orifice plates or structured in-line packings. The high gas velocity relative to the liquid in the venturi and cocurrent mixers atomizes the liquid, while the high turbulence promotes contacting of the phases. Venturi units are used mostly for dust collection only and can capture very fine particles. However, they usually require a cyclone separator to remove the liquid from the treated gas and may even need further mist eliminating devices.

While these categories roughly divide the various scrubber types, there is considerable overlap among the categories.

13.3.1 Fundamentals

According to Semrau (1984), a particulate scrubber consists of two parts. The first is a contactor stage in which a spray is generated and brought in contact with the particulate-laden gas stream. The second is an entrainment separation stage where the spray and deposited particles are separated from the cleaned gas. These two stages can be separate or physically combined. The spray may be generated by the flow of the gas itself in contact with the liquid, by spray nozzles, by a motor-driven mechanical spray generator, or by a motor-driven rotor through which both the gas and liquid pass. The separation can be accomplished using inertial separators such as cyclones or various forms of impingement separators. These devices can remove essentially all of the droplets of the size produced by the scrubber.

The mechanism of particle collection employed by the spray droplets in most scrubbers is (1) inertial impaction and interception for particles above 0.3 μm in diameter and (2) Brownian diffusion for particles smaller than 0.3 μm (Calvert, 1977). Inertial impaction occurs when the particle-containing gas stream flows around a liquid spray droplet. The inertia of the particles tends to make them continue to flow toward the droplet as the gas flow is deflected, and some of them will strike the liquid surface and be collected. Thus, inertial impaction describes the effects of small-scale changes in flow direction. The inertia of a particle depends on both its size and its density. Therefore, the ease of collecting a particle also depends on these factors. They have been lumped together to give a new parameter, the aerodynamic particle diameter, which is shown in Eq. 6 (Calvert, 1977).

$$d_{pa} = d_p (\rho_p C')^{1/2} \tag{6}$$

where:

d_{pa} = aerodynamic particle diameter
d_p = physical particle diameter
ρ_p = particle density
C' = dimensionless Cunningham correction factor

Fortunately, most methods for measuring particle size actually determine the aerodynamic diameter. Therefore for inertial impaction, it is not necessary to know the paticle density or actual physical particle size.

Brownian diffusion takes over for very small particles. It occurs when particles are small enough to respond to collisions with gas molecules, and thus begin to act like gas molecules in their own right. This starts to occur as the particles become less than 0.1 μm in diameter. Brownian motion causes the particles to diffuse randomly through the gas until they strike a spray droplet and are collected on it. Collection efficiency for particle impaction *decreases* with particle size. However, the collection efficiency for Brownian diffusion *increases* as particle size is reduced. Since these two mechanisms are the main ones operative in most scrubbers, we see a deep minimum in collection efficiency for particles of about 0.3 μm in diameter.

The term *cut diameter* is used to describe the capability of a particle scrubber. This is the aerodynamic particle diameter that will be collected at 50% efficiency. The particle collection efficiency increases sharply with particle diameter. Therefore, to a good approximation it can be assumed that all particles larger than the cut diameter will be collected and all particles smaller will be completely passed.

All scrubbers show a correlation between power consumption and collection efficiency. Increased efficiency requires an increase in power consumption. Furthermore, the required power consumption also increases as the particle size to be collected is decreased. The correlation of collection efficiency with power input is with the power dissipated in the gas-liquid contacting, i.e., the "contacting power." Extensive studies of this correlation have shown that the collection efficiency of a scrubber on a given particulate-containing feed gas is dependent essentially on only the contacting power. It is affected to only a minor degree by the size or geometry of the scrubber or by the way in which the contacting power is applied. Senkau (1984) defines contacting power as the power per unit of volumetric gas flow rate that is dissipated in gas-liquid contacting and is ultimately converted to heat. Thus, only the power actually dissipated in vapor-liquid contacting counts when determining contact power. Other power losses due to mechanical equipment inefficiencies, kinetic energy changes in the gas flow, or pressure drops not caused by gas-liquid mixing do not count as contacting power. In the simplest case, where all the contacting energy is obtained from the gas flow in the form of pressure drop, contacting power is equivalent to the friction loss across the wetted equipment. This is the effective friction loss, or gas contacting power.

For scrubbers which do not receive all of their contacting power from the gas phase, the power input from other sources is not directly measurable but must be estimated. Such estimates indicate that contacting power from various sources is additive in its relation to scrubber performance. Gas phase contacting power may be calculated from the effective friction loss by Eq. 7.

$$P_G = 0.1575 F_E \qquad (7)$$

where:

P_G = gas contacting power, hp/1000 ft^3/min

F_E = gas friction, inches of water

Similarly, liquid contacting power is based on the power input from a liquid injected through a hydraulic spray nozzle. It is approximately equal to the product of the nozzle feed pressure and the volumetric liquid rate. The liquid contacting power is then calculated from Eq. 8.

$$P_L = 0.583 p_F Q_L / Q_G \tag{8}$$

where:

P_L = liquid contacting power, hp/1000 ft^3/min
p_F = pressure, psig
Q_L = liquid flow rate, gal/min
Q_G = gas flow rate, ft^3/min

The correlation with efficiency is based on total contacting power, which is the sum of P_G, P_L, and P_M, where P_M is any mechanical power that may be supplied by a power-driven rotor.

Figure 13.10 shows a log-log plot of cut diameter versus power input for a variety of scrubbers.

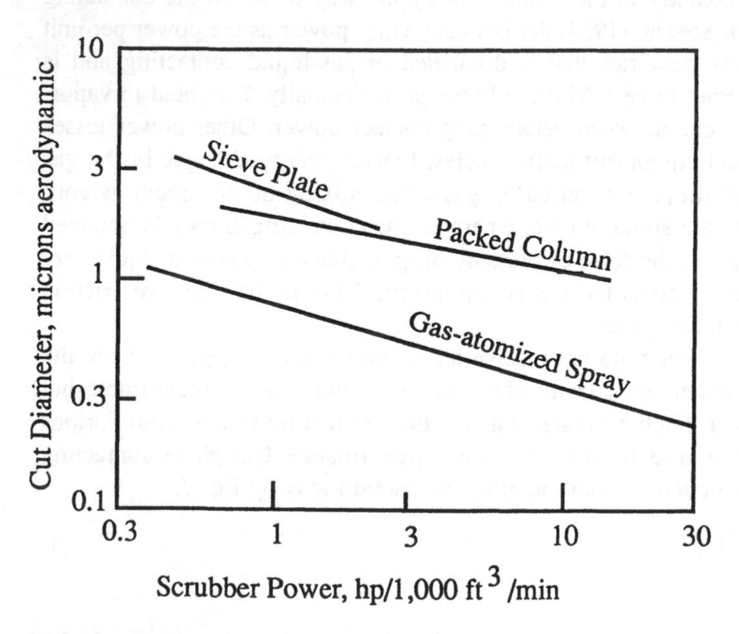

Figure 13.10 Illustration of cut diameter versus power input plot for scrubbers.

13.3.2 Specific Applications

As mentioned above, scrubbers share with absorbers the ability to remove soluble gases and vapors from a gas stream. Therefore, scrubbers are used in most of the same applications described for absorbers—removal of a large variety of toxic, odorous, or environmentally unacceptable gaseous materials from gases being vented to the atmosphere or recycled to the workplace. The sources of these vent or stack gases include many types of chemical plants, sewage plants, manufacturing plants, and utilities. However, in addition to these absorption-type applications, there are many applications which depend on the ability of scrubbers to remove particulate materials from gas streams. The targets of these particulate scrubbing applications are dusts and mists. They need to be removed from gases vented to the atmosphere to eliminate unsightly smoke and visible plumes as well as to eliminate release of environmentally damaging liquids and solids into the surroundings. In addition, valuable particulate products can be recovered from gaseous process streams by scrubbing.

Some examples of particulate scrubbing applications are:

- Workplace ventilation systems that use scrubbing to clean air for recycling by removing oil mists and dusts from grinding and metalworking operations
- Removal of acid mists from vent gases from acid manufacturing plants
- Removal of dusts from exhaust gases resulting from size reduction, calcining, and smelting operations
- Collection of fly ash from stack gases in power generation
- Removal of visible plumes from exhaust or stack gases from industrial plants

In most of these applications the scrubbing solvent is water and occasionally a water solution.

13.3.3 Equipment

The types of equipment used for scrubbing vary according to whether the primary purpose of the scrubbing is the collection of particles or the absorption of soluble solutes from the gas stream. However, good gas-liquid contacting is required in either case, just as with absorption and stripping. Thus, all scrubbers can collect particles and gases to some extent and the equipment is much like that used for absorption. Sizes can range from small local units placed near the source of the particle-laden stream up to very large units designed to handle the entire exhaust gas stream from a large industrial plant. Venturi scrubbers with capacities up to 250,000 ft^3/min per individual unit can be built, while some low-energy spray towers used in the mineral industries have double that capacity. Packed towers have been designed to scrub 120,000 ft^3/min of flue gas. Calvert (1977) has given a good description of the various types of scrubbers and his descriptions will be largely followed here.

The types of scrubbers he discussed are:

- Plate
- Massive packing
- Fibrous packing
- Preformed spray
- Gas-atomized spray
- Centrifugal
- Baffle- and secondary-flow
- Impingement-and-entrainment
- Mechanically aided
- Moving bed
- Combinations of the above

Plate scrubbers are essentially the same as plate-type distillation towers. They consist of a vertical tower with one or more horizontal vapor-liquid contacting plates spaced vertically inside the tower. The gas enters the bottom of the tower and the liquid is introduced on the top plate. From there, the liquid flows down through the tower across each plate in succession as a shallow liquid pool. The gas flows upward through perforations in each plate and bubbles through the liquid layer on each plate in succession until it exits at the top of the column. The plates may be perforated sieve plates, bubble cap plates, valve trays, or slotted trays. The type of tower is named for the type of trays inside (for example, a bubble-cap tower). Collection efficiency increases as size of the gas bubbles is decreased.

Massive packing scrubbers are packed tower scrubbers like those used in absorption and distillation. They can be packed with manufactured materials like saddles, rings, or other shapes specially designed to maximize gas-liquid contacting. On the other hand, they can be simply beds of crushed stone. The flow pattern of gas and liquid can be countercurrent, cocurrent, or crossflow. With mist collection and drainage, there may even be no additional flow of liquid required. Collection efficiency increases as packing size is decreased. However, massive packing is subject to plugging by solids and may need frequent removal from the tower for cleaning.

Fibrous packing consists of beds of fiber made from plastic, glass, or steel. The fibers are small in diameter but strong enough to support the collected particles without matting together. The particles collect on the fibers and are then washed off by the flowing scrubber liquid. Flow can be countercurrent, cocurrent, or crossflow in the same manner as with massive packings. Collection is by inertial impaction on the fibers as the gas flows around the fibers. Collection efficiency increases as fiber diameter decreases.

Preformed spray refers to a spray that has been formed by liquid flow through a nozzle before coming in contact with the gas. The droplet size

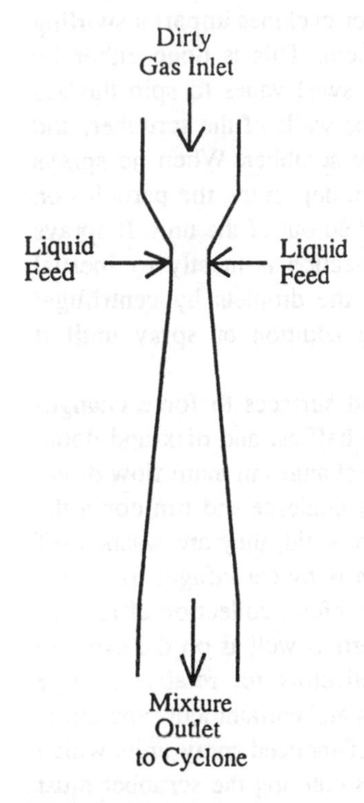

Figure 13.11 Typical example of a venturi scrubber.

distribution is determined by the nozzle design and liquid properties independently of gas flow. The gas is then fed through the spray in a contacting chamber designed to optimize good contacting of liquid and gas. Again, flow can be countercurrent, cocurrent, or crossflow. Particle collection is by inertial impaction on the droplets and an optimum droplet size often exists. Efficiency can be improved by increasing spray nozzle pressure and liquid to gas ratio.

Gas-atomized spray scrubbers use the moving gas stream to atomize the liquid and then accelerate the droplets. Typical devices are the venturi scrubber and various designs of orifice scrubbers. Liquid may be introduced in various places where the high gas velocity will form droplets from the liquid and pull them into the gas flow. With venturi scrubbers the liquid is usually introduced at the throat of the venturi scrubbers as illustrated in Figure 13.11. Particle collection then occurs by inertial impaction as the gas flows around the droplets. Efficiency increases with gas velocity and liquid to gas ratio. However, it is usually best to increase liquid to gas ratio first to achieve a given cut diameter.

Centrifugal scrubbers in the form of cylinders or cyclones impart a swirling or spinning motion to the gas passing through them. This is done either by tangential introduction of the gas or by providing swirl vanes to spin the gas inside the cylinder. Liquid can be added to wash the walls of the scrubber, and sprays can be used for gas-liquid contacting in the scrubber. When no sprays are used, collection is achieved by centrifugal force, depositing the particles on the inside wall of the cyclone from where it is washed out of the unit. If sprays are used to increase gas-liquid contacting, the collection is mostly by inertial impaction on the spray droplets with removal of the droplets by centrifugal force. Collection performance improves with the addition of spray until it matches that of a pre-formed spray scrubber.

Baffle- and secondary-flow scrubbers use solid surfaces to force changes in gas velocity or flow direction. Louvers, zig-zag baffles, and disk-and-donut baffles are examples of target surfaces that produce changes in main flow direction. If the particles being collected are liquid, they coalesce and run down the target surface into a liquid sump. If the particles are solid, they are washed off the target surfaces intermittently. Particle collection is by centrifugal force due to the change in direction of the main gas flow. Therefore, collection efficiency depends on the radius of curvature of the flow pattern as well as on the expanse of the baffle surface. They are only efficient collectors for relatively large particles and, as such, are mainly used as precleaners and entrainment separators.

Impingement-and-entrainment scrubbers are self-induced spray units which consist of a shell that contains liquid so that the gas entering the scrubber must impinge on the liquid surface and skim over it to reach a gas exit duct. This is illustrated in Figure 13.12. Some liquid droplets are formed at the surface and carried into the gas where they act as the particle collection and mass transfer surfaces. The gas exit duct is usually designed to change the direction of flow of the gas-liquid mixture flowing through it to reduce droplet entrainment. Particle collection is attributed to inertial impaction and appears to be comparable to that of an atomized scrubber operating at the same pressure drop.

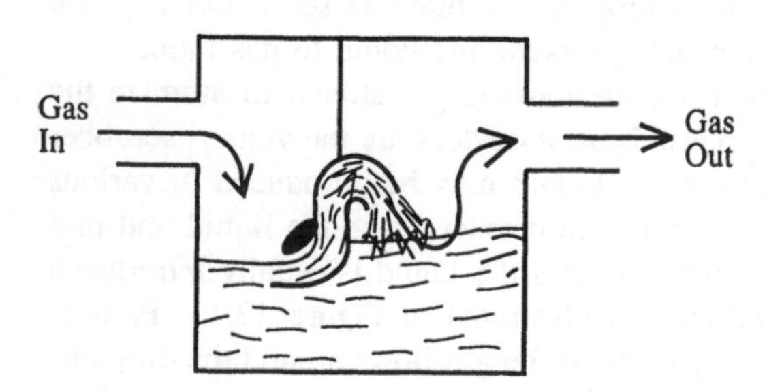

Figure 13.12 Illustration of an impingement-and-entrainment scrubber.

Mechanically aided scrubbers use a motor-driven device between the inlet and outlet of the scrubber body. Frequently, this is a fan used to move air through the scrubber. Particles are collected by impaction on the fan blades. Liquid is usually fed to the hub of the fan where part of it is atomized by impact with the fan blades. The rest runs over the blades washing off the collected particles. As it leaves the blades, it atomizes also. The liquid droplets are recaptured by the fan housing and drain into a sump. Particle collection mechanisms, in order of importance, are inertial impaction on liquid droplets, inertial impaction on the fan blades, and centrifugal deposition in the housing.

Moving bed scrubbers can be either fluid beds or less violently agitated beds which still move and mix under the influence of the gas flow. The contacting vessel holds a support grid on which the movable packing is placed. The gas flows upward through the bed, and liquid is sprayed into the bed either up from the bottom or down onto the top surface of the bed. Particle collection is from inertial impaction on atomized liquid and on the packing elements. Moving bed scrubbers provide both good particle collection and good mass transfer characteristics for absorption. The movement of the bed keeps the packing clean and reduces solids deposition problems.

Combinations of these scrubber designs have been made in many different variations, which will not be described here. Furthermore, designs combining other driving forces such as electrostatics have also been developed to produce effective hybrid types.

Some vendors of various types of scrubbers are given in Tables 13.1, 13.2, and 13.4.

13.3.4 Common Problems and Troubleshooting

The problems encountered with wet scrubbers usually show up as high utility requirements or low collection efficiencies. They can also have problems with corrosion and erosion depending on the nature of the solvents or materials being recovered. The erosion usually occurs with recycling of liquid streams used to recover solid particles. Gilbert (1977) has discussed the common problems for packed towers and low-energy and high-energy scrubbers. We will summarize his discussion here.

Packed Towers

The major problems with packed towers are the same as for absorbers and distillation columns, that is, high pressure drop, flooding, entrainment, foaming, channeling, and poor liquid distribution. The high pressure drop and attendant flooding are caused by excessive flow rates of either gas or liquid or both, or by partial plugging of the packing with solids. They can be readily monitored by installing a differential pressure manometer or gage across the packed bed.

Periodic cleaning of the packing as well as reducing flow rates can help eliminate these problems. If reduction in flow rates is undesirable, changing to a more open packing with lower pressure drop may also prevent flooding.

Entrainment can be caused by excessive gas flow, very fine particles in the feed gas, and faulty scrubbing liquid distributors, which make fine mists at the top of the packing. Entrainment can be greatly reduced or eliminated by installing a mesh bed between the top of the packing and the gas outlet. This bed can be continuously or periodically washed to remove solid particles from the mesh. Foaming is largely a property of the liquid and its surface tension and viscosity. It can usually be reduced by the addition of small amounts of chemicals which collect at the surface and alter the surface tension of the liquid. Furthermore, packed towers are somewhat more forgiving than plate towers with respect to foaming.

Channeling of the liquid phase occurs when the liquid is not distributed well through the cross-section of the packing, but rather flows down in preferred channels which bypass contacting with the gas phase. This can be caused by poor liquid distribution at the top of the packing. In addition to ensuring that good liquid distribution is taking place at the top of the packing, it may be necessary to redistribute the liquid at several points between the top and bottom of the packed bed.

Low-Energy Scrubbers

Low-energy scrubbers are typified by ejector venturis, mechanically assisted units, and spray columns. They do not normally have difficulty with solid plugging due to their open design. However, they may have abrasion problems due to solids buildup in the liquid in the recycle loop. This is caused by high velocities of the liquid slurry through distribition nozzles. Abrasion may be minimized by increasing the liquid make-up rate and thus lowering the solids concentration in the recycle liquid. It may also be decreased by removing a concentrated slurry in the recycle loop.

The gas-liquid separators conventionally used with low-energy scrubbers are usually gravitational or impingement designs. The gravity settling chamber is the simplest of these but is only useful for very coarse liquid droplets entering the separator. It is often supplemented by a simple stack of parallel plates to reduce the vertical distance the particles must settle before collecting on the surfaces. These units must be built for easy cleaning to remove collected solids.

The impingement separators can use designs with mesh mist eliminators or chevron plates. Flow through these units creates a tortuous path causing the droplets to impact on the mesh or chevron plates, where they coalesce and run off to a sump. These units are less susceptible to plugging, but still require periodic cleaning to remove solids.

Low-energy scrubbers can generally improve poor performance by increasing pressure drop through the unit. This is usually done in venturi scrubbers by increasing the gas velocity through the throat of the venturi.

High-Energy Scrubbers

High-energy designs frequently consist of a high-energy venturi coupled with a cyclone. These relatively open units do not usually have plugging problems. The only place plugging is likely to be encountered is in the recycle liquid distributors. High pressure drops indicate either gas or liquid flow rates higher than design. With these high-energy designs, low presssure drop, which tends to reduce collection efficiency, is more troublesome than high pressure drop. In this case a final polishing step is added after the cyclone to eliminate any mist carryover at the lower cyclone velocity. This can be a mesh pad or chevron baffles, which are often included in the original design if flow rates are expected to fluctuate widely.

High-energy scrubbers are also more likely to encounter the problem of re-entrainment common to electrostatic precipitators. Re-entrainment occurs when the flowing gas has a sufficiently high local velocity to pick up and carry out of the scrubber liquids and solids that had previously been collected on a solid surface. It can be avoided by washing the collecting surface with liquid as particles are being collected and by avoiding high local gas velocities at the collecting surfaces.

13.3.5 Economics

The capital cost of scrubbing systems depends on the size and complexity of the system, the materials of construction, and the auxiliary equipment needed for operation. Usually two or three theoretical stages are used and the common construction materials are stainless alloys, polypropylene, PVC, or fiberglass-reinforced plastic (FRP). The auxiliaries include any pretreatment needed for preparing the feed gas and any posttreatment of the scrubber liquid prior to disposal. Auxiliaries also include heat exchangers, ductwork, and instrumentation for both analysis and control of operation.

The type of scrubber needed is determined by the particle size distribution in the feed gas as well as whether the particles are liquid or solid. Costs generally vary widely with the particle size that needs to be removed from the gas and generally increases as the size of the particles to be captured decreases. Best cost estimates can be obtained by contacting a vendor at early stages in planning.

The operating costs are energy related and cover the cost of operating fans and pumps to overcome gas and liquid pressure drops through the system. The energy costs can be approximated from the pressure drop by assuming the theoretical power required is 0.16 hp/1000 ft^3/min of gas for each inch of water pressure drop. The actual power requirement is double this, assuming a 50% fan and motor efficiency. Again, contacting a vendor to get realistic values is important.

13.4 SUMMARY

Absorbers, strippers, and scrubbers play an important role in minimizing air and water pollution while at the same time recovering valuable materials. They are basically gas-liquid contactors equipped to separate the two phases after thorough contacting is achieved. Scrubbers and absorbers find most of their use in preventing air pollution, while strippers are more useful in preventing water pollution. While these are often the most visible uses, many industrial and mineral industry operations use them to recover valuable materials for recycling or sale.

REFERENCES

Calvert, S. (1977). How to Choose a Particulate Scrubber, *Chemical Engineering*, August 29, 1977: 54–68.

Diab, S. and Maddox, R. N. (1982). Absorption, *Chemical Engineering*, December 27, 1982:38.

Edwards, W. M. (1984). Mass Transfer and Gas Absorption, *Perry's Chemical Engineers' Handbook*, 6th ed. (R. H. Perry and D. W. Green, eds.), McGraw-Hill, New York, p. 14-7.

Gilbert, W. (1977). Troubleshooting Wet Scrubbers, *Chemical Engineering*, October 24, 1977: 140–144

Semrau, K. T. (1984). Gas-Solids Separation, *Perry's Chemical Engineers' Handbook*, 6th ed. (R. H. Perry and D. W. Green, eds.), McGraw-Hill, New York, pp.20-75 to 20-121.

14

Adsorption, Ion Exchange, and Reversible Reactions with Solids

14.1 INTRODUCTION

The adsorption, ion exchange, and reversible chemical processes discussed in this chapter are based on reversible selective interactions of gases or liquids with solids. They are very important to hazardous waste minimization. This is because of their effectiveness at recovering hazardous materials when they are present at low concentrations in the feed stream. The hazardous materials may be recovered and recycled because of their value, or may be recovered to remove them from either a process stream to be recycled or a waste stream to be discarded. Typical feed streams are industrial process streams, wastewater streams, or gas streams destined for venting to the atmosphere. The material to be removed and/or recovered from the feed stream is captured by the solids and then removed from the feed stream along with the solids. The feed stream may be either liquid or gaseous. The solids may be amorphous or crystalline materials chosen for their selectivity for either non-polar, polar, or ionic molecules. Typical amorphous materials are activated carbons, activated aluminas and clays, silica gels, and polymer-based ion exchange resins. Typical crystalline materials are natural and synthetic zeolites and molecular sieves.

Contacting of the solids with the feed stream is achieved either by passing the feed stream through a captive bed of solids or by dispersing the solids in the feed. The dispersed solids are then removed from the feed by settling, centrifugation, filtration, and the like. Coarse granular solids are used when contacting is done in fixed beds, while finer powdered materials can be used when the solids are dispersed in the feed and then recovered. Typical examples of processes using fluid interactions with solids are (1) the use of granular activated carbon in fixed beds as an adsorbent to remove organics from wastewater streams, (2) the use of powdered activated carbon, again as an adsorbent, in biological water treating plants to improve removal of organic materials, (3) the use of ion exchange resins to remove metals from aqueous waste streams, and (4) the use of adsorbents to recover organic solvents from ventilation exhaust leaving printing, painting, or dry cleaning operations.

The processes using solids to recover materials from feed streams are usually cyclical. The cycle contains (1) a contacting period followed by separation of the solids from the feed, and (2) regeneration and/or reactivation of the solids for further use in a subsequent cycle. The time span of the cycle depends on the concentration of material to be removed from the feed stream and the capacity of the solid for that material. Low concentrations of material to be removed from the feed and high capacities of the solids for this material can lead to long on-stream times. This may make regeneration of the solids uneconomical compared to discarding them. However, these are also the exact conditions which make separations using solids look economically attractive in the first place. The choice of regeneration versus disposal of spent solids is made based on economic analysis of these alternatives.

The reversible interactions of liquid or gaseous molecules with solids can be conveniently divided into three classes depending on the forces causing the organics to stick to the solids. These classes are (1) adsorption, both physical and chemical, (2) ion exchange of either or both cations and anions, and (3) reversible chemical reactions such as complexing and chelation. They will be discussed in that order in this chapter.

14.2 ADSORPTION

Adsorption is the name given to the tendency for a solid surface to be covered with one or more layers of molecules attracted to the surface from its gaseous or liquid surroundings. If the attraction between the solid surface and the adsorbed molecules is purely physical, the adsorption is called physical adsorption and is readily reversible. An example of physical attraction forces is the van der Waals forces that cause vapors to condense into liquids. On the other hand, if the attractive forces are chemical in nature and result in the formation of a chemical bond between the solid adsorbent and the adsorbed molecules, the adsorption is called chemical adsorption or chemisorption. In this latter case the removal of the chemisorbed material to regenerate the adsorbent is more difficult and may require chemical techniques to reactivate the spent adsorbent. For this reason most adsorbents are operated as physical adsorbents with regeneration by reducing pressure, raising temperature, or flushing with solvents. However, even physical adsorbents are often slowly deactivated over many adsorption/regeneration cycles by chemisorptions due to unknown trace contaminants in either the adsorbent or the feed stream. At some specific deactivation level, the adsorbent must be either reactivated or discarded.

Two general types of adsorbents are commonly used. The first, typified by activated carbons, has a low-energy surface and is selective for non-polar molecules. These can be hydrocarbons and other organic compounds of low polarity such as ethers, ketones, esters, and halogenated hydrocarbons. This type of adsorbent is useful for recovery of organic materials from wastewaters and exhaust gases. The other type, which consists of activated clays, aluminas, and silica gel, has a higher energy surface that is more selective for polar molecules. These materials are usually not effective for water solutions because the water selectively covers all the adsorption sites to the exclusion of everything else. However, they do find use for removing polar impurities such as water, oxidation products, or naturally occurring impurities like sulfur and nitrogen compounds from organic process streams. They are also useful for cleaning slightly contaminated waste streams to permit recycling of the major components.

Zeolites and molecular sieves are special cases which depend on pore size and structure, i.e., access to the internal surface, for their selectivity rather than on adsorbent surface energy. They can be made from specially treated carbons as well as from natural or synthetic crystalline materials.

14.2.1 Fundamentals

The primary factors that make adsorption practical as a separation process are (1) the ability to manufacture adsorbents having high surface areas per unit weight and (2) surfaces which are very selective for the materials to be removed from the feed stream. The surface area depends on the adsorbent structure. The selectivity depends on the nature of the adsorbing surface and the strength of its interactions with adsorbed molecules.

Adsorbent Structure

The total amount of feed adsorbed is a function of the available surface area of the adsorbent and can be calculated for an adsorbed layer one molecule deep, a monolayer, from the surface area and the area occupied by a single molecule. Clearly it is desirable to provide high surface area adsorbents to get high capacity for adsorbed material. One way to get increased surface area is to

Table 14.1 Surface Area of Solid Spherical Sand Particles

Particle diameter, cm.[a]	Surface area, m^2/g[b]
1	0.00023
0.1 (1 mm)	0.0023
0.01	0.023
0.001	0.23
0.0001 (1 μm)	2.3

[a] 2.54 cm = one in.
[b] 1 m^2/g = 4887 ft^2/lb.

reduce the size of the adsorbent particles. However, as is shown in Table 14.1, the surface area does not become very large until particle sizes in the micron range are obtained.

A better way to achieve high surface area in adsorbents is to prepare the adsorbent in a "microporous" form, but in relatively large particle size. Such adsorbents can have surface areas on the order of 100 to 1000 m^2/g in large particle sizes. Both activated carbons and solid oxide adsorbents are prepared in this form, which is illustrated by Figure 14.1.

The structure illustrated in Figure 14.1a shows a pore network where the diameter of the pores gradually decreases, starting with large pores at the surface of the particle and ending with tiny pores in the interior of the particle. It is clear from this diagram that only very small molecules can penetrate the structure to the ends of the small pores. Intermediate-size molecules compete with small molecules for surface area in the intermediate size pores, and large molecules compete only in the pores of larger size. Thus, size fractionation occurs in the adsorbent because of the different amounts of surface area available to different sized adsorbing molecules. If the feed contains a range of solute sizes that are to be removed by adsorption, this pore structure is desirable in the adsorbent. However, if the solute is of large uniform molecular size, it is desirable to maximize the surface area available in large pores.

In contrast to Figure 14.1a, which shows a smooth decrease in pore size with length, the structure shown in Figure 14.1b shows some "ink bottle" pores, where the pores increase and decrease in size on the way from the surface of the particle to the end of the pore. In this case, a lot of the surface area measured with small molecules is not available to molecules larger than the constrictions in the pores. This effect is carried to the extreme in the pores of molecular sieves which are illustrated in Figure 14.1c. Here the crystalline solid has a uniform-size opening to a chamber where the surface area and pore volume are concentrated. The size of the opening is determined by the atomic structure of

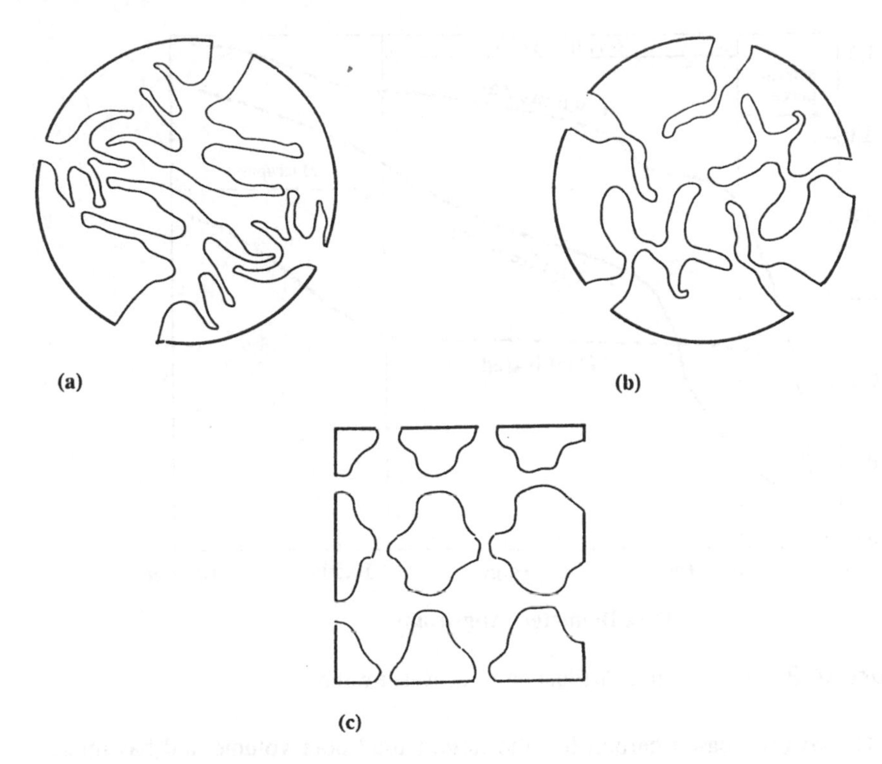

Figure 14.1 Porous structure of adsorbents. (a) Decreasing pore diameter, (b) ink bottle pores, (c) molecular sieve pores.

the crystal and can be controlled over a range of diameters from 3 to 8.5 Å by selection of the cations used in the synthesis of the crystals. These crystals exclude molecules of a size larger than the openings in the crystals and have thus acquired the name *molecular sieves*. The crystals of molecular sieves are usually too small to use as is for adsorption processes and are usually pelleted with a large pore oxide binder to get particles large-enough for convenient handling.

The pore structure of solid materials can be characterized by measurement of either surface area distribution according to pore size or pore volume distribution according to pore size. Typical pore volume distributions are shown in Figure 14.2 for activated carbons made from coal, lignite, and coconut shells.

The pore sizes of the carbons in Figure 14.2 range from about 10 Å up to tens of thousands of Angstroms. Pores smaller than about 40 Å are called *micropores* and pores larger than about 5000 Å are called *macropores*. Those in the range between these two extremes are called *intermediate pores*.

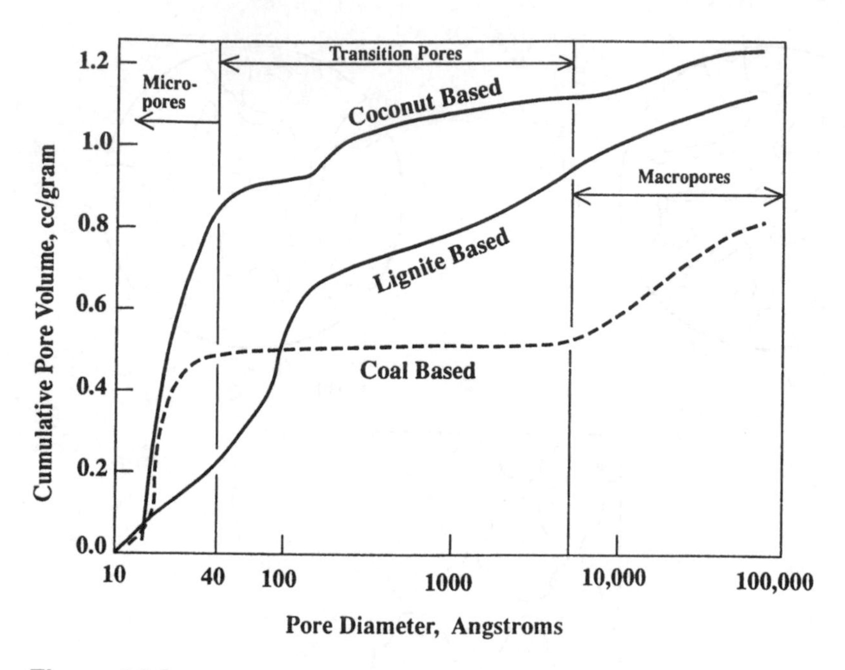

Figure 14.2 Pore volume distribution of activated carbon.

The coconut-based carbon has the largest total pore volume and has most of its pores in the micropore region. This makes it particularly valuable for the adsorption of small molecules from gaseous feed streams. These molecules can penetrate to the interior of the carbon and condense to a liquid, filling the small pores. The condensation is called capillary condensation and occurs in pores below a certain size. This critical size is determined by the partial pressure of the component in the gas phase divided by its saturation pressure at the same temperature. The saturation pressure is the vapor pressure at which the component would condense in the gas phase at the same temperature. Thus, the small pores fill first at low concentrations of vapor in the gas phase and larger pores fill progressively as concentration in the gas phase increases. The largest pores fill only when the concentration is high enough that the partial pressure reaches the saturation pressure. Thus, in operation the largest pores are rarely filled. However, they provide a low resistance pathway to the intermediate and micropores and are particularly important during regeneration. The optimum carbons for removing contaminants from gases have most of their pore volume in the micropore range.

For liquid phase adsorption, there is no capillary condensation involved. The adsorption is controlled by surface forces and access to the surface. Thus, the optimum carbons for liquid phase adsorption require pores sufficiently large

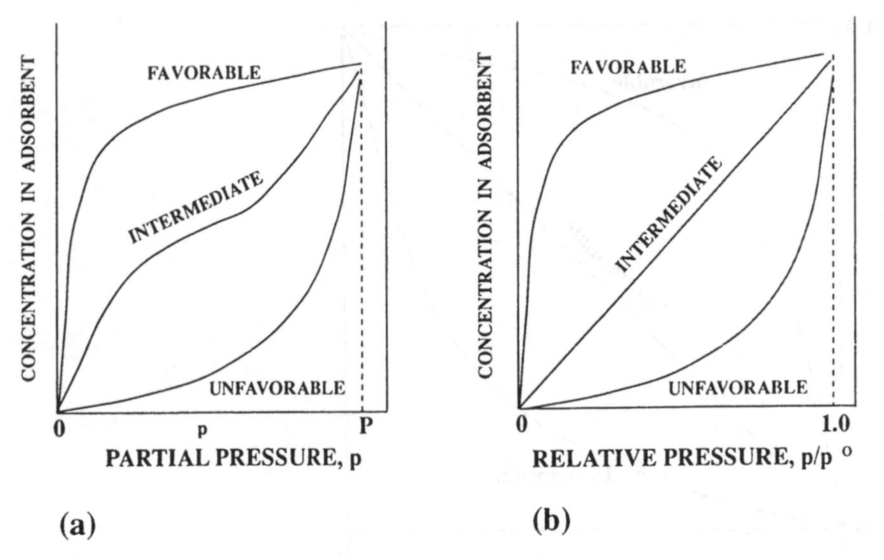

Figure 14.3 Adsorption isotherms for (a) gases and (b) vapors on activated carbon.

to let the molecules reach the internal surface but small enough to provide adequate surface area for adsorption. Smaller pores give higher surface areas for a given pore volume. From the above discussion, it is clear that coconut- or coal-based carbons are better for adsorbing gases, and lignite-based carbon is better for treating liquid streams.

Adsorption Isotherms

Adsorption is a dynamic competition among the molecules in the feed for a position on the surface of the adsorbent. The more strongly adsorbed molecules will displace the more weakly adsorbed molecules from the surface of the solid until an equilibrium is established with the composition of the feed stream. The relationship between the amount of material adsorbed and its concentration in the feed in equilibrium with the adsorbent is called an adsorption isotherm. Isotherms are illustrated in Figure 14.3 for a single component in a gaseous feed.

With gaseous feeds the concentration in the feed is represented by either the partial pressure of the adsorbable component in the feed or the ratio of the partial pressure of the adsorbed component to its saturation pressure at the same temperature. The first type of isotherm shown in Figure 14.3a is used for adsorption of fixed gases and the second type shown in Figure 14.3b is used with condensible vapors such as water, gasoline, and solvents, i.e., volatile organic compounds or VOCs.

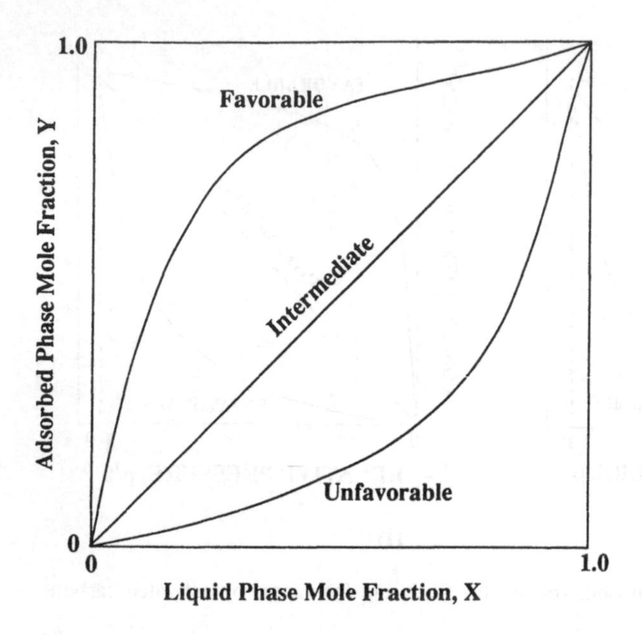

Figure 14.4 Adsorption isotherms for liquid phase adsorption.

For liquid feeds the measure of concentration can be any convenient unit such as weight percent, parts per million, mole percent, or mole fraction. However, it must be remembered that with a liquid feed there is always a competition between the solute and the carrier liquid for a position on the surface of the adsorbent. The equilibrium between the adsorbed material and the external liquid can be described by isotherms of the type shown in Figure 14.4, which plots the composition of the adsorbed phase, Y, against the composition of the liquid phase, X, in equilibrium with it.

The shape of the isotherms may vary from concave downward through linear to concave upward. The concave downward isotherms are very desirable because they give higher solid loadings at low effluent concentrations. This gives better adsorbent capacity while at the same time giving good removal of contaminant from the feed. Concave upward isotherms do not give good contaminant removal at high solids loadings. In other words, the equilibrium liquid phase must contain a lot of adsorbable material to load up the adsorbent. Therefore, if suitable adsorbents having concave downward isotherms are available, they are to be preferred. Valenzuela and Myers (1989) give the adsorption isotherms for a large number of gases, gas mixtures, and liquid mixtures in a variety of adsorbents.

Selectivity and Competition for Adsorption Sites

There is always competition among the various components of the feed to occupy the surface of the adsorbent. Even when there is only one component present in the feed other than the carrier solvent, there is competition for available surface between that component and the carrier solvent. With two or more components present in the carrier, they all compete with each other and with the carrier for space on the surface. Under these conditions the equilibrium between the phases is plotted on a carrier-free basis, with the adsorbed phase as the Y axis and the solvent-free carrier phase composition as the X axis. This method of handling equilibrium relationships is discussed in Chapter 2. To get reliable selectivity data for a given adsorbent with a specific feedstock, it is necessary to do laboratory testing of the adsorbent with the actual feedstock to be used. In this way unexpected effects of unknown trace impurities can be anticipated for the full-scale application.

Heat of Adsorption

When a molecule is adsorbed on the surface of a solid, the reaction forming the adsorption bond gives off heat. The magnitude of this heat release increases with the strength of the bond formed. Thus, the heat release is smallest for weak physical adsorption and rises to a maximum for strong chemisorption. This heat is usually readily removed in liquid phase adsorptions where the heat-carrying capacity of the liquid is large. However, for vapor phase adsorptions the heat-carrying capacity of the gas phase is much lower than that of a liquid and supplementary heat removal may be needed. The heat must be removed from the bed or the bed temperature can rise to a point where either further adsorption will not take place or, worse yet, a fire may be started in the adsorber bed.

Bed Performance

When a fluid feed is passed through a freshly regenerated fixed bed adsorber, a concentration profile is rapidly established over a portion of the bed near the fluid inlet. The upstream end of the concentration profile has the same composition as the feed, while the downstream end has a concentration representative of equilibrium with fresh adsorbent. This is illustrated in Figure 14.5 by the bold concentration curve. The width of the band of adsorbent in which the concentration profile occurs is called the active part of the bed and depends inversely on the number of theoretical adsorption stages per unit length of the adsorber bed. For a large number of stages per foot, the gradient is very sharp; while for small numbers of stages per foot, the profile can extend over a long distance in the bed. Sharp bands require rapid equilibration between the feed and the adsorbent and result from readily available surfaces and many theoretical stages. This is usually provided by small particle size. These concentration gradients are also illustrated in Figure 14.5.

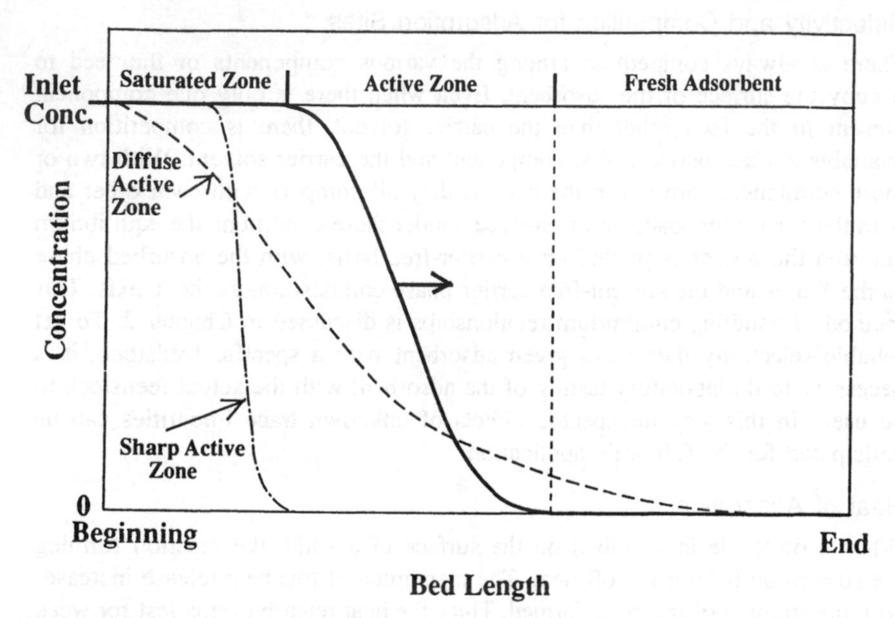

Figure 14.5 Concentration profiles in adsorbent beds.

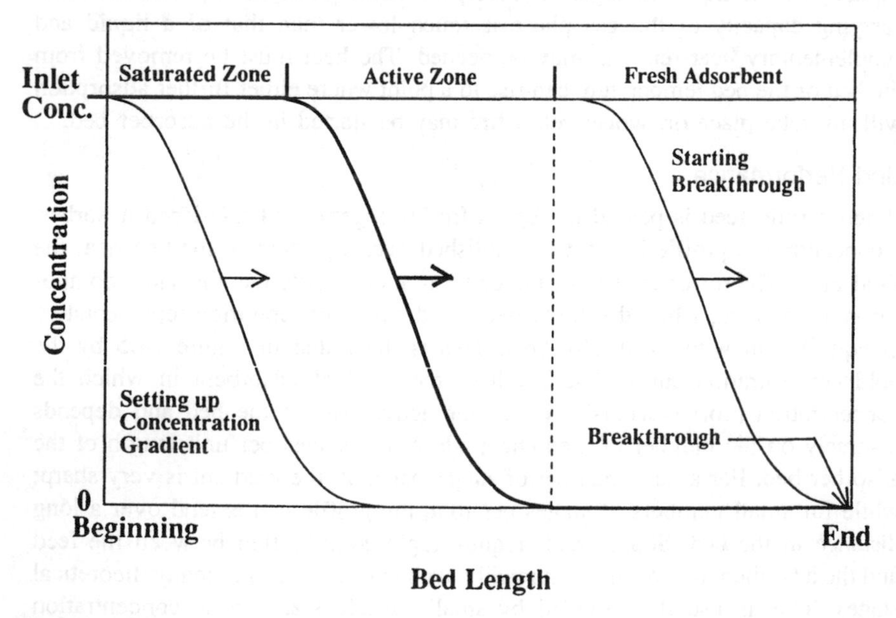

Figure 14.6 Movement of concentration profiles through an adsorbent bed.

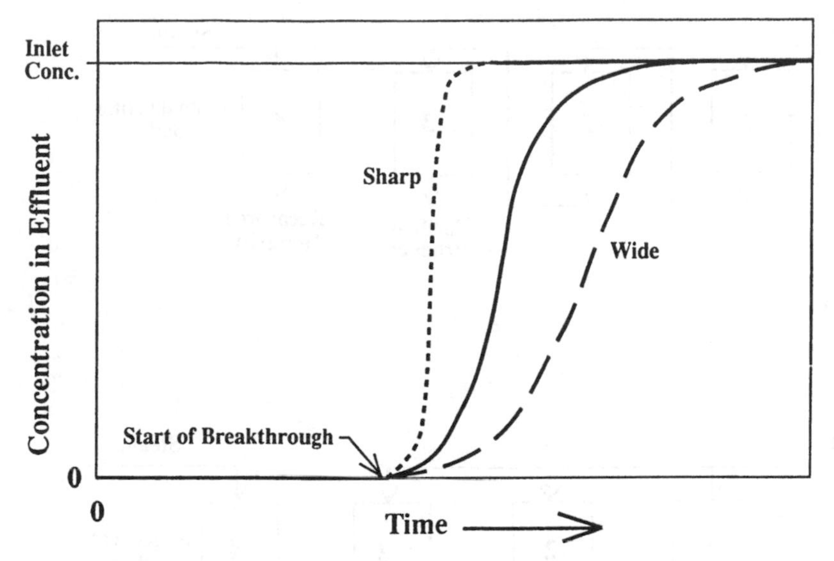

Figure 14.7 Breakthrough curves in adsorption.

As the bed of adsorbent gradually uses up its capacity to hold adsorbed material, the concentration gradient moves through the bed until it arrives at the exit end of the bed. This is shown in Figure 14.6.

As the concentration profile moves out of the bed, the concentration of feed impurity in the exit stream begins to increase with time, giving what is called a *breakthrough curve*. These breakthrough curves are illustrated in Figure 14.7.

Like the concentration profiles, the breakthrough curve is very sharp for large numbers of stages per foot and much wider for small numbers of stages per foot. To get good adsorbent capacity as well as good purity of the effluent stream, it is desirable to have a sharp breakthrough curve.

Process Configurations

Fixed Beds: There are many different process configurations used for adsorption separations. The simplest is flow of a feed stream through a single fixed bed of adsorbent. The feed is usually passed downflow through the adsorber to prevent the adsorbent from lifting from the bed at high fluid velocities and being entrained out of the vessel with the exit stream. With a single fixed bed in batch operation, flow is started to a bed of fresh or newly regenerated adsorbent and continued until breakthrough occurs in the effluent stream from the adsorber bed. At this point the flow of feed is stopped and the adsorbent is regenerated, usually in situ. If interrupting the flow to achieve regeneration is undesirable, more than one adsorbent bed can be provided so that as one bed

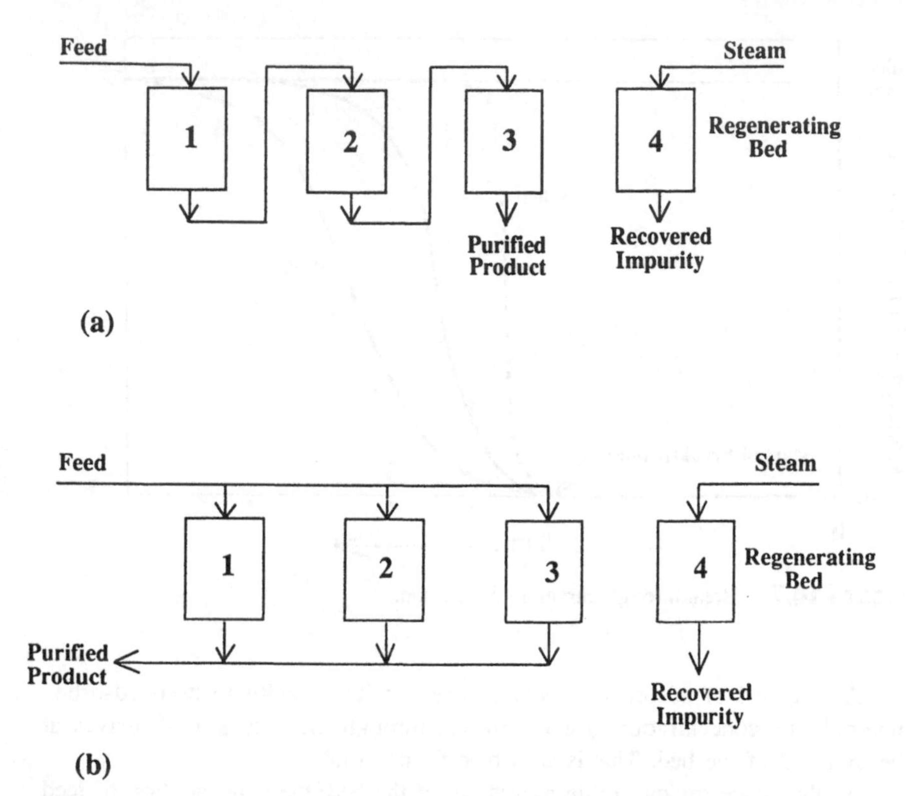

Figure 14.8 (a) Series and (b) parallel bed arrangements.

is taken offline for regeneration, another regenerated adsorbent bed can be put on-stream. The number of beds needed depends on how long it takes for breakthrough in a bed versus how long it takes to regenerate a bed. Multiple beds can be connected in parallel or in series with switching of feed to a regenerated bed each time breakthrough occurs. If regeneration takes a long time relative to the adsorption part of the cycle, several beds may be on regeneration for each one in the adsorption part of the cycle. However, if regeneration takes only a short time, most of the adsorber beds can be operated on-stream in parallel while one bed is being regenerated. These configurations are shown in Figure 14.8. Three beds are shown on-stream, while one is being regenerated. The piping is normally done so that each bed can be regenerated separately while the other beds are on-stream.

The series arrangement of adsorbent beds gives higher mass loadings of the adsorbent for a given purity of effluent. However, the parallel arrangement gives much larger flow rates for a given amount of adsorbent. A design com-

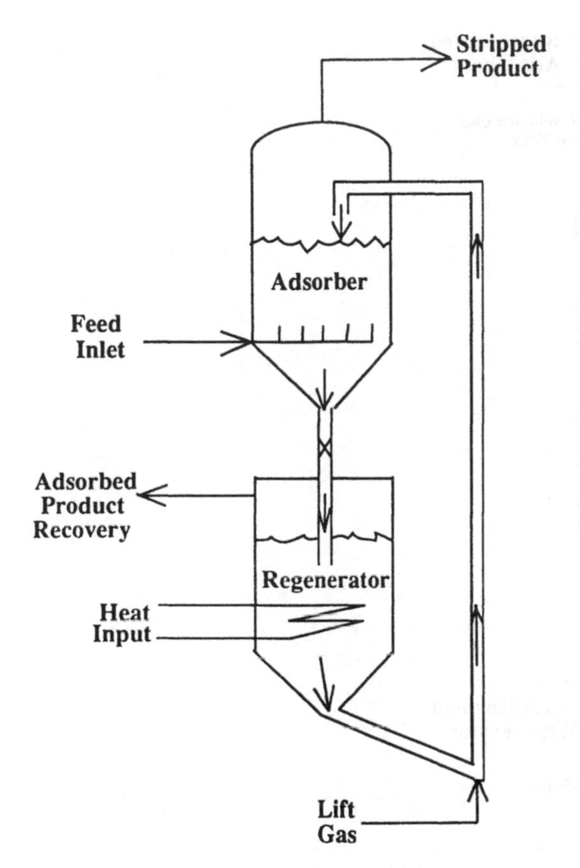

Figure 14.9 Fluid bed adsorber-regenerator.

promise is frequently made by using a combination of series and parallel configurations in which sets of parallel beds are connected in series to get the efficiency of series flow along with the capacity of parallel flow.

Other Adsorber Designs: A variety of designs have been developed over the years in an attempt to get true countercurrent flow in the contacting of fluid feeds with solid materials. These usually hinge on ingenious ways to move solids through the solid-fluid contacting apparatus. For example, the fluidized bed unit shown in Figure 14.9 takes advantage of the flow properties of fluidized solids to build an adsorber bed and a regenerator zone into a single piece of apparatus able to accept continuous gaseous feed. The solids are pumped to the top of the vessel by an air or inert gas lift and fall by gravity through the

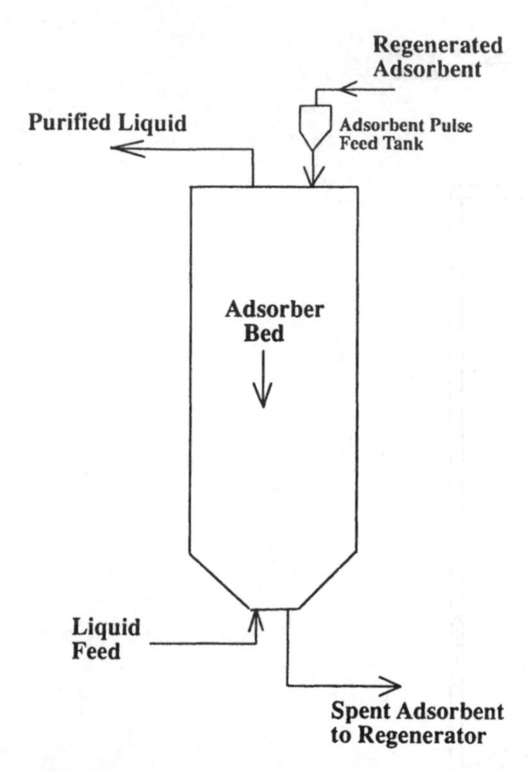

Figure 14.10 Pulsed bed adsorber.

adsorber bed and thence into the lower part of the apparatus where the solids are regenerated using heat.

A second design for moving solids through an adsorber bed is shown in Figure 14.10 and finds its applications principally with liquid feeds. This design also makes use of gravity flow to pass solids down through the adsorption bed to where the bottom portion of the bed is periodically removed for regeneration. Fresh or regenerated adsorbent is added at the top of the column to maintain the bed depth and the amount of adsorbent inventory.

In addition to the countercurrent solids flow designs, a crossflow design has also been tested at the pilot plant level for removing volatile organics from air. Larsen and Pilat (1991) have shown an observed collection efficiency of 90% for removing ethanol vapors from air using a crossflow moving bed panel filter of activated carbon. A schematic of their apparatus is given in Figure 14.11. This device has a thermal regenerator for the adsorbent built into the bottom of the apparatus and also uses an air lift to recycle regenerated adsorbent

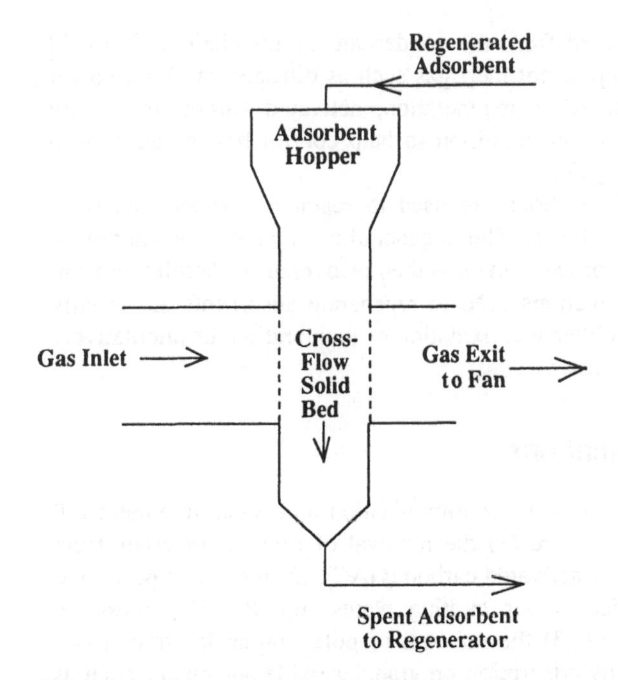

Figure 14.11 Schematic of crossflow VOC adsorber.

to the top of the moving adsorbent bed. The flow of solids is due to gravity and the rate is set by the speed of an auger removing them from the regenerator section and feeding them into the air lift.

Regeneration Techniques

Spent adsorbents are usually regenerated by either steam, hot gas, or vacuum stripping if the adsorbed material is sufficiently volatile. However, for less volatile adsorbates, they can be removed by (1) desorption with solvents which are effective in competing for the adsorption sites, thus removing the adsorbate from the surface and dissolving it, (2) combustion, pyrolysis, or other chemical reactions of the adsorbed material which remove it by conversion, and (3) biological conversion of the adsorbed material. The second and third techniques do not recover the adsorbates as such, so if their value is sufficient a regeneration technique can be used which will recover them intact.

Generally, steam, hot gas, or vacuum regeneration are used to regenerate adsorbents for vapor phase feeds. The adsorbate is recovered by condensing it out of the regeneration gas, and in the case of steam the entire regeneration effluent can be condensed. Phase separation of the water and organic layers permits recovery of the organic layer by decanting. If the organic compounds are soluble in water they can be recovered from the condensate by distillation. To avoid the additional distillation step, a hot inert gas such as nitrogen can be used for stripping instead of steam. After regeneration, activated carbon is usually brought back on-stream in a wet condition to help control heat of adsorption during the next adsorption cycle.

Solvents and chemical reactions are used to regenerate adsorbents from operation with liquid phase feeds. The regeneration solvents are chosen to displace and dissolve the adsorbate, which is then recovered by distillation from the solvent. The chemical reactions used to regenerate adsorbents are usually oxidation reactions such as either wet oxidation or incineration, or alternatively they can be pyrolysis reactions.

14.2.2 Specific Applications

The applications of adsorption to waste minimization and waste treatment fall into three major classes. These are (1) the removal of organic materials from gases or liquids using granular activated carbon (GAC), (2) the use of powdered activated carbon in biological water treating plants (the PACT process) to improve organic removal, and (3) the removal of polar impurities from gases or non-polar liquid streams by adsorption on granular oxide adsorbents such as activated aluminas, clays, silica gel, and molecular sieves. These groups are discussed below.

GAC Processes

Granular activated carbon is the form of microporous carbon normally used in fixed bed or moving bed adsorptions applied to liquid or gaseous feeds. The granules are usually in the size range from 8 to 40 mesh and are made either by pelleting or extruding finer carbons or by grinding and sieving larger particle size carbons.

Organics from Air: This application is a solution to the air pollution problem caused by the drying of solvent-based inks and paints by evaporation of the carrier solvent. The ventilation air from the application stalls and the drying chamber is pulled by exhaust fans through beds of activated carbon to remove the solvent vapors from the air. Typical solvents recovered in this way are hydrocarbon solvents and paint thinners for installations handling up to 500,000 ft^3/min.

Organics from Wastewater: Probably the largest single use of adsorption is the removal or recovery of organic materials from wastewaters using activated carbon. The activated carbon has been used to pretreat feeds to biological oxidation sewage plants, as an additive during biological oxidation, and as a posttreatment for final polishing of the effluent from biological oxidation. The latter use first went commercial in sewage treatment plants and was known as *advanced* or *tertiary treatment.*

PACT Process

The PACT process, described by Berrigan (1991), aerates wastewater in the presence of powdered activated carbon and biomass so that adsorption on the carbon of some impurities and biological oxidation of other impurities take place simultaneously. Each process enhances the effectiveness of the other. The effluent from the aeration is pumped to a sedimentation tank where the bio sludge and the powdered carbon settle together, clarifying the wastewater. If necessary, more than one stage of aeration and sedimentation can be used to get a crystal clear effluent. The carbon-containing sludge can be incinerated with heat recovery in multiple hearth furnaces.

One of the major benefits of this system is that it can be added to a wastewater treating plant that already uses biological oxidation without addition of new vessels. Essentially all that is needed is a way to add the powdered carbon to the inflowing stream at the right place. The PACT process can be used as a pretreatment for industrial wastewaters, for treatment of landfill or other leachates, or as an improvement to biological oxidation–based sewage treatment plants.

The main application of the PACT process is in combination with biological oxidation processes for wastewater treating. However, the use of powdered activated carbon as an adsorbent for organic contaminants from wastewaters in the absence of biomass is also an effective application of adsorption in waste minimization. The carbon is recovered from the treated effluent by gravity separation, filtration, or centrifuging, and can be either regenerated or incinerated.

Oxide Adsorbent Processes

The oxide adsorbents most frequently used are activated natural or synthetic clays and aluminas, silica gel, and synthetic molecular sieve zeolites. Like the activated carbons they have microporous structures which give surface areas of from about 100 to 800 m^2/g and pore volumes from 0.2 to 0.4 cm^3/g. In contrast to the activated carbons, which are selective for non-polar molecules, these materials are selective for more polar molecules, such as water; sulfur compounds; nitrogen compounds; and low molecular weight alcohols, ketones, esters, and organic acids. In addition, the zeolites are also selective by size and shape. This makes zeolites effective for gas phase separations based on either (1) molecular sieving by molecular size or (2) on different diffusion rates through the holes in the zeolites for different components in the feed.

Like activated carbons, the oxide adsorbents are used with either liquid or vapor feed streams and the particle sizes and pore size distributions are optimized for the particular application being considered.

Drying: A major application of oxide adsorbents is the drying of either liquid or gaseous feeds. The adsorbents used are molecular sieves, activated aluminas, and silica gel, which all have a strong affinity for water and selectively remove it from either vapor streams or non-polar liquid streams. Drying applications listed by Keller, Anderson, and Von (1987) include removal of water from air, argon, helium, hydrogen, methane, chlorine, hydrogen chloride, sulfur dioxide, and gaseous fluorocarbons. Removal of water from gases allows them to be refrigerated to low temperatures without frost formation and consequent plugging of equipment. This is important where air-operated controllers are used outdoors or in cold boxes. Drying potentially corrosive gases such as hydrogen chloride, chlorine, sulfur dioxide, and halogenated hydrocarbons and fluorocarbons prevents them from reacting with the water to form their corrosive acid components. Finally, the presence of traces of water can poison catalysts or alter the effectiveness of catalytic reactions.

Recycling Industrial Oils and Solvents: Many industrial oils become degraded in use by the accumulation of water and/or highly polar reaction products from heating or oxidation. This make them unsuitable for their original use. Examples of such oils are insulating oils, heat-treating quench oils, and cooling oils for metalworking. These oils can usually have their impurities removed by passing them through a column of silica gel, activated alumina, activated bauxite, or molecular sieves. The purified oil can then be recycled to either the original use or a less critical use. When the adsorbent is spent, it can be regenerated for a new adsorption cycle, usually by thermal stripping or oxidation.

The same adsorption techniques used to clean oils for recycling can be used with other relatively non-polar liquids such as halogenated hydrocarbons and liquid fluorocarbons. The main requirement is that the impurities be more highly polar than the base fluid.

Equipment

The equipment used for adsorption is basically that required to contact a solid with a gas or liquid and then separate the solid from the unadsorbed fluid. When its adsorption capacity has been exhausted, the solid is regenerated to reactivate it for a new adsorption cycle. Most adsorption plants use multiple downflow fixed beds connected either in series or in parallel. They are equipped with the necessary valves, piping, and controllers to permit operation with one or more of the beds in the regeneration mode while the rest are on-stream. These accessories consume an appreciable portion of the capital investment.

Special equipment such as the fluid bed and pulsed bed units must be custom designed and fabricated for the specific application being considered. Table 14.2 lists some vendors of adsorbents and adsorption equipment.

Table 14.2 Vendors of Adsorption Equipment

Adsorption Systems, Inc. P.O. Box 387 T Millburn, NJ 07041 (201) 762-6304	Airco Generon Systems 12941 I-45 N. Houston, TX 77060 (713) 873-0123
Air Products & Chemicals 7201 Hamilton Blvd. Allentown, PA 18195 (215) 481-4911	Amcec Corporation 2625 Butterfield Road Oak Brook, IL 60521 (708) 954-1515
American Norit Company, Inc. 420 Agmac Ave. Jacksonville, FL 32205 (800) 641-9245	Barnebey & Sutcliffe Corp. P.O. Box 1516 Columbus, OH 43216 (614) 258-9501
Calgon Carbon Corporation P.O. Box 717 Pittsburgh, PA 15230 (412) 787-6700	Carbtrol Corporation 39 Riverside Avenue Westport, CT 06880 (800) 242-1150
Ensol, Inc. 4322 Woodlake Drive Bakersfield, CA 93309 (805) 322-3872	Filter Flow Technology 3027 Marina Bay Dr., Suite 110 League City, TX 77573 (713) 334-6080
Hoyt Corporation 251 Forge Road Westport, MA 02790 (800) 343-9411	Kreha Corp. of America 420 Lexington Ave. New York, NY 10170 (212) 867-7040
Met-Pro Corporation Systems Division 160 Cassell Rd., P.O. Box 144 Harleysville, PA 19438 (215) 723-6751	Sorbent Control Technologies 200 N. Spring St. Elgin, IL 60120 (708) 695-2900
Tigg Corporation P.O. Box 11661 Pittsburgh, PA 15228 (412) 563-4300	Union Carbide Corporation Linde Division 39 Old Ridgebury Rd. Danbury, CT 06817 (203) 794-2000
Westates Carbon, Inc. 2130 Leo Avenue Los Angeles, CA 90040 (213) 722-7500	Zimpro Passavant Environmental Systems, Inc. 301 West Military Rd. Rothschild, WI 54474 (715) 359-7211

14.2.3 Common Problems and Troubleshooting

The most common problems encountered with adsorption separations are listed here and discussed in this section.

- Channeling of fluid feed
- Plugging of adsorbent bed
- Heat release
- Regeneration of adsorbent

Channeling of Fluid Feed

Channeling of a liquid or gaseous feed through a fixed bed occurs when the feed finds one or more paths of lower resistance through the adsorbent bed. It flows selectively through these channels, thus bypassing the bulk of the adsorbent. This results in diluting the product with untreated feed, and thus limits both the product purity that can be achieved and the capacity of the adsorbent bed.

Channeling is rarely encountered with gaseous feeds and when it occurs it is due to non-uniformity of the packing of the particles in the fixed bed. This non-uniformity can result from too wide a range of particle sizes, i.e., too many fines, in the adsorbent charge. It can also result from formation of fines due to particle motion during operation at high upflow gas velocities, from dirt or particulates in the feed stream, or from bridging of particles during filling of the bed with adsorbent. The remedies are (1) to clean the feed stream of particulates, (2) to back-flush the adsorber bed to blow out fines, (3) to dump the bed to screen out fines, and (4) to use a narrower size range of larger particles.

Channeling is more frequently encountered with liquid feeds than with gaseous feeds. It is usually caused by poor feed distribution to the bed, air pockets in the bed, or partial plugging with fines. Fines can be removed from the feed by filtration or from the bed by back-flushing, and their production can be minimized during operation by downflow of the feed through the bed. Liquid feed distribution is enhanced for downflow operation by using grids or rings perforated with multiple inlet nozzles to spray the liquid feed on the top of the adsorbent bed. Alternatively, the adsorbent bed could be run flooded by holding the liquid level above the top of the adsorbent bed. Air pockets can be avoided by initially filling the bed with liquid from the bottom, thereby displacing the air with liquid from the bottom up and avoiding formation of air pockets.

Plugging of Adsorbent Bed

As mentioned, plugging of the adsorbent bed can occur as a result of (1) dirt or particulate fines in the feed, (2) fines in the adsorbent charge, or (3) fines generated by attrition of the adsorbent during operation. This is particularly a potential problem with downflow liquid feeds, where the fines cannot be carried out of the bed effectively with the liquid effluent. To remedy bed plugging, the feed can be prefiltered or a back-flushing capability can be built into the adsorption apparatus. This upflow back-flushing is done at high liquid velocities and loosens up the bed to permit flushing out of the bed the small particulate matter that is plugging the bed. When the bed has been stripped of the fine particulates, it is returned to downflow operation on the feed stream. Other remedies are removal of the adsorbent for screening to remove the fines or replacement of the adsorbent with more attrition-resistant particles of larger size.

Heat Release

Heat release due to either heat of adsorption or the catalytic oxidation of volatile organic chemicals (VOCs) and the resulting warming of the adsorbent bed is usually a problem only with vapor phase feeds. This is because liquid feeds usually have sufficient heat-carrying capacity to remove the heat as it is generated and carry it out of the bed. In contrast, vapor feeds do not have much heat-carrying capacity and consequently the bed heats up, thereby reducing the adsorption capacity of the bed and in the extreme presenting a potential fire hazard. Microporous solid adsorbents are normally good thermal insulators and thus maximize the heating in the bed by their prevention of efficient heat loss to the surroundings. The potential for a fire is greatest when recovering VOCs from air with an activated carbon bed. This is because the bed itself is combustible and the air in the feed supports the combustion once ignition occurs. With liquid feeds there is no air to support combustion, and with oxide adsorbents there is insufficient combustible material in the bed to generate a very serious fire.

The best way to prevent excessive temperature rise in the bed during gas phase adsorption is to limit the concentration of adsorbable materials in the feedstock to less than about 1%. This can be done by either condensing out excess volatile organics from the feed or by diluting the feed with recycled product gas to lower VOC concentration. With the lower concentration of adsorbable material in the feed, the heat can be carried out with the product gas with only an acceptable small rise in temperature through the bed.

Heat release is at its maximum with fresh or newly activated or regenerated adsorbents. This is because adsorbents in this highly active condition react most rapidly with the feed contaminants using the highest energy adsorption and catalytic sites. Zanitch (1979) has listed the following procedures applicable to vapor phase adsorption in carbon beds to minimize the likelihood of a bed fire. These procedures also provide early detection and quenching of any fire encountered.

- A virgin bed should be steamed before the first adsorption cycle. The residual condensate will remove heat by vaporization of the water.
- The bed should not be left dormant except in the regenerated state, i.e., with no combustible organics on the adsorbent.

- Design parameters should avoid high inlet concentrations and low flow rates.
- Instruments should monitor temperature rise across the bed and outlet CO/CO_2 content and signal an alarm at the first sign of decomposition of organics in the bed.
- A water deluge safety system should be provided to cool the bed by spraying water on it if an excursion from normal operation occurs.

Regeneration of Adsorbent

Activated carbon adsorbents *for vapor phase operation* are usually regenerated by steaming, vacuum, hot gas stripping, or a combination of vacuum and hot gas stripping. Steam is usually preferred because it heats the bed rapidly and can be readily separated from many recovered organics by condensation and decanting. The excessive activity of freshly regenerated adsorbent can be moderated by leaving a small amount of condensate in the pores after cooling. Over many cycles of adsorption and regeneration, the carbon may lose capacity due to buildup of impurities which are not removed on regeneration. At this point the carbon must be either discarded or reactivated by heating in a rotary or multihearth furnace in the presence of steam to pyrolize or volatilize strongly adsorbed material. The char residue is then activated at 1200 to 1900°F.

Activated carbon adsorbents *for liquid phase operation* are regenerated by thermal, biological, chemical, hot gas, or solvent techniques. The materials to be recovered from the adsorbent are usually of higher molecular weight than those on vapor feed adsorbents. Therefore more severe regeneration techniques are needed and the recovered product must be sufficiently valuable to justify recovery rather than simply discarding the spent adsorbent to fuel use.

Oxide adsorbents are stable to air at much higher temperatures than activated carbons and thus can frequently be reactivated by burning off carbonaceous residues left after steam stripping. The temperature of the burning must be controlled to prevent loss of microporous surface area and thus adsorption activity. This is usually done by limiting oxygen supply during burning.

Economics

The economics of adsorption processes are very specific to the particular application being considered and utility costs at the particular site where the equipment is to be installed. However, some generalizations can be made regarding costs for an adsorption installation. First, the size of the installation depends on feed flow rate, adsorbent capacity, and adsorbate concentration in the feed. The flow rate determines the diameter of the vessels, the adsorbent capacity determines how much adsorbent is needed, and the feed concentration determines how long the adsorption cycle may go before regeneration is required.

The nature of the feedstock determines the materials of construction required and thus the cost of process piping, vessels, and so on.

14.3 ION EXCHANGE

Ion exchange is an effective process for converting undesirable electrolytes into less offensive materials. It is most commonly used either to recover valuable materials from aqueous process streams or to remove toxic materials from wastewater streams prior to disposal. Solid synthetic ion exchange resins are usually used in fixed beds in much the same manner as adsorbents in adsorption plants. By exchanging positive hydrogen ions for positive metal ions and negative hydroxyl ions for the negative anion of the metal compound, the net effect is the replacement of the metal compound by water. This can effectively demineralize an aqueous process stream by replacing various electrolytes with pure water. It is often used for that purpose in treating wastewaters.

14.3.1 Fundamentals

Ion exchange depends for its effectiveness on the general property of electrolytes (acids, bases, and salts) in solution to dissociate into ionic form, as illustrated in Eq. 1.

$$AB \leftrightarrows A^{+} + B^{-} \tag{1}$$

In Eq. 1 the neutral molecule AB splits into a positive ion A^+ and a negative ion B^-. The positive ions (cations) can be exchanged with other positive ions and the negative ions (anions) can be exchanged with other negative ions using either solid or liquid ion exchange materials. The solid ion exchangers are porous resins or gels in the form of polymer beads which operate in processes much like adsorption. The liquid ion exchangers are reactive solvents and operate in processes much like liquid-liquid extraction. Naturally enough, the materials used to exchange cations are called cation exchangers and those used for anions are anion exchangers. Ionization can occur in certain other solvents as well as in water and, thus, non-aqueous ion exchange is also practiced, although on a much smaller scale.

Resin Structure

Most ion-exchange resins are made of either styrene divinylbenzene, acrylate divinylbenzene, or methacrylate divinylbenzene three-dimensional copolymer networks. The divinylbenzene acts as a crosslinking agent to give structural stability to the copolymer, much in the same manner that vulcanization stabilizes rubber. The active exchange sites are made by grafting appropriate functional

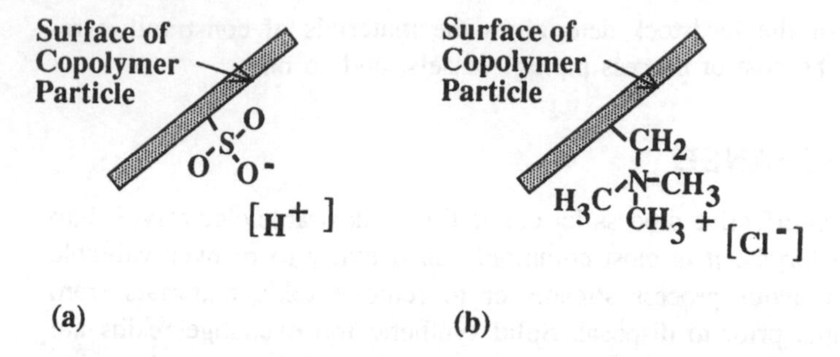

Figure 14.12 Strong acid and base ion exchange sites. (a) Sulfonic acid site, strong acid cation exchanger, (b) quaternary ammonium site, strong base anion exchanger.

groups onto reactive regions of the copolymer. For example, with the styrene copolymer, strong acid sites are made by reacting sulfuric acid with the aromatic rings of the copolymer. This generates the active sulfonic acid site shown in Figure 14.12a. Corresponding strong base sites result from grafting a quaternary ammonium salt onto the aromatic ring as shown in Figure 14.12b. In each case, the exchangeable ion is shown in brackets.

Weak acid sites are obtained by grafting carboxylic acid functionality onto the styrene rings and weak base sites result from grafting tertiary amine functionality onto the aromatic rings. This is illustrated in Figure 14.13.

The ionic functionality provides both the ion exchange sites and the ability to swell the resin with water. This swelling provides mobility of ions inside the resin particle and is controlled by both the degree of crosslinking in the resin

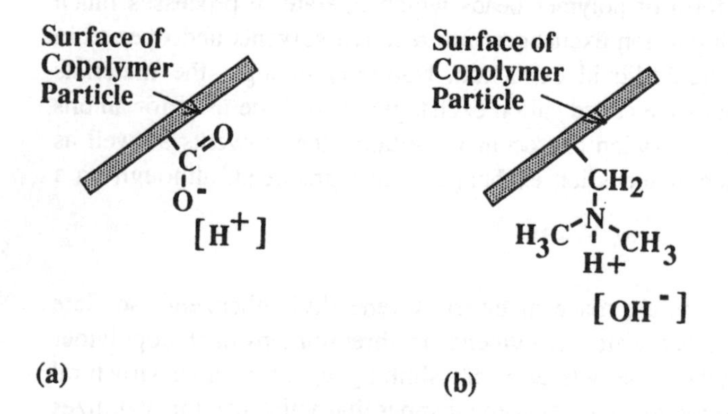

Figure 14.13 Weak acid and base ion exchange sites. (a) Carboxylic acid, weak acid cation exchanger, (b) tertiary amine, weak base anion exchanger.

and the amount of ionic functionality introduced into the resin. Resin swelling decreases with increased crosslinking and increases with increased functionality.

The swollen ionic gel provides the second phase to equilibrate with the surrounding fluid. The ion exchange sites alter the ratio of different types of cations or anions in the gel from that in the surrounding fluid and thus provide a basis for a separation. By using a mixture of cationic and anionic exchange resins, both ions in an electrolyte can be removed from the surrounding fluid.

Exchange Reactions

A typical cation exchange reaction is shown in Figure 14.14a for the exchange of a nickel cation for two protons. This removes the metal from the surrounding fluid and replaces it with two hydrogen ions. Correspondingly, Figure 14.14b shows the anion exchange reaction of a nitrate anion with a hydroxyl anion. The overall effect of passing a stream containing nickel nitrate through both cation and anion exchange resin beds is to remove the nickel nitrate and replace it with water.

When the resin beds are spent because all the exchange sites are full of the ion being removed, they can be regenerated. Sulfuric acid can be used to put protons back on the cation exchange bed and sodium hydroxide restores the hydroxyl ions to the anion exchange bed. The metal ions are recovered in the acid regeneration solution as metal sulfates, and the anions are recovered in the

$$\text{-}SO_3^- H^+,\ \text{-}SO_3^- H^+ \quad \text{plus} \quad Ni^{++} \longrightarrow \text{-}SO_3^-,\ \text{-}SO_3^-\ Ni^{++} \quad \text{plus} \quad 2\,H^+$$

(a)

$$\text{-}N(CH_3)_4^+(OH)^- \quad \text{plus} \quad NO_3^- \longrightarrow \text{-}N(CH_3)_4^+\,(NO_3)^- \quad \text{plus} \quad (OH)^-$$

(b)

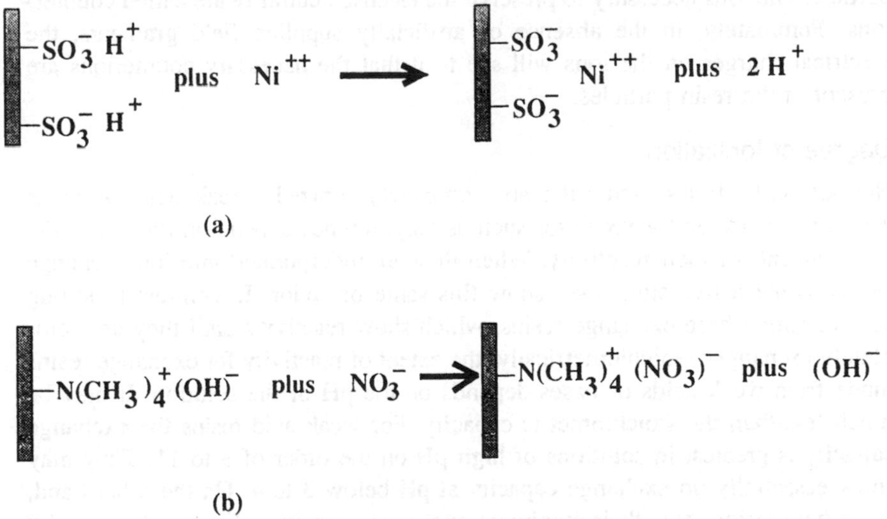

Figure 14.14 Ion exchange demineralization of nickel nitrate. (a) Exchange of nickel, (b) exchange of nitrate.

caustic solution as sodium salts. The capacity of the resin before regeneration increases with the number of active exchange sites in the resin. Typical values for capacity are on the order of 5.0 millimoles per gram (mmol/g) of dry resin. When swollen with water and packed into a resin bed, the bed has a capacity of about 1 to 1.4 eq./L of bed volume. Thus, 1 L of bed volume will remove about 1 equiv. wt of ions, that is, 1 ionic weight divided by its valence. This amounts to cation capacities of 64/2 = 32 g/L of bed for copper, 65.4/2 = 32.7 g for zinc, 108/1 = 108 g for silver, and 58.7/2 = 29.4 g for nickel. Corresponding anion capacities are 62/1 = 62 g/L of bed for nitrate, 35.5/1 = 35.5 g for chloride, and 96/2 = 48 g for sulfate ion. The larger the ionic weight and the lower the valence of the ion, the greater the ion exchange bed capacity. To convert capacity to pounds per cubic foot of resin bed, multiply the value in grams per liter by 0.062. Thus, a bed capacity of 32 g/L of copper is 1.98 pounds of copper per cubic foot of resin.

Electric Neutrality

An important factor often overlooked in considering ion exchange is the concept of electrical neutrality. That is, for every positive ion in the system, there must be a negative ion and vice versa. Furthermore, the positive and negative ions must be distributed together throughout the system to prevent the formation of undesired electrical field gradients. This means that for every cation entering a resin particle, a corresponding anion must also enter the particle to preserve electric neutrality. The same is true for each anion entering an anion exchange particle. The ions necessary to preserve the electric neutrality are called counterions. Fortunately, in the absence of artificially supplied field gradients, the electrical charges on the ions will see to it that the necessary counterions are present in the resin particles.

Degree of Ionization

Not all acids, bases, and salts are completely ionized. Weak acids such as carboxylic acids and weak bases such as alkyl amines depend on the pH of the environment for their reactivity. When they are incorporated into ion exchange resins as the active sites, they show this same behavior. In contrast to strong acid or strong base exchange resins, which show reactivity until they are completely exchanged stoichiometrically, the extent of reactivity for exchange resins made from weak acids or bases depends on the pH of the solution. It may be much less than the stoichiometric capacity. For weak acid resins the exchange capacity is greatest in solutions of high pH on the order of 9 to 11. They may show essentially no exchange capacity at pH below 3 to 4. On the other hand, weak base resins show their maximum exchange capacity at lower pH, generally in the region from 3 to 7. They may have essentially no exchange capacity at pH of 9 and above. The primary benefit of weak ion exchange resins is selectivity among ions when the feed contains a variety of potentially reactive ions. This is in contrast to strong exchange resins where all ions of a particular charge will react with the appropriate exchange resin, giving less ion selectivity. An additional benefit of weak exchange resins is their ability to be regenerated with essentially a stoichiometric amount of regeneration solution. The strong exchange resins may require several times the stoichiometric amount of regeneration solution to be fully regenerated.

14.3.2 Process Configurations

The process configurations used for ion exchange are very similar to those used for adsorption. This is to be expected because both processes involve contacting of a solid with a fluid followed by separation of the solid and fluid phases.

Fixed Bed Units

In its commonest form, the ion exchange process uses a fixed bed of ion exchange resin though which the liquid feed is passed until the composition of the effluent begins to change, i.e., until breakthrough of feed components. When only cations or only anions are to be exchanged, a single bed of exchange resin can be used. In this case, the feed is interrupted when breakthrough occurs and the bed is regenerated. If it is undesirable to interrupt the feed, a pair of ion exchange beds can be used with one bed being regenerated while the other bed is on-stream. This technique is illustrated in Figure 14.15 and provides a continuous stream of treated feed by switching the beds back and forth from on-stream to regeneration.

If both anions and cations are to be exchanged, as in demineralization, two ion exchange beds can be used in series. More preferably, a mixed bed of cation and anion exchange resins is used. Again, if continuous feed is required, two parallel trains of mixed bed systems are required so that one bed can be regenerated while the other is on-stream. One train of the mixed bed system is shown in Figure 14.16.

The sequence of operation of the mixed bed cycle is (1) on-stream, (2) backwash, (3) regeneration of cation resin, (4) regeneration of anion resin, and (5) remixing of bed. During the backwash, the mixed resin bed is fluidized by the upflowing wash water and then stratified into separate layers of cation and anion exchange beds when allowed to settle. This segregates the mixed resins so that they can be regenerated separately. To improve the sharpness of the regeneration step, an inert resin of intermediate density can be used in the mixed bed. This provides a band of inert resin between the exchange resins when they are segregated by the backwash. This minimizes the washing of the resin beds with the wrong regenerant. The addition of water during regeneration keeps the regenerant from entering the wrong bed.

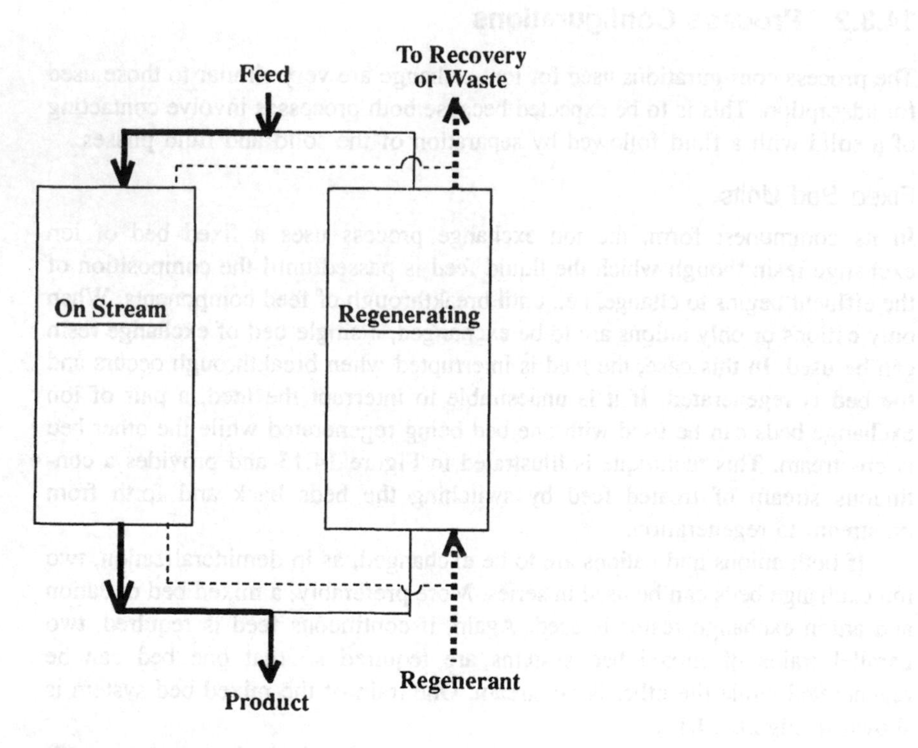

Figure 14.15 Duplex ion exchange system.

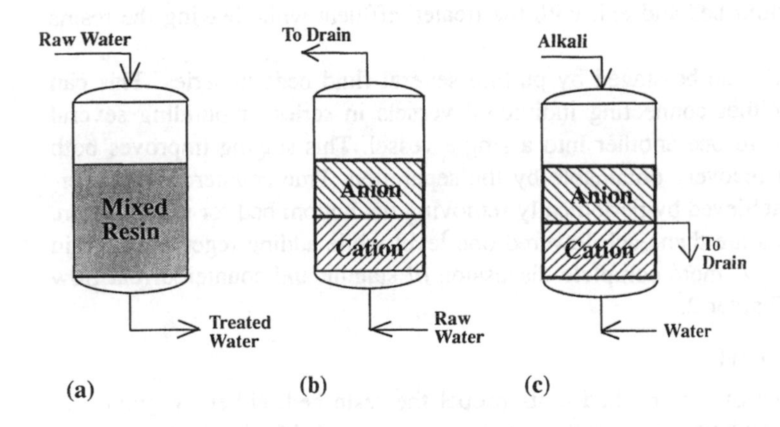

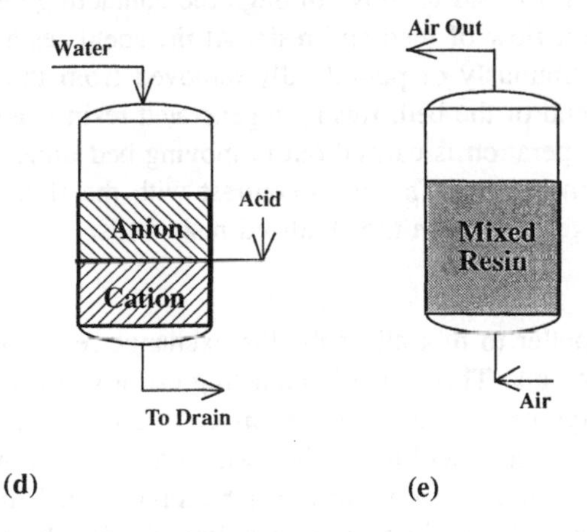

Figure 14.16 Mixed bed ion exchange system. (a) On-stream, (b) backwash, (c) regenerating anion resin, (d) regenerating cation resin, (e) remixing bed.

Fluid Bed Units

While fixed bed units are generally used with clear (i.e., filtered) feedstocks free of particulates, fluid beds can be used to treat feedstocks containing appreciable amounts of particulates. This is often done in the mining industry. However, the particulates in the feed must be of such size and density that they will pass through the fluid bed and exit with the treated effluent while leaving the resins in the vessel.

Fluid beds can be staged by putting several fluid beds in series. This can be done by either connecting individual vessels in series or building several fluid beds above one another into a single vessel. This staging improves both the yield and recovery obtainable by the separation. True countercurrent operation can be achieved by periodically removing the bottom bed for regeneration. The other beds are then each lowered one level while adding regenerated resin to the top bed. A more complete discussion of staging and countercurrent flow is given in Chapter 2.

Moving Bed Units

Recent designs of moving bed units propel the resin bed, either by gravity or mechanically or both, through the ion exchange vessel. Moving bed units take advantage of the ability of the resin bed to move through the contacting vessel to get continuous countercurrent flow of feed and resin. At the spent resin end of the bed, resin is either continuously or periodically removed from the bed for regeneration. At the other end of the bed, freshly regenerated resin is added to the bed. The ion exchange operation is carried out in moving bed units with the beds in the slumped or dense form. This is in contrast with the fluid bed units in which the ion exchange is done in the fluidized resin beds.

Stirred Tanks

Stirred tanks use a mixing impeller to mix all of the ion exchange resin in the tank with the liquid feed in the tank. This type of contacting can give only one batch theoretical contacting stage per vessel. However, by connecting a number of stirred tanks in series with periodic settling of the resin and counterflow of the treated liquids, multistage countercurrent flow can be simulated. This is much the same as simulating it with multiple fluid beds. Periodically the resin in one of the tanks must be taken off the line and regenerated, freshly regenerated resin must be put on-stream, and feed and product lines must be moved to accommodate the new sequence of on-line vessels.

Continuous Deionization

A recent development in ion exchange is the introduction of truly continuous deionization by the Ionpure Technologies Corporation (Wilkins and McConnelee, 1988). In this process the resin bed does not move or need separate

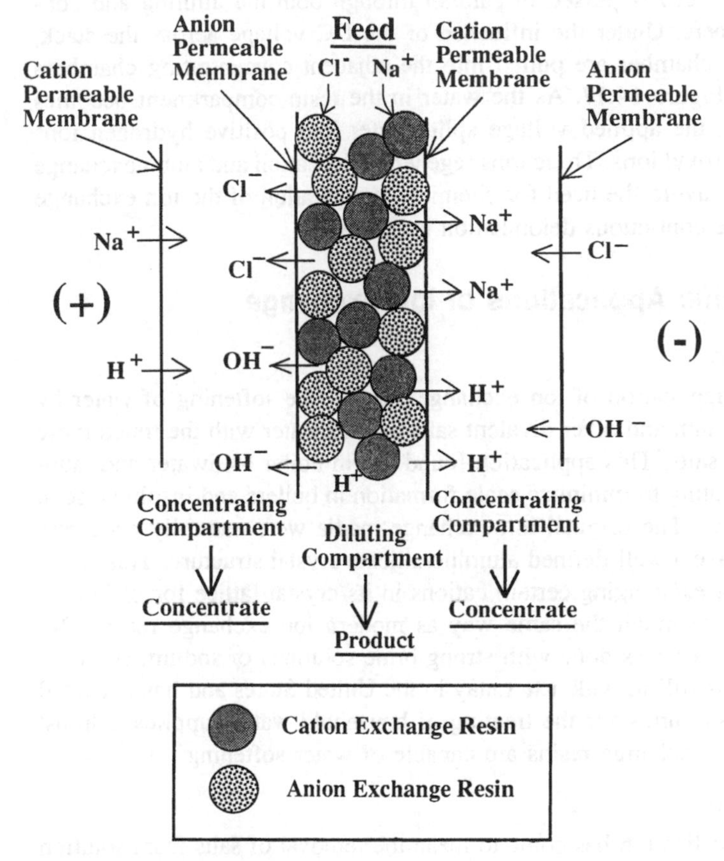

Figure 14.17 Continuous deionization.

regeneration. This is accomplished by using ion exchange technology coupled with electrodialysis technology as illustrated in Figure 14.17. Electrodialysis is discussed in Chapter 8.

This process uses many parallel pocess flow chambers much like a dialysis stack with mixed bed deionization or diluting compartments sandwiched between the ion-concentrating compartments. The deionization compartment has a cation exchange membrane on one side and an anion exchange membrane on the other side. The basic repeating unit is called a cell pair, just as with electrodialysis, and in this case consists of one resin chamber and one ion-concentrating chamber. Stacks consisting of many cell pairs in parallel are used to achieve high-volume throughputs. The stack has a DC voltage applied across it to provide both ion mobility and continuous resin regeneration.

The aqueous feed is passed in parallel through both the diluting and concentrating chambers. Under the influence of the DC voltage across the stack, ions in the resin chamber are pulled into the adjacent concentrating chambers as illustrated in Figure 14.17. As the water in the resin compartment becomes depleted of ions, the applied voltage splits water into positive hydrogen ions and negative hydroxyl ions. These ions regenerate the cation and anion exchange resins in situ and avoid the need for chemical regeneration of the ion exchange resins. Thus, true continuous deionization is achieved.

14.3.3 Specific Applications of Ion Exchange

Water Softening

The first major application of ion exchange was for the softening of water by replacing the calcium and other divalent salts in hard water with the much more tractable sodium salts. This application found use in boiler feedwater and laundry feedwater treating to minimize scale formation in boilers and insoluble soap scums in laundries. The original ion exchange media were naturally occurring zeolites which have a well-defined alumino-silicate crystal structure. This structure is capable of exchanging certain cations in its crystal lattice for cations in aqueous solution in much the same way as modern ion exchange resins. Regeneration of the bed was done with strong brine solutions of sodium chloride. This application is still in wide use today in the United States and has extended into small domestic units for the treating of household water supplies. Almost all modern cation exchange resins are capable of water softening.

Demineralization

The term demineralization has come to mean the removal of salts from solution by exchanging them for water. This is done by exchanging the cation for a proton and the anion for a hydroxyl ion. This can be done using a mixed bed of cation and anion exchange resins or by passing the feed successively through cation and anion exchange beds.

The applications of demineralization range from the preparation of ultrapure water as a feedstock to the treatment of effluents and wastewaters before disposal. An increasing use is for treating of groundwater for the removal of trace toxic materials.

Ultrapure water is used for (1) high pressure boiler feed, (2) electronic manufacturing applications, and (3) preparation of high-purity solutions in the medical, pharmaceutical, and chemical manufacturing industries. At the outlet end of processing, demineralization is used either to remove toxic materials from the waste stream so that it can be safely discarded, or to recover valuable materials for either recycling to the process or sale for their economic value. It is of particular use in recovering radioactive salts from mildly radioactive aqueous streams generated in the nuclear industry.

Metals Recovery

Both precious metals and baser metals are effectively recovered from either aqueous waste streams or aqueous process streams by ion exchange. Advantage has been taken of this fact both in treating of wastes from the metals-forming and photographic industries. Furthermore, ion exchange is used in the processing of metal-containing streams in the mining industry where aqueous reagents are sometimes used to recover metals by hydrometallurgy. Typical metal recovery operations using ion exchange are (1) recovery of silver from photographic solutions, (2) recovery of copper, nickel, chromium, platinum, silver, and gold from plating rinsewaters, (3) recovery or purification of metals from hydrometallurgy solutions, and (4) recovery or purification of uranium, thorium, and rare earths in the nuclear industry. The use of ion exchange in these applications appears to be growing.

Process Tuning

Ion exchange is frequently used as an intermediate step in industrial processes either to improve the overall process or to improve the final product quality. For example, in the sugar industry cation exchange is used to remove calcium and magnesium salts from the sugar syrup by exchanging them with sodium before evaporation. This prevents scaling in the evaporators. In addition, deionization of the sugar solution improves yield. The color of the final sugar product is also improved by using ion exchange resins to decolorize the sugar solution before crystallization.

Another example of this type of application is the adjustment of pH in the preparation of various foods. Either acidic or basic functionality can be added to or removed from the process stream to alter the taste of the product. Thus, the tartness of various products, such as fruit juices and tomato sauces, can be controlled. Ion exchange resins are finding increasing use as catalysts because of their easy removal from the products of the catalytic reaction. For example, many reactions conventionally catalyzed by liquid acids are now catalyzed using cation exchange resins in the acid form. Typical of the acid catalyzed reactions are esterifications, isomerizations, reactions of alcohols with olefins to form ethers as in the manufacture of methyl-*t*-butyl ether (MTBE), hydration of olefins, and dehydration of alcohols.

Purification

The purification of products by ion exchange is very widespread and is common in the manufacture of reagent chemicals, pharmaceuticals, medicines, and biologiical products, to name a few. For example, used glycol and other solvents can have ionic corrosion products removed by ion exchange. Recirculated chilled water systems which contain dichromates as corrosion inhibitors can have the chromium salts removed before disposal of the water to the sewer. Analytical reagents can have ionic impurities removed by ion exchange, thereby increasing their purity. Finally, in the manufacture of medicines and pharmaceuticals, ion exchange is used to recover the active ingredients from fermentation broths or to separate desirable products from undesirable by-products. These are just a few of the many applications of ion exchange for product purification.

14.3.4 Equipment

The types of equipment used in ion exchange are very much like those used in any solid-liquid contacting operation, such as for example, adsorption. That is, fixed beds, fluid beds, moving beds, and stirred tanks are all used. Multiple beds can be connected in various configurations to provide multi-staging and simulate true countercurrent operation, as well as to allow regeneration of part of the system while the rest is still on-stream. Some of the major suppliers of either ion-exchange equipment or ion-exchange services are listed in Table 14.3.

14.3.5 Common Problems and Troubleshooting

Some of the common problems encountered with ion exchange are much like those of adsorption discussed earlier in this chapter. These are channeling and bed plugging. The solutions to these problems are the same as previously discussed for adsorption. Heat release is not as much of a problem because exchange reactions are more thermally neutral than adsorption reactions. However, in addition to the problems in common with adsorption, ion exchange operations can run into other difficulties such as absorption of organics into the resin beads or poor resin stability.

Absorption of organic materials occurs because they can dissolve in the polymer structure of the resin beads. This can be either good or bad depending on how the resin is being used. The resins can show some selectivity for non-ionic materials in the feed, thus removing organics as well as electrolytes from wastewaters. On the other hand, they may be released on regeneration, thus contaminating the recovered product with organics. If the recovered product is the high value product, this contamination is undesirable. The remedy is to use a resin that does not dissolve the organics as readily (i.e., more highly crosslinked or with higher ionic content) or to remove the organics prior to ion exchange.

Poor resin stability can be caused by (1) mechanical attrition, (2) cracking and spalling due to swelling and shrinking of the resin beads as the nature of the attached ion changes due to ion exchange, or (3) chemical reaction with strong chemical agents in the feed stream. The solution to these problems is

Table 14.3 Vendors of Ion Exchange Equipment

Culligan Commercial/Industrial Systems 1 Culligan Parkway Northbrook, IL 60062 (800) 451-3260	Hydro Group, Inc. Environmental Products Div. 97 Chimney Rock Rd. Bridgewater, NJ 08807 (201) 563-1400
Ionics, Inc. P.O. Box 99 W Bridgeville, PA 15017 (412) 343-1040	Ionpure Technologies Corp. 10 Technology Drive Lowell, MA 01851 (508) 934-9349
Kinetico Engineered Systems P.O. Box 127 10845 Kinsman Rd. Newbury, OH 44065 (216) 564-5397	Klenzoid Equipment Co. P.O. Box 44 T Wayne, PA 19087 (800) 345-6365
Lancy International, Inc. 181 Thorn Hill Road Warrendale, PA 15086-7527 (412) 772-0044	Memtek Corporation 28 Cook Street Billerica, MA 01821 (508) 667-2828
Mobile Water Technology 2070 Airways Blvd. P.O. Box 14867 Memphis, TN 38114 (901) 744-1142	Rohm & Haas, Dept. TR Independence Mall West Philadelphia, PA 19105 (800) 338-1205
Trionetics, Inc. 2023 Midway Drive Twinsburg, OH 44087 (216) 425-2846	

clearly to use more resistant resin beads or to eliminate the offending chemicals from the feedstock.

14.3.6 Economics

The economics of ion exchange as a separation process depend strongly on resin costs, plant size, degree of automation required, regenerant costs, and costs of disposing of regenerant wastes. Strong acid resins are the least expensive, with weak acid and both strong and weak base resins being several times more expensive. The working life of the resin must also be considered in calculating the operating cost of the resin.

The size of the plant and the degree of automation needed set the capital costs for a given installation, while operating costs also include labor and utilities. This makes operating costs site-dependent. For an installation driven by product quality improvement, the value added by ion exchange must be greater than the cost of running the ion exchange process. However, if the installation is driven by waste disposal prohibitions, the costs of ion exchange must be less than the costs of alternative legal disposal or waste reduction methods. This latter area of application is becoming increasingly important as costs of waste disposal are increasing.

14.4 REVERSIBLE CHEMICAL REACTIONS

Reversible chemical reactions can be extremely selective for the separation of materials on the basis of their chemical reactivity. An example is the removal of acid gases such as carbon dioxide or sulfur dioxide from flue gases from combustion. When the flue gas is scrubbed with an aqueous alkaline solution, these acid gases react with the solution to form soluble salts which stay dissolved in the liquid phase. The non-reactive components of the flue gas are not dissolved and continue up the stack. The separation factor between reactive and non-reactive components in the feed is theoretically infinite and in practice achieves values approaching 100. That is, one theoretical stage of separation will separate a 50:50 feed mixture into a 99% pure product mixture. There are many types of reversible chemical separations, one of the oldest being the separation of various organic molecules by extraction with sulfuric acid of different strengths.

The material used as the reactive separating agent can be used either as a solid or a liquid. If it is applied as a solid to form a solid product, the process is very similar to adsorption and ion exchange. In this case the reagent is usually supported on a solid carrier to get high surface area. It is also frequently incorporated into a polymer resin in much the same way as the active sites in an ion exchange resin. However, it can also be used as a simple solid reagent where it is insoluble in the feed stream and forms a reaction product which is also insoluble in the treated product stream. Examples of this latter operation are the use in the past of (1) a bed of solid sodium hydroxide pellets to remove odorous mercaptans from gasoline and heating oil and (2) the use of a bed of sponge iron to react with sulfur compounds from town gas made from coal.

When the chemical reagent is applied as a liquid or solution to form a product soluble in the reagent, the separation process is very similar to solvent extraction when treating a liquid, and absorption or scrubbing when treating a gas. Examples of these processes are (1) recovery of acetic acid from dilute aqueous solutions by liquid extraction with solutions of reactive amines dissolved in water-insoluble hydrocarbons and (2) scrubbing of all sorts of flue gases.

14.4.1 Fundamentals

Equilibrium

The reversible reactions used for separation processes can be described by Eq. 2 where one molecule of reactive component in the feed reacts with n molecules of reagent to form one molecule of product.

$$\text{Reactant} + n \times \text{reagent} \leftrightarrows \text{product} \tag{2}$$

This reaction can be described for liquid phase reactions by an equilibrium constant, K_e, as shown in Eq. 3.

$$K_e = \frac{[\text{product}]}{[\text{reactant}]\,[\text{reagent}]^n} \tag{3}$$

The values in brackets are the concentrations, or more precisely the activities, of the various components in the liquid phase. The exponent, n, is the number of molecules of reagent required to react with one molecule of the desired reactant to be removed. The value of K_e at reaction conditions must be large enough to achieve the desired degree of completeness of the reaction to remove the reactant from the feedstock. If the feedstock is a gas rather than a liquid, it is still the concentration of the reactant dissolved in the liquid phase that enters the equation. This concentration is directly related to the composition of the gas phase through a proportionality constant known as Henry's law constant. Thus, for scrubbing operations, the composition of the tail gas is controlled by liquid phase equilibria, and high equilibrium constants lead to good cleanup of the gas phase.

In contrast to liquid phase reagents, solid reagents have a constant activity. Thus, if both the reagent and the reaction product are solids, Eq. 3 reduces to Eq. 4 with all the proportionality constants included in the overall equilibrium constant.

$$[\text{Reactant}] = K_c \tag{4}$$

This equation shows that if the reaction is allowed to come to equilibrium, the composition of the effluent stream will be constant until all the solid reagent is consumed. This is typical of many processes which treat liquids with solids.

On the other hand, if the reactant is in the gas phase and the reagent and reaction product are solids, then Eq. 3 reduces to Eq. 5, again with all the proportionality constants included in the overall equilibrium constant.

$$p_r = K_p \tag{5}$$

p_r is the partial pressure of the reactant in the gas phase and k_p is the pressure equilibrium constant. Eq. 5 shows that a solid reagent treating a gaseous reactant to form a solid product will reduce the partial pressure of the reactant in the effluent gas to a constant value until the solid reagent is used up. The values of the constant compositions determined by Eqs. 4 and 5 are thus a function of temperature.

Types of Practical Reversible Chemical Reactions

In order for a chemical reaction to be adequately reversible to become the basis for a separation process, the strength of the bond formed by the reaction must be within certain limits. It is usually stronger than Van der Waals attraction between molecules and weaker than the covalent bonds that hold organic molecules together. Typical of such bonds are those of chelates, pi complexes, and hydrogen bonded complexes.

Chelates: Chelating reagents are so named because they involve more than one bonding site and reminded the early researchers of the claws of a crab, or chela. They generally react with metals such as copper, nickel, zinc, and mercury to form five-membered rings with the metal bound to two or more reactive sites. The active sites of chelates are usually electron donors such as nitrogen, phosphorus, arsenic, antimony, oxygen, sulfur, selenium, and tellurium. These elements bond with metal ions eager to accept their electrons. A typical chelating structure is that of ethylene diamine which is illustrated in Figure 14.18 with the chelating bonds shown as dotted lines.

Pi Complexes: Pi complexes are similar to chelates in that an electron donor is involved. However, instead of the electrons being provided by the lone pair on, for example, a nitrogen or phosphorus atom, they are now donated by an organic pi bond. Typical organic electron donors from pi bonds are olefins, aromatics, and nitriles. They complex readily and reversibly with copper and silver monovalent ions. Figure 14.19 shows a pi complex of copper with ethylene.

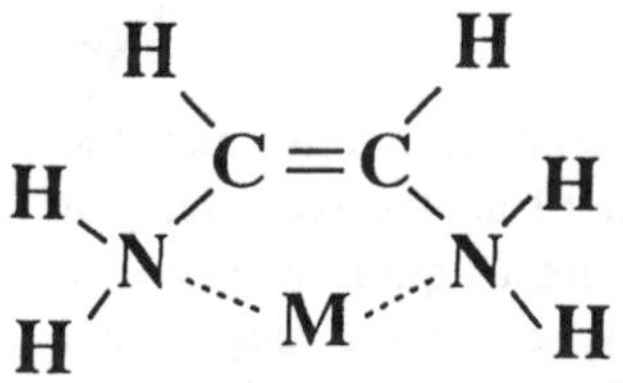

Figure 14.18 Ethylenediamine chelate of a metal. M, metal.

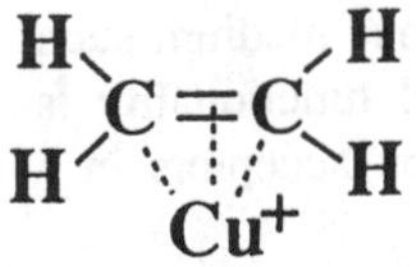

Figure 14.19 Copper pi complex of ethylene. Cu^+, monovalent copper.

Hydrogen Bonding: Hydrogen bonding occurs when a hydrogen atom is attracted to two different electron donors, thus forming a weak bond between the two donor molecules. Water is a great example of hydrogen bonding because the network formed by the hydrogen bonds is the reason water has such a high boiling point. With a molecular weight of 18, water would boil a lot nearer the boiling point of methane if it weren't for the hydrogen bonding. Conceptually, the hydrogen bonding of water is illustrated in Figure 14.20. The hydrogen bonds are shown as the dotted lines.

While water is a major example of hydrogen bonding, many other organic molecules have hydrogen bonds. Again the bonding is between a hydrogen atom and a pair of electron donors. Thus, the same donor atoms found in chelates can have hydrogen bonds when they are in organic molecules. Thus, organic acids, alcohols, amines, and other electron donors can all be involved in hydrogen bonding. To make a solid separating agent that functions through hydrogen bonding, the appropriate functionality must be built into the solid. For example, either a hydrogen bond acceptor or a hydrogen bond donor can be built into a solid polymer matrix. It will then hydrogen bond to its opposite functionality in a non-aqueous system. That is, an electron donor, such as an organic nitrogen compound grafted to a polymer matrix will react through hydrogen bonding

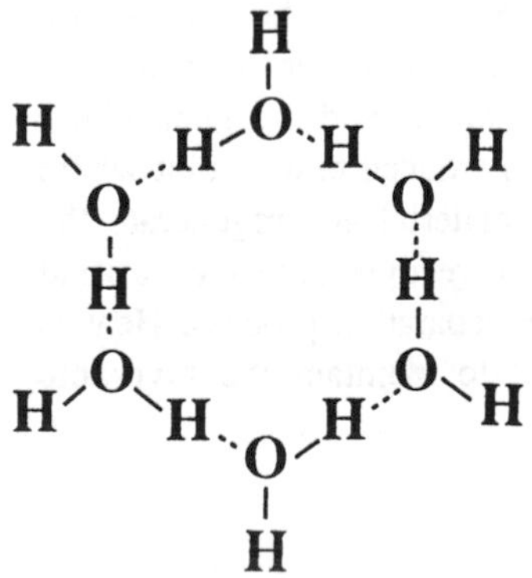

Figure 14.20 Hydrogen bonding in water.

with alcohols, organic acids, water, and so on, in a non-aqueous medium such as hydrocarbons. Similarly, if the acid, alcohol, or hydroxyl functionality is bound to the solid, it will react with amines and other proton acceptors in a non-aqueous medium through hydrogen bonding.

Reaction Rates

Reaction rates with solids vary from very slow to very fast. The slow reactions occur when diffusion in the solid is the limiting factor because diffusion in solids is very slow. However, if the solid is porous, as with most adsorbents, the surface reaction rates can be very fast. Capacity is then limited by the amount of the total solid that is available as surface molecules. For this reason solid reagents are frequently dispersed on a high surface area inert solid to maximize the amount of the reagent that is available.

When the reaction products remain on the surface after reaction, they can form a protective coating which keeps the solid and fluid reagents from coming in contact with each other. When this is so, the reaction stops after the surface is coated. However, when the reaction products are continuously removed from the solid surface by being either volatile or soluble in a liquid feed, the reaction will continue until one of the reagents is used up. Thus the solid reagent cannot be regenerated in the treating vessel, but must be recovered from the fluid stream leaving the reactor vessel and then regenerated and returned to the reactor.

Regeneration

The regeneration step is an important part of the use of reversible chemical reactions as separation processes, just as it is with ion exchange or adsorption. However, regeneration of chemical reagents usually requires more energy because the bonding is usually stronger than adsorption bonding and it doesn't have the energy balance obtained by exchange reactions. It is normally reversed thermally to recover the reagent and the purified reacted product.

Regeneration of a solid reagent is accomplished by raising the temperature of the solid addition compound to a level where the equilibrium constants in Eqs. 4 and 5 are favorable for the decomposition of the adduct. Then either pressure or concentration of the adducted material is reduced to the point where the complex decomposes to release the adducted material and regenerate the solid reactant. These materials are then separated to recover both the purified product and the reactive solid for recycling to the separation process. Heat is provided during the decomposition of the adduct to maintain the favorable equilibrium temperature for regeneration.

14.4.2 Specific Applications

The most frequent applications of reversible reactions of solids involve either a solid reagent dispersed on an inert carrier, or a chemical functionality built into a solid polymer matrix. In most cases, either of the reactants can be immobilized on the solid to recover the other one from a liquid or gaseous feedstock. For example, electron donors can be immobilized to recover electron acceptors, or electron acceptors can be immobilized to recover electron donors. An example of this type is the recovery of metals (electron acceptors) using immobilized organic nitrogen bases and the recovery of organic nitrogen bases (electron donors) using immobilized metals.

Some applications of reversible reactions of solids are:

- Recovery of copper, nickel, cobalt, or iron from aqueous waste streams using a chelating resin such as Dowex A-1, which contains iminodiacetate groups bound to a polystyrene matrix
- Treatment of process gases to remove reactive toxic materials such as (1) carbon monoxide with supported cuprous salts or (2) hydrogen sulfide with iron oxide
- Recovery and purification of olefins and diolefins from hydrocarbon vapor mixtures using porous solid cuprous chloride
- Decontamination of aqueous waste streams containing radioactive metals by using solid chelates
- Recovery of volatile toxic components from vent gases using silver supported on carbon

14.4.3 Equipment

The equipment used for reversible chemical reactions of solids with fluids is much like that for any other process which contacts fluids with solids. Fixed beds, moving beds, and fluid beds are used with either liquid or gaseous feeds. In addition, stirred tanks are also used with liquid feeds. These units are operated in much the same way as described earlier for adsorption and ion exchange. Appropriate reactor vessels can be obtained from custom fabricators and the units can be designed by most process engineering firms.

14.4.4 Common Problems and Troubleshooting

The problems encountered with separation processes depending on reversible reactions of solids are much the same as those of adsorption on solids, that is, channeling, bed plugging, removing reaction heat, and adequacy of regeneration. The first two are mitigated in the same way as for adsorption. However, reaction heat cannot be effectively removed by dilution of the feed because dilution reduces the concentration in the feed and thus the recovery obtainable by the chemical reaction. A better way is to remove the heat by heat exchange. This is relatively simple for stirred tanks, fluid beds, and moving beds, but relatively difficult for fixed beds. Thus, if heat release is large, it may dictate the design of the reaction vessel and the type of contacting process used. One solution for fixed beds is to provide external heat exchange on the process fluid which is withdrawn, cooled, and returned at frequent points along the fixed bed.

An additional concern with reversible reactions in bulk solids is the change in solid structure as the reaction and regeneration cycles occur. This can cause breakdown of the solid structure and excessive attrition and fines formation. The problem can be remedied by preparing the solids in an attrition resistant porous binder which allows access to the reactive solids but prevents attrition and solids breakdown.

14.4.5 Economics

Like adsorption, the economics of separation processes based on reversible chemical reactions depend on the size of the installation, which determines the feed flow rate, the capacity and cost of the solid reagent, the concentration of reactant in the feed, the operating conditions of temperature and pressure needed to provide adequate recovery of the desired product, and the frequency of regeneration needed. The equilibrium constant for the reaction determines the operating conditions needed for both reaction and regeneration of the solid as well as the recovery that can be obtained at a given temperature. In general, if the equilibrium constant is adequate, the process is less expensive for low concentrations of reactant material in the feed.

14.5 SUMMARY

Separation processes based on adsorption, ion exchange, or reversible reactions of solids all have in common the interaction of liquid or gaseous fluids with solids. This interaction can be made very specific by selecting the proper solid. Furthermore, more than one type of interaction can be built into a single solid, making it selective for more than one type of component. However, all the solids used in these processes have limited capacity before they must be regenerated. Thus, they find their most effective applications where the concentration of the reactive component in the feedstock is low. This makes them particularly effective for hazardous waste streams, where low concentrations of toxic components are frequently encountered and must be removed before discarding the waste.

The most frequently used technique for contacting the solids with fluid feed streams is with fixed beds. These have the advantage of providing many theoretical stages of contacting in a single vessel. However, they suffer the disadvantage that they must be removed from the process line for regeneration. This is usually overcome by running two or more beds in parallel so that one bed can be regenerated while the others are on-stream. Moving bed and fluid bed designs, with continuous regeneration of the solids, have also been developed and are particularly useful for large-scale applications.

These processes are already well established for removing volatile organics from waste gas streams as well as inorganic and organic contaminants from liquid aqueous waste streams. It is expected that they will continue to hold an important place in waste treatment.

14.6 REFERENCES

Berrigan, J. K., Copa, W. M., and Randall, T. L. (1991). Biophysical Treatment of Wastewaters from the Refining Industry Using the PACT System, Presented at A.I.Ch.E. Summer National Meeting, August 18–21, 1991.

Heilshorn, E. D. (1991). Removing VOCs from contaminated water, *Chemical Engineering*, February, 1991:121.

Hutchins, R. A. (1980). Liquid phase adsorption: maximizing performance, *Chemical Engineering*, February 25, 1980:101.

Keller, G. E. II, Anderson, R. A., and Von, C. M. (1987). Adsorption, *Handbook of Separation Process Technology* (R. W. Rousseau, ed.), Wiley, New York, p. 649.

Larsen, E. S. and Pilat, M. J. (1991). Design and testing of a moving bed VOC adsorption system, *Environmental Progress*, *10*(1):75.

Parmele, C. S., O'Connell, W. L., and Basdekis, H. S. (1979). Vapor phase adsorption cuts pollution, recovers solvents, *Chemical Engineering*, December 31, 1979:59.

Valenzuela, D. P. and Myers, A. L. (1989). *Adsorption Equilibrium Data Handbook*, Prentice Hall, Englewood Cliffs, NJ.

Wilkins, F. C. and McConnelee, P. A. (1988). Continuous deionization in the preparation of microelectronics-grade water, *Solid State Technology*, August, 1988: 87–92.

Zanitsch, R. (1979). "Solvent Recovery from Low Concentration Emissions," Seminar on Emission Control for Publication, Packaging, and Specialty Gravure Printers, sponsored jointly by Gravure Research Institute and Gravure Technical Association, Louisville, KY, September 26, 1979.

15

Solvent Extraction, Solvent Precipitation, and Leaching

15.1 INTRODUCTION

Solvent extraction and solvent precipitation are commonly used either to recover valuable materials or to remove contaminants from aqueous or non-aqueous waste streams. Leaching is solvent extraction applied to solid feeds. They are usually physical processes whose selectivity is dependent on the relative solubilities of the feed components in a chosen solvent. For example, extraction with a liquid organic solvent such as kerosene, which is essentially insoluble in water, can be used to remove polar organic compounds such as phenols, organic acids, and ketones, etc., from industrial wastewaters. This makes the wastewater more suitable for disposal and at the same time recovers valuable chemicals for reuse or sale. The solvents used for extraction and solvent precipitation can be either conventional liquids below their critical point or supercritical fluids operating at conditions above the critical point of the pure solvent. Most, if not all, of the separations that can be made with supercritical fluid solvents can also be made with liquid solvents. However, in some specific applications there are enough operating benefits to favor supercritical fluids. Examples of these benefits are (1) lower operating temperature for heat-sensitive materials,

(2) ease of solvent recovery by pressure reduction or temperature increase, and (3) faster phase separation due to lower viscosities and higher density differences between phases. The trend is toward increased use of supercritical fluid extraction solvents in recent years. However, their use still represents only a small fraction of the total solvent extraction operations in existence.

The interaction between solvent and solute can be purely physical intermolecular attraction, called London or dispersion forces, or can have some additional chemical forces superimposed on the physical attraction. These chemical forces increase the solvent selectivity for the materials that interact chemically with the solvent. They are the basis for extracting polar organics, metals, and ionic materials from aqueous streams. Examples of such interactions are ionic bonding, hydrogen bonding, and chemical complexing. The extraction processes using ionic interactions are frequently called liquid ion exchange.

In contrast to liquid extraction, which extracts a dissolved material from a liquid feed into a solvent, solvent precipitation works by dissolving an antisolvent into the liquid feed. This results in the formation of a second liquid or solid phase insoluble in the feed. The same fundamentals governing solubility of liquids in each other apply to both solvent extraction and solvent precipitation. However, the solubility properties of solvents and antisolvents are quite different, as will be discussed later in this chapter.

Solvent extraction is usually applied to liquid feeds. It is then called liquid-liquid extraction or supercritical fluid extraction. However, it can also be applied to solids and sludges to remove soluble liquid or solid impurities. This permits both the recovery of valuable materials and the regeneration of solid adsorbents and catalysts. This type of extraction is usually referred to as leaching. In this chapter, solvent extraction will be discussed from the perspectives of both its liquid extraction applications and its leaching applications.

15.2 LIQUID-LIQUID EXTRACTION

Liquid-liquid extraction is the commonest of the solvent extraction processes. It is usually the second choice, after distillation, when considering alternatives for separating large volumes of liquid mixtures. If the feed mixture contains only a small amount of impurity and the impurity has a higher boiling point than the rest of the mixture, liquid-liquid extraction can even become preferable to distillation on an energy consumption basis. This is because distillation would have to vaporize almost all of the feed to recover the small amount of impurity. Extraction could directly remove the minor component. Liquid extraction is widely used in the petroleum refining, chemical, and nuclear industries for the purification of aqueous and non-aqueous liquid process streams. However, its current use in hazardous waste minimization applications is largely confined to removal and recovery of organic materials from aqueous waste streams. Its potential for removal of undesirable impurities from organic streams to make them recyclable has not yet been widely recognized. Future applications in waste minimization seem quite likely.

15.2.1 Terminology

The diagram shown in Figure 15.1 gives the basic terminology for liquid-liquid extraction. This diagram illustrates operation with a solvent less dense than the feed. With a solvent more dense than the feed, the solvent and feed inlets would be exchanged and the extract and raffinate exits would be exchanged.

The streams entering the extraction unit (or extractor) are the feed and solvent, while the streams leaving the extractor are the extract and raffinate. These streams are defined as follows:

Feed: The liquid mixture or solution to be separated.

Solvent: The liquid solvent which selectively extracts one or more components from the feed.

Extract: The solvent-rich phase containing the components extracted from the feed.

Raffinate: The refined product from which materials have been extracted—from the French word for refine, *raffiner*.

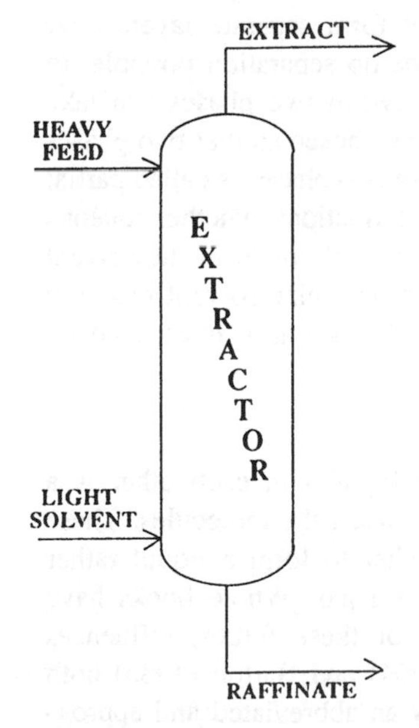

Figure 15.1 Basic terminology for liquid-liquid extraction.

There are several other terms to describe operation inside the extractor. These are:

Overall interface: The level-controlled interface between settled phases. It is usually held at either the top or bottom of the extraction column.

Continuous phase: The phase that is continuous from the controlled interface to the opposite end of the extraction column.

Dispersed phase: The phase that is broken up into droplets or filaments to provide the surface area for mass transfer in the mixing zones.

If the overall interface is controlled at the top of the column, the heavy phase is continuous and the light phase is dispersed. If the overall interface is controlled at the bottom of the column, the light phase is continuous and the heavy phase is dispersed.

15.2.2 Fundamentals

Liquid-liquid extraction requires the existence of two liquid phases between which all of the components in the system can distribute at equilibrium. For example, if a wastewater feed and an organic solvent are mixed thoroughly and then allowed to settle, they form two liquid phases in equilibrium. The impurities in the wastewater as well as the water itself will distribute between the two layers. The organic solvent will also distribute between the two layers. However, if the organic solvent and the wastewater do not form separate layers after mixing (i.e., are completely miscible), there will be no separation possible. In the absence of a second phase, no distribution between two phases can take place. Thus, solvents for treating wastewater must be chosen so that two phases will form after mixing and settling. This formation of two phases is called partial miscibility and is a requirement for all liquid-liquid extractions, whether aqueous or non-aqueous. It is desirable to have high solubility of solute in the solvent but low solubility of solvent in the feed mixture to minimize solvent recovery costs. Choice of solvents to achieve partial or total miscibility is dictated by solubility theory.

Solubility Theory

The solubility of non-electrolytes (i.e., non-ionic liquids) in each other is a function of the strength of the attractive forces between the molecules. These forces hold the individual molecules closely together to form a liquid rather than permitting them to be widely separated as in a gas. Whole books have been written on how *cohesive energy*, the sum of these forces, influences solubility. For example, Hildebrand and Scott (1964) and Barton (1983) both discuss this subject in detail. We will present only an abbreviated and approximate version to illustrate how solubility is affected by the nature of the solute and solvent. The goals are (1) to achieve some intuition about how to choose solvents and (2) to learn how to control the solubility of various materials in the solvents during extraction operations.

A good measure of the attractive forces between molecules (i.e., the cohesive energy) is the amount of energy required to pull the liquid molecules apart to a distance where the attractive forces are essentially zero. For liquids well below their boiling points, this is closely approximated by the energy of vaporization because the molecules in the vapor are about 1000-fold less concentrated than they are in the liquid. The cohesive energy per unit volume of liquid is called the cohesive energy density and is obtained by dividing the molar energy of vaporization by the molar volume of the liquid, as shown in Eq. 1.

$$\text{C.E.D.} = \frac{\text{molar energy of vaporization}}{\text{molar volume}} \tag{1}$$

where:

C.E.D. = cohesive energy density

molar energy of vaporization = internal energy change for vaporizing one molecular weight of liquid

molar volume = volume of one molecular weight of liquid

The internal energy change, ΔE, is obtained from the heat of vaporization, ΔH, by correcting for the energy of expanding the vapors from one atmosphere to zero pressure. This correction is made using the perfect gas law, i.e., by RT, where R is the gas constant (1.99) and T is the absolute temperature in degrees Kelvin (298°K for 25°C). Thus:

$$\Delta E = \Delta H - RT \tag{2}$$

Cohesive energy densities blend on a volume fraction basis. Thus, the value for mixtures is the sum of the volume fractions of each component times its cohesive energy density. This is shown in Eq. 3 for a mixture of A and B.

$$\text{C.E.D.}_M = \text{VF}_A + \text{C.E.D.}_A + \text{VF}_B \times \text{C.E.D.}_B \tag{3}$$

where:

C.E.D. = cohesive energy density

VF = volume fraction, the volume of component divided by total volume

Subscripts M, A, and B refer to mixture, component A, and component B, respectively.

Equation 3 shows that the cohesive energy density of a given solvent, A, can be raised or lowered by adding a second miscible solvent, B. Solvent B can have either higher or lower cohesive energy density than A. The mixed homogeneous solvent then behaves like a single pure solvent having the same cohesive energy density as the mixture.

Hildebrand and Scott (1964) define a parameter for describing solubility relationships for non-polar, non-associating systems. This parameter, the square root of the cohesive energy density, has become known as the Hildebrand solubility parameter, represented by a greek delta (δ). The value of δ can be determined directly from the energy of vaporization, if it is known, or calculated from the heat of vaporization by correcting for the pressure-volume product of the vapor. This is illustrated in Eq. 4.

$$\delta = \sqrt{\text{C.E.D.}} = \sqrt{\frac{\Delta E_V}{V_M}} = \sqrt{\frac{\Delta H_V - RT}{V_M}} \tag{4}$$

where:

δ	=	solubility parameter, Hildebrands
C.E.D.	=	cohesive energy density, cal/cm^3
ΔE_V	=	energy of vaporization, cal/g mole
V_M	=	molar volume, cm^3/g mole
ΔH_V	=	heat of vaporization, cal/g mole
R	=	gas constant, 1.99
T	=	absolute temperature (°C + 273)

The values of the solubility parameter can be given in any system of units as a square root of energy per unit volume. We will use the original Hildebrand units of square root of calories per cubic centimeter, which have been named Hildebrands. The values of solubility parameter for liquids at 25°C (77°F or 298°K) range from 5 to 6 for fluorocarbons and from 6 to over 10 for hydrocarbons. The hydrocarbons start with paraffins at the low end and increase with cyclization, unsaturation, aromaticity, and molecular weight. Organic chemicals other than hydrocarbons vary from 9 to 15, and water has the maximum solubility parameter of 23.4. Table 15.1 gives selected solubility parameter values for a variety of organic chemicals and solvents. These values were taken from tables compiled by Seymour (1984).

The solubility parameter quantifies the old rule of thumb for predicting solubility of materials in each other, i.e., like dissolves like. If two materials have essentially the same solubility parameter exclusively due to dispersion forces, they have the same attractive forces between molecules of like and unlike kind. Thus, they have no preference between mixing with molecules of their own kind and molecules of the other material. The result is that they mix in all proportions with each other and do not form two liquid phases. On the other hand, if the attractive forces between molecules are very different for the two

Table 15.1 Selected Hildebrand Solubility Parameter Values for Liquids at 25°C

Liquid	Solubility parameter, Hildebrands
Halogenated hydrocarbons	
Perfluoroheptane	5.8
Perfluoremethylcyclohexane	6.0
Carbon tetrachloride	8.6
Freon 1,1,2	7.8
Perchloroethylene	9.3
Hydrocarbons	
n-Hexane	7.3
n-Heptane	7.5
Dodecane	7.8
Cyclopentane	8.7
Cyclohexane	8.2
Methylcyclohexane	7.3
1-Hexene	7.4
1-Octene	7.6
1,5-Hexadiene	7.7
Benzene	9.1
Toluene	8.9
o-Xylene	9.0
Tetralin	9.5
Decalin	8.8
Naphthalene	9.9
Anthracene	9.9
Phenanthrene	9.8
Ethers	
Diethyl ether	7.4
Diisopropyl ether	6.9
Furan	9.4
Tetrahydrofuran	9.1
Ketones	
Acetone	9.9
MEK, (Methyl ethyl ketone)	9.3
Cyclohexanone	9.9

Table 15.1 (Continued)

Liquid	Solubility parameter, Hildebrands
Esters	
Ethyl acetate	9.1
n-Butyl formate	8.9
Amyl acetate	8.5
Alcohols	
Methanol	14.5
Ethanol	12.7
Cyclohexanol	11.4
Selected solvents	
Acetonitrile	11.9
Aniline	10.3
Phenol	11.8
N-Methyl-2-pyrollidone	11.3
Furfural	11.2
Quinoline	10.8
Dimethyl formamide	12.1
Inorganic compounds	
Ammonia	16.3
Water	23.4

materials, the one with the higher attractive forces, and thus higher solubility parameter, prefers to stay with its own kind. It resists making internal space for molecules of the other kind. This leads to complete separation of the two materials into separate phases, each of which contains essentially none of the other material, as with gasoline and water. Clearly a full spectrum of intermediate conditions exists, depending on how great a difference exists in the solubility parameters for the individual components.

Figure 15.2 is a sketch using data from Martin et al. (1981) to illustrate the general effect of the solubility parameter on solubility of solutes in solvents. The bell-shaped curve, centered at a solubility parameter of 9.9 (that of naphthalene), gives the solubility of naphthalene in a variety of solvents having various solubility parameters ranging from 5 to 15. The upper curve is for the solubility of naphthalene in hydrocarbons and halogenated solvents, while the lower curve

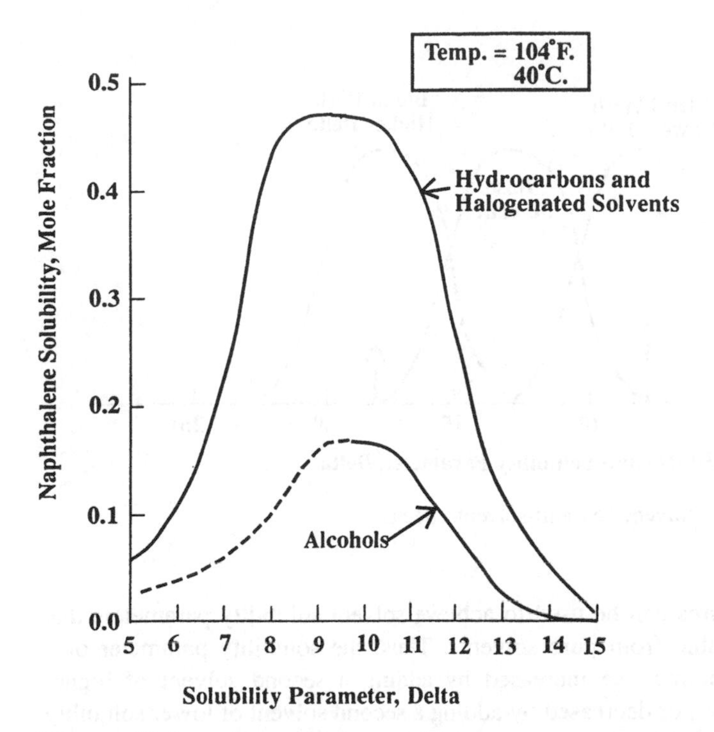

Figure 15.2 Effect of solvent solubility parameter on naphthalene solubility.

is for low molecular weight alcohols. The maximum solubility occurs when a solute has the same solubility parameter as the solvent. However, mutual solubility reduces to essentially zero as the solubility parameter of the solvent becomes more than five units different from that of the solute in either direction. The height of the bell curve at its maximum depends on how difficult it is for the solute to force its way into the solvent. If the cohesive energy density of the solvent is low and consists entirely of dispersion forces, as is the case for hydrocarbons and halogenated solvents, it does not take much energy to disrupt the attraction between solvent molecules. Furthermore, most of this disruption is compensated for by the new attraction between solvent and solute. Under such conditions the maximum solubility is high. If, on the other hand, the solvent molecules are held together strongly by additional forces of polarity, hydrogen bonding, or chemical interactions, it takes more energy to make a "hole" for the naphthalene in the solvent. This extra energy is not compensated for by new interactions between the naphthalene and the polar molecules. As a result the maximum solubility for naphthalene is lower in the alcohols than in the non-polar hydrocarbons.

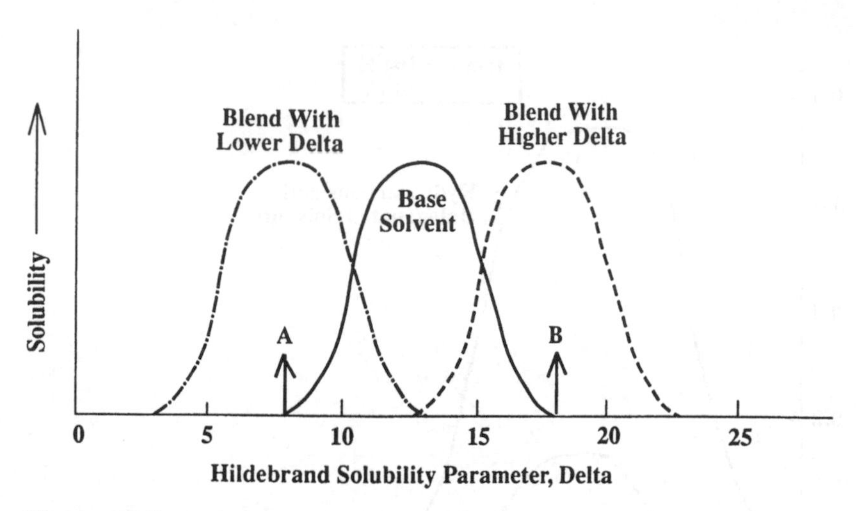

Figure 15.3 Pro-solvent and anti-solvent effects.

Solvent mixtures can be used to achieve solvent solubility parameters that may not be available from pure solvents. Thus, the solubility parameter of a given base solvent may be increased by adding a second solvent of higher solubility parameter, or decreased by adding a second solvent of lower solubility parameter. The second solvent is called either a pro-solvent or anti-solvent, depending on whether it causes solubility to increase or decrease. This is illustrated in Figure 15.3.

Figure 15.3 illustrates that when the solubility parameter of the solute A is lower than that of the base solvent, addition of a second solvent of lower solubility parameter than the base solvent will give a solvent blend with decreased solubility parameter. The characteristics of this solvent blend are shown by the dot-and-dash bell to the left of the base solvent. This solvent blend will increase solute solubility by decreasing the difference in solubility parameter between the solvent blend and solute A. This represents pro-solvent character for the second solvent additive. However, if the base solvent has a solubility parameter lower than that of the solute (B), the addition of the second solvent will increase the difference in solubility parameters between solvent blend and solute. This will decrease solubility of B. This is anti-solvent character for the second solvent. Similarly, the addition of a high-delta second solvent to the basic solvent gives the bell curve shown as the dashed line. In this case the second solvent is an anti-solvent for A and a pro-solvent for B.

Looking at the bell curves another way, as in Figure 15.4, the center of the bell can be looked on as the solubility parameter of a solvent, and the base of the bell is the range of solubility parameters of solutes which will dissolve in

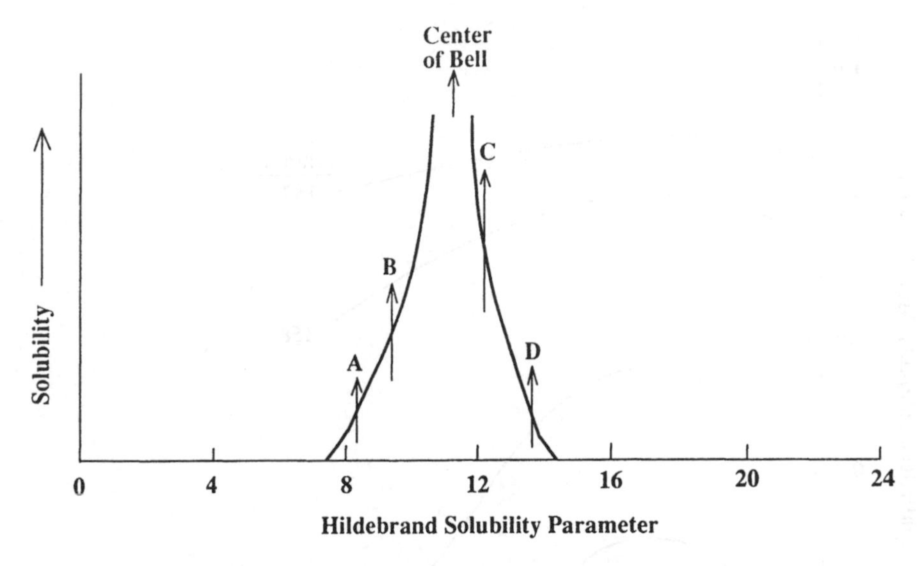

Figure 15.4 Solubility of various solutes in a given solvent.

the solvent. The sides of the bell represent the solubilities in the solvent of solutes A, B, C, and D of different solubility parameters. The same rules of solvent blending just discussed also apply to this diagram. Adding a second solvent of higher solubility parameter will move the bell to the right and increase solubility for solutes C and D, which have higher solubility parameters than the solvent. Correspondingly, the solutes A and B, with lower solubility parameters than the solvent, will have their solubilities decreased by adding the second solvent. Similarly, moving the bell to the left increases the solubilities of A and B, while decreasing the solubilities of C and D. The concepts of Figure 15.4 are useful for deciding what solutes are likely to dissolve in a given solvent or solvent blend.

By combining the effects of solvent blending with choice of extraction temperature, almost any desired solubility can be achieved. The effect of increasing temperature is twofold. First, the absolute level of solubility of solute in solvent increases, and, second, the bell-shaped solubility curves get much wider so that larger differences in solubility parameter can be tolerated while still achieving a desired solubility. This is illustrated in Figure 15.5.

Since Hildebrand's and Scott's work, the concept of how molecular attraction influences solubility has been extended to all liquid systems. The overall solubility parameter has been subdivided into the various different types of attractive forces that can exist between molecules. The cohesive energy provided by the various types of forces is still additive to give the overall cohesive energy.

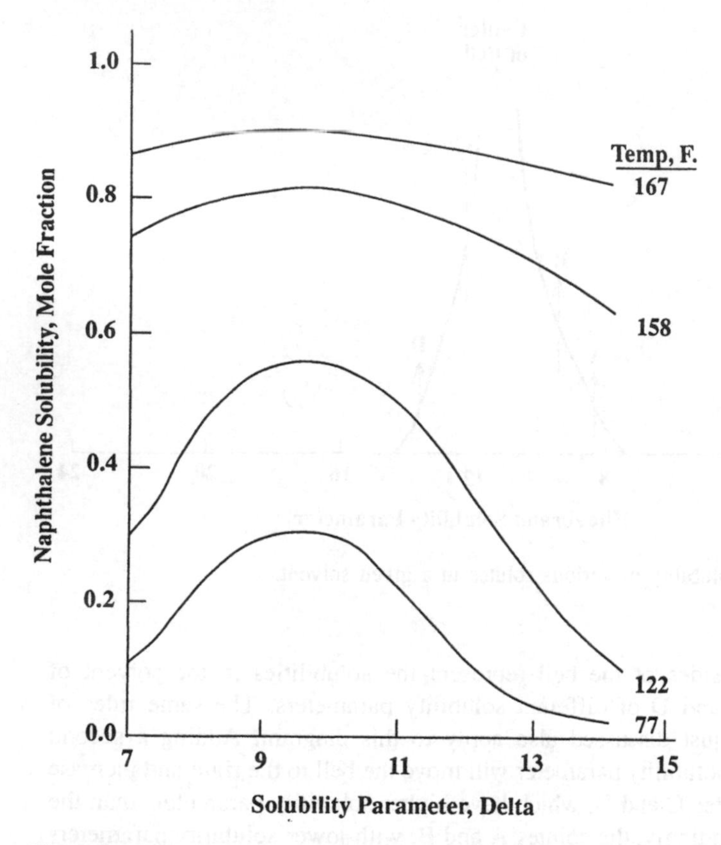

Figure 15.5 Effect of temperature on naphthalene solubility in various solvents.

Therefore, the squares of the solubility subparameters based on the different types of forces are also additive to give the square of the overall solubility parameter. The number of types of forces considered by different authors varies from two to five, with three, due to Hansen (1967), being most popular. The Hansen parameters are illustrated in Eqs. 5 and 6.

$$\text{C.E.} = \Delta E_V = \Delta E_D + \Delta E_P + \Delta E_H \tag{5}$$

where:

C.E. = cohesive energy
ΔE_V = molar energy of vaporization
ΔE_D = cohesive energy due to dispersion forces
ΔE_P = cohesive energy due to polar interactions

ΔE_H = cohesive energy due to hydrogen bonding

$$\delta_O^2 = \delta_D^2 + \delta_P^2 + \delta_H^2 \tag{6}$$

where:

δ = solubility parameter

Subscripts O, D, P, and H refer to overall, dispersion, polar, and hydrogen bonding contributions, respectively.

Solubility can be enhanced if both the overall and the subparameters are matched for the solvent and solute. For example, if one has a choice between two solvents with the same overall solubility parameter to dissolve a given solute, the solvent most closely matching the non-dispersion cohesive forces of the solute will dissolve more of the solute. Where only dispersive forces are involved, a match of overall solubility parameter gives complete miscibility in all proportions. Detailed discussion of this is beyond the scope of this book, but is given in Barton (1983). For our purposes the overall solubility parameter will be confined to dispersion forces and complexing forces. Complexing forces is a catch-all name for all chemical forces beyond the physical dispersion forces.

Selectivity

The second major factor important to solvent extraction is selectivity of the solvent. This is given in the form of a separation factor, as was discussed in Chapter 2. If only a single solute is involved, as with the extraction of phenol from a wastewater stream to make the wastewater disposable or recyclable, a simple equilibrium distribution constant for phenol between the extracting solvent and water describes the extraction selectivity. The extent of separation is then determined graphically by the solvent to feed ratio, the number of extraction stages, and the feed composition. However, for more complicated systems, where two or more solutes are being separated from each other, the separation factor or selectivity is determined by the relative distribution constants for the individual components between the two phases calculated on a solvent-free basis.

If the separation factor or distribution constant is very large, the desired separation can often be made in a single stage consisting of a mixer and settler. More typically, liquid extractions employ multiple contacts with countercurrent flow of the liquid phases between stages.

Types of Operation

Batch extractions where the feed and solvent are placed in a mixing vessel, thoroughly mixed, and then allowed to settle into two quiescent liquid layers are more commonly used for treating small volumes. The solvent layer containing most of the solute to be removed from the feed is separated from the residual feed layer. If necessary, a second batch of fresh solvent is contacted with the

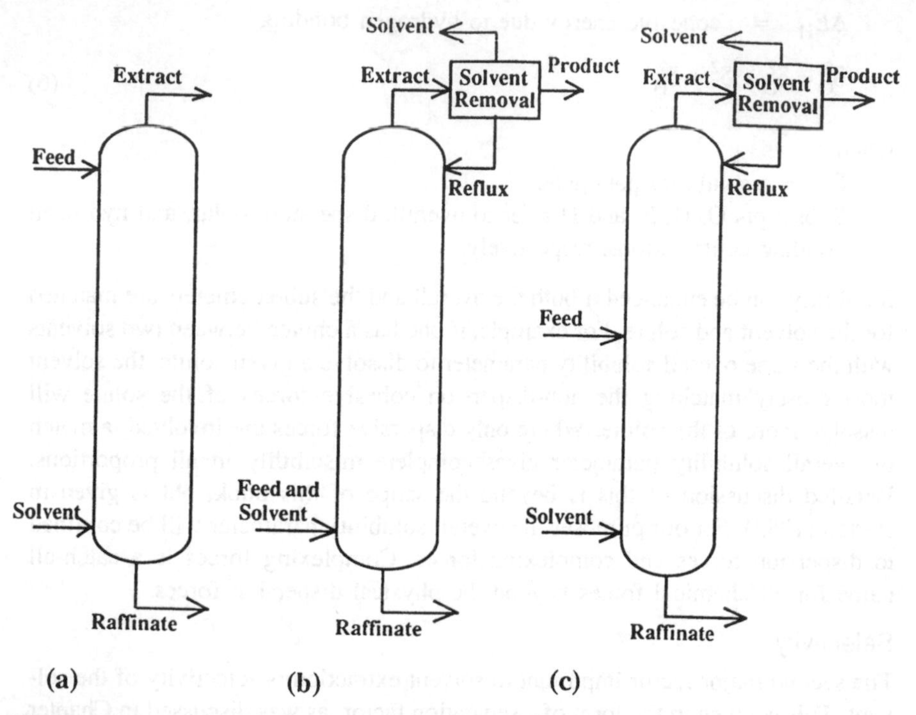

Figure 15.6 Various types of liquid-liquid extractions. (a) Stripping, (b) enriching, (c) compound operation.

remaining feed layer to remove more solute. This is called batch stripping and can be repeated as many times as necessary to strip the feed of solute to a level where it can be recycled or safely discarded.

For large volumes of feed it is usually better to do the extraction continuously in a multistage extraction apparatus of one of the types shown in Figure 15.6. If the purpose of the extraction is to remove solute from the feed stream, the operation is called *stripping*. For the stripping operation, the feed and solvent are fed at opposite ends of the apparatus and flow countercurrently through a series of mixing and settling zones. The mixing and settling zones are sufficient in number to provide the required degree of separation. If the purpose of the extraction is to recover the solute in high purity, the operation is called *enriching*. For enriching, reflux is probably required at the enriching end of the extractor to provide adequate solute product purity. The recovered solvent is recycled to solvent feed. For *compound operation*, where both stripping and enriching are combined in the same apparatus, the feed is introduced at some intermediate point between the ends of the extractor. Reflux is provided at the enriching end of the apparatus. The sketches shown in Figure 15.6 illustrate these types of operation and they are discussed in more detail in Chapter 2. The various kinds of apparatus used commercially are described in Section 15.2.4.

Solvent Recovery

Solvent recovery is almost always an integral part of every solvent extraction operation. It is usually done by distillation or a secondary extraction step. However, sometimes a suitable solvent can be obtained from some waste stream destined for destruction in a way that is compatible with disposal of the solute to be extracted. This can eliminate the need for solvent recovery and save a lot of money. For example, if a waste fuel stream is destined for fuel blending for on-site use, it could be used to extract organic oxygenated solvents such as phenols, alcohols, and ketones from wastewater and then be sent to fuel blending. Thus, the solvent would be free and would not have to be repurified as part of the extraction operation. Furthermore, the fuel value of the recovered solvents would also be used. The innovative use of available waste streams in this manner can often kill two birds with one stone and help make waste management pay.

15.2.3 Specific Applications

Wastewater Extraction with Organic Solvents

The major application of liquid-liquid extraction to hazardous waste minimization is the stripping of organic solutes from process wastewaters. This makes the stripped wastewater suitable for disposal or reuse and recovers the organic solutes for purification and recycle or sale. The solvents used for wastewater extraction are usually chosen for their low solubility parameters to make them insoluble in water but still effective solvents for the wastewater solute. Such materials are either hydrocarbons or simple, relatively non-polar oxygenated compounds such as ethers, ketones, or esters. Typical solutes that are already being recovered in this way are phenols, chlorophenols, and acetic acid.

The *phenols* are found in wastewater streams from petroleum refining, petrochemical production, phenolic resin production, and coal conversion. For recovery of phenols from wastewaters, the Phenosolvan process, developed by Lurgi, uses diisopropyl ether as solvent, and the Chem-Pro process is reported to use methyl isobutyl ketone as solvent. Both of these processes are effective and are in commercial use.

The *chlorophenols* are found in the wastewaters from wood preserving, primarily as pentachlorophenol but also as lesser quantities of the other chlorophenols. Wallin et al. (1981) showed in the laboratory that 97% removal of pentachlorophenol could be achieved by single-stage batch extraction using No. 2 fuel oil as solvent at a solvent to wastewater ratio of 1 to 4. This degree of extraction could be increased to 99% under the same conditions by using a mixed solvent of No. 2 fuel oil containing 10% of amyl alcohol still bottoms. Number 2 fuel oil is a hydrocarbon stream boiling a little higher than kerosene made up mostly of paraffin, naphthene, and single-ring aromatic molecules. It is often called home heating oil. The composition of the amyl alcohol still bottoms was not given, but is assumed to be mostly oxygenated compounds having a higher solubility parameter than the fuel oil. This would lead to increased solubility of the pentachlorophenol in the solvent and thus account for the better separation results with the mixed solvent.

Acetic acid is found in a variety of industrial wastewaters and can be recovered by liquid-liquid extraction using ethyl acetate or ethyl acetate/benzene mixtures as solvent. Glitch Package Plants (Parsippany, NJ) has designed acetic acid recovery plants for a variety of industries using a combination of solvent extraction and distillation as the recovery method.

Extraction of Liquid Organic Feeds

The liquid-liquid extraction of liquid organic feeds is a very common process in the petroleum and petrochemical industries. It is used for either the purification of products by removal of contaminants or for recovery of a valuable product from a mixed feed. The solvent usually has a higher solubility parameter than the feed to provide partial miscibility with the feed, i.e. to generate two phases. The solvent then selectively extracts the higher solubility parameter and lower molecular weight components of the feed. However, extraction of organic feeds is just starting to find application in the management of organic hazardous wastes. The first important application is the removal of PCBs from transformer oils. This has been studied by Union Carbide, which piloted a process using dimethyl formamide as solvent, and by the Electric Power Research Institute, which developed a process based on diethylene glycol monomethyl ether (methyl carbitol) as solvent. Both of these processes use continuous countercurrent multistage extraction to recover PCBs for destruction by incineration and to reduce their concentration in the transformer oil to acceptable levels.

It is expected that liquid-liquid extraction will find further application to organic feedstocks for the removal of toxic materials and the recovery of valuable materials for recycle or sale.

15.2.4 Equipment

The apparatus used for liquid-liquid extraction always includes a mixing device to thoroughly contact the two liquid phases and a relatively quiescent settling zone where the two phases are allowed to disengage and form two separate liquid layers. One mixing zone combined with one settling zone comprises a single stage of liquid-liquid extraction. While batch extractions are usually carried out in a single-stage unit, most continuous extractions are done in multistage units with counterflow of the phases between stages. A wide variety of devices have been invented which differ in the way the phases are mixed, the design of the settling zones, and the way the stages are laid out to get countercurrent flow. A good summary of these designs with the advantages and disadvantages of each is given by Cusack and Fremeaux (1991) as part 2 of a three-part series published in *Chemical Engineering* entitled "A Fresh Look at Liquid-Liquid Extraction." Part 1 by Cusack et al. (1991) covers extraction systems, and Part 3 by Cusack and Karr (1991) covers extractor design and specification.

The types of equipment available include (1) batch mixing tanks, (2) countercurrent mixer-settler units, (3) countercurrent columns containing a variety of internals where the mixing energy is provided by the density difference between the flowing phases, (4) countercurrent columns where the mixing energy is mechanically supplied by paddles, rotating disks, vertically reciprocating plates, vibrating plates, or fluid pulses, and (5) centrifuges where both the mixing energy and enhanced gravity for better settling are provided by the centrifugal forces of a specially designed centrifuge. These units all have advantages and disadvantages and best results occur when a vendor runs a pilot test for the specific application and specifies the proper equipment for the specific application.

Major vendors of liquid-liquid extraction equipment are listed in Table 15.2.

15.2.5 Common Problems and Troubleshooting

The successful operation of a liquid-liquid extraction process involves a series of compromises among competing effects. For example, solubility in the solvent must be high enough to require a minimum amount of solvent circulation, but not high enough to lose solvent selectivity or make settling difficult by reducing density difference between the phases. Mixing of the phases must be adequate to provide good stage efficiency, but not so severe that settling and entrainment are problems. Feed rate must be high enough to provide good capacity for the extraction unit, but not so high that flooding ruins the separation. These are now discussed.

Mixing and Flooding

Many extraction columns use the density difference between the phases to provide the mixing energy. The density difference drives the flow of the fluid phases through the extraction column internals, such as packings and orifice plates. In such cases, it is possible to pump feed and solvent into the column faster than they can flow through the column internals under the influence of density difference between the phases. Under these conditions the phases being fed to the column must exit by the path of least resistance. This usually means

Table 15.2 Vendors of Liquid-Liquid Extraction Equipment

APV Chemical Machinery, Inc. 1000 Hess Street Saginaw, MI 48601 (517) 757-1300	Artisan Industries 73 T Pond Street Waltham, MA 02154 (617) 893-6800
Critical Fluid Systems (Supercritical) 500 W. Cummings Park Suite 6600 Woburn, MA 01801 (617) 937-0800	Fluitron (Supercritical) 30 Industrial Drive Northamton Industrial Park Ivyland, PA 18974 (215) 355-9970
Koch Engineering Co. P.O. Box 8127 Wichita, KS 67208 (316) 832-5110	Mixing Equipment Co., Inc. 138 Mt. Read Blvd. Rochester, NY 14603 (716) 436-5550
Otto H. York Co., Inc. P.O. Box 3100 Parsippany, NJ 07064 (800) 524-1543	Xytel Corporation 801 Business Center Drive Mount Prospect, IL 60056 (708) 299-9200

that the light phase pumped into the bottom of the column exits with the heavy phase, and the heavy phase being pumped into the top of the column exits with the light phase. This bypassing of the effective extraction region of the column can occur with either or both phases and leads to catastrophic loss of extraction efficiency. This condition is called flooding. The remedy is to lower the pump rates on the feed and solvent streams to levels where the internal flow under the influence of density difference between the phases is adequate to prevent flooding.

The same type of flooding can also occur with columns using mechanical energy to provide the phase mixing in each stage. The optimum degree of mixing is that which provides a stage efficiency of 80 to 90%, i.e., 0.8 to 0.9 theoretical stages per actual stage. More intense mixing raises the pressure drop through the mixing zone by creating a finer dispersion and thus leads to flooding. In this case the condition can be corrected by lowering either the flow rates and the mixing intensity or both.

Entrainment and Settling

A less catastrophic form of bypassing occurs when the phases do not completely separate in the settling zones of each stage. The droplets of heavy phase still dispersed in the light phase flow with the light phase either to the next stage or out of the column. Similarly, light phase droplets can be entrained in the heavy phase. The effect of entrainment is to lower stage efficiency by diluting each phase with the composition of the alternate phase. Entrainment can have a variety of causes, including (1) inadequate settling time due to excessive feed rates, (2) overmixing, which generates too fine a dispersion, (3) inadequate density difference between the phases due to high mutual solubility, (4) high viscosities of the phases being contacted, which causes slow settling, and (5) the presence of surfactants and dirt, which increases the difficulty of settling by stabilizing the dispersion or forming an emulsion. The obvious remedies are to cut down flow rates, reduce mixing intensity, reduce mutual solubility, switch to a less viscous solvent, and remove dirt and emusifiers from the feed streams.

Solubility and Loss of Second Phase

With non-aqueous systems, the solubility of the phases in each other can vary greatly as the solvent becomes more enriched with the more soluble components of the feed. An attendant danger is that the mutual solubility of the phases becomes so great that the phases become totally miscible and only one phase remains. Clearly, at this point no further enrichment of the solvent phase is possible. The addition of an antisolvent, usually water, to the more polar solvent phase at intermediate points in the extractor can maintain the presence of two phases and allow higher purity solutes to be recovered. Solubility can also be increased by adding prosolvents to the solvent feed and then controlled by the addition of antisolvents at intermediate points in the extraction. A typical example is the use of methylamine as prosolvent and water as antisolvent with liquid ammonia as the main solvent. This can give complete solubility control in the extraction of aromatic hydrocarbons from petroleum. Similarly, both ammonia and water can be used as antisolvents when extracting oils with methylamine.

15.2.6 Economics

The costs associated with installing and operating a liquid-liquid extraction process are unique to the individual application contemplated. However, some general guidance can be given regarding the sensitivities of costs to various options.

The capital and operating costs of liquid-liquid extraction are very closely tied to the choice of solvent for the extraction. For example, if the solvent is available as a waste stream from other operations, there is essentially no cost associated with solvent losses to the raffinate. Furthermore, if the extract and the solvent are destined for incineration, there is no solvent recovery required from the extract phase. This combination requires no capital or operating cost for solvent recovery and no operating cost for solvent losses. Clearly, this is a very unusual case, but one worth seeking.

A more typical liquid-liquid extraction requires solvent recovery from both the extract and raffinate streams. Under these conditions choice of solvent determines both how complicated the solvent recovery process needs to be and how much solvent must be recovered. The total solvent circulation rate is determined by the amount of feed which must be removed as extract, its solubility in the extract phase, and the amount of solvent leaving in the raffinate. High solubility in the extract phase reduces solvent circulation rate, and high solubility in the raffinate raises it. A compromise between the loss of selectivity at high extract solubilities, which requires more stages in the extractor at increased capital cost, and the reduction in solvent recovery costs achieved with higher extract solubilities has led to an optimum extract solubility usually in the range of 15 to 30 volume percent.

The cost of solvent recovery also depends on the technique used to recover the solvent. Distillation is the commonest solvent recovery technique and it is least expensive when the solvent and the solute have widely different boiling points. In such cases a simple flash distillation or fractionation with only a few stages at low reflux ratio should be adequate. This can give recovered solvent at adequate purity for recycling to the extractor. However, in all cases using distillation for solvent recovery, at least one vaporization of all the solvent used is required. Thus, solvents with lower heats of vaporization (i.e., hydrocarbons, fluorocarbons, and ethers) require less energy for solvent recovery.

Methods for solvent recovery other than distillation are also used in special cases. Such cases include extraction of heat-sensitive materials which degrade on distillation or with solvents which form azeotropes in distillation. The solvent recovery alternatives inlude secondary extraction of the extract phase with a more specific solvent which could not be used as a primary solvent because it is soluble in the feed. An example of this is the use of a caustic solution for the re-extraction of phenols from a hydrocarbon solvent used to recover phenols from wastewater. The phenols are then recovered from the caustic by neutralization. The use of such techniques is usually more expensive than solvent recovery by distillation due to the added complexity of the process. However, in some specific cases such techniques may be the optimum overall process.

The cost of the solvent is also an important factor, other things being equal, in the economics of an extraction operation. There will always be some solvent makeup requirement to compensate for solvent losses, even with the best solvent recovery systems. The cost of this solvent makeup is directly proportional to solvent price. Furthermore, the cost of the original solvent inventory is also directly proportional to solvent price and needs to be considered in optimizing the choice of extraction solvent.

15.3 SUPERCRITICAL FLUID EXTRACTION

Supercritical fluid extraction is the use as an extraction solvent of a material at a pressure and temperature above its vapor-liquid critical point. The process is very similar to liquid-liquid extraction. The critical point of the solvent is the temperature above which it cannot be liquefied, no matter how high the applied pressure. The corresponding vapor pressure at the critical temperature is called the critical pressure and is usually between 500 and 1200 psig for light hydrocarbon gases, light freons, and carbon dioxide. The viscosity and diffusion properties of a supercritical solvent are much like those of a gas. However, the density, and thus the molecular interactions leading to solute solubility, are more like those of a liquid. These molecular interactions achieve the solubilization of high molecular weight solutes into a supercritical gas phase at temperatures well below the boiling point of the solute. This permits use of the supercritical gas as an extraction solvent for high boiling point heat-sensitive materials. It should be remembered that while this discussion is limited to the *extraction* application of supercritical fluids, once the solute is in the supercritical carrier phase, it can be treated by adsorbents, separated by chromatography, and subjected to other separation processes.

Most, if not all, of the separations possible with supercritical gas solvents could also be made with liquid solvents either well below or in the vicinity of their critical points. What is it then that drives the increasing application of supercritical gas extraction? It is the easier and less expensive solvent recovery, coupled with essentially complete solvent removal from the products, because of (1) the availability of non-toxic supercritical gas solvents such as carbon dioxide, (2) the ability to process essentially non-volatile materials at moderate temperatures, and (3) the low phase viscosities and high density differences between the phases, which enhance effective mass transfer and phase disengagement. The solubility of solutes in the supercritical fluid increases as the density of the supercritical fluid is increased. Thus, operation at high pressure gives high solubilities in the supercritical solvent. Reduction of pressure on the supercritical gas phase decreases solubility and precipitates out part of the solute. This phenomenon can be used to perform intermediate fractionation of solute by reducing pressure in steps during solvent recovery. The supercritical solvent can be recovered from both the extract and raffinate merely by lowering pressure below the critical and letting the gaseous solvent flash off. Thus, by choosing a material having a low critical point, an effective extraction solvent of low viscosity, high solubilizing power, rapid mass transfer, and easy solvent recovery is obtained. The penalty for these benefits is the requirement for (1) operation at moderate to high pressure in pressure vessels and (2) a compressor in the solvent recycle loop. Table 15.3 gives the critical temperatures, pressures, and densities of some major candidates for supercritical fluid solvents.

Table 15.3 Critical Constants of Potential Supercritical Solvents

Solvent	Critical constants		
	Temperature, °F	Pressure, psia	Density, g/cm³
Ethylene	50	742	0.22
Chlorotrifluoromethane	84	561	0.58
Carbon dioxide	88	1073	0.46
Ethane	90	717	0.21
Propylene	197	676	0.23
Propane	206	617	0.22
Dichlorodifluoromethane	233	582	0.56
Ammonia	270	1639	0.24
n-Butane	307	529	0.23
n-Pentane	387	485	0.23
Isopropyl alcohol	456	779	0.27
Methanol	465	1157	0.27
Ethanol	470	928	0.28
Benzene	551	701	0.30
Water	706	3211	0.32

Relatively mild operating temperatures are possible for supercritical extractions using light hydrocarbons, light freons, or carbon dioxide as solvents. These range from 50 to about 250°F at pressures from 600 to several thousand psig. The use of water as a supercritical solvent requires temperatures above 705°F and pressures above 3200 psig. Thus, supercritical water finds more use as a medium for incineration by oxidation than as a solvent for extraction without conversion.

15.3.1 Fundamentals

The dramatic increase with increasing pressure of the solubility of solutes in gases and vapors above the critical temperature begins as the pressure approaches the critical pressure and extends well above the critical pressure of the gas. This is illustrated for naphthalene in ethylene in Figure 15.7, drawn from data taken from Diepen and Scheffer (1948). Ethylene has a critical temperature of 50°F and a critical pressure of 727 psig.

At a constant temperature of 77°F, the vapor pressure of solid naphthalene is constant. Thus, its equilibrium concentration in the vapor phase decreases as total pressure is increased up to the region of the critical pressure. However, in

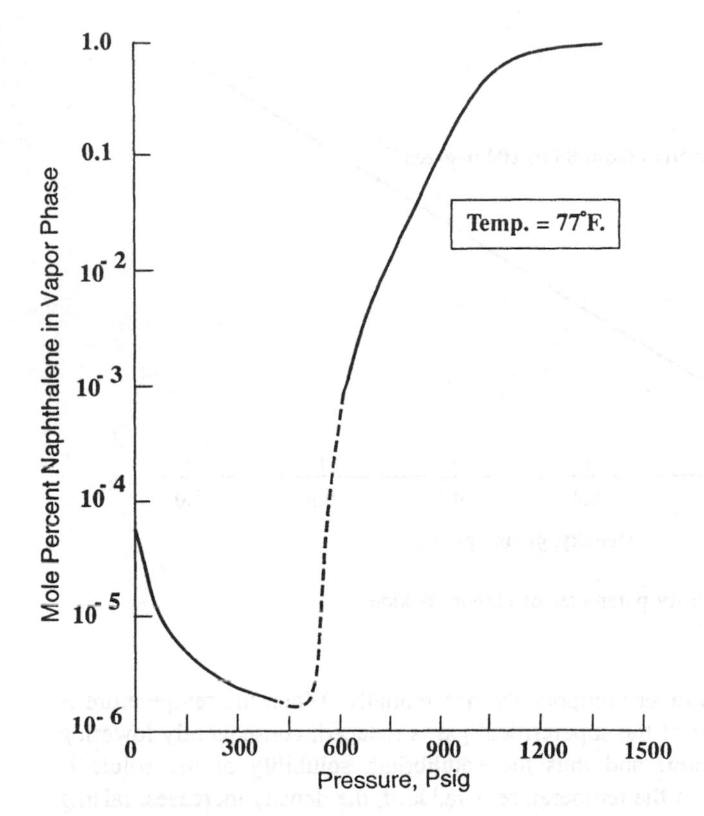

Figure 15.7 Solubility of naphthalene in ethylene vapor.

the vicinity of the critical pressure the trend changes sharply. As pressure is further increased, the equilibrium concentration of naphthalene in the vapor phase rises to levels more typical of liquid solubility, that is, to 1.0 mol % of naphthalene in ethylene, or 4.4 wt % naphthalene in the supercritical phase. This value is about 10,000 times higher than the value predicted by the perfect gas laws and is due to the interactions between the dense gas and the solute molecules.

As is illustrated in Figure 15.8, the solubility parameter for carbon dioxide is a function of density only over a temperature range from slightly below the critical to 16°F above the critical. Temperature and pressure affect the solubility of the solute in the supercritical gas phase through their effects on the density of the compressed gas and thus on the solubility parameter of the dense gas. In general, the solubility parameter of dense gases is lower than that of most solutes it is used to extract. Thus, when extracting mixtures, it solubilizes the

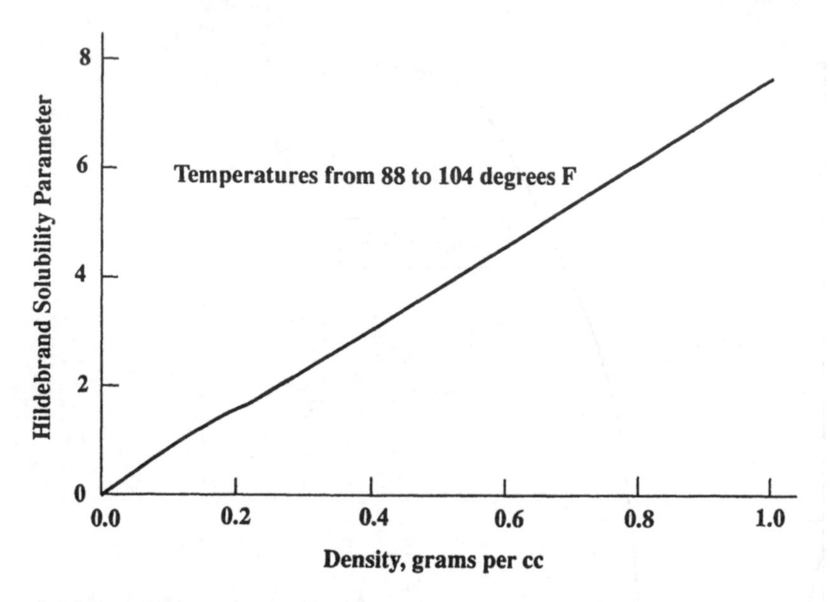

Figure 15.8 Solubility parameter of carbon dioxide.

lower solubility parameter components preferentially. When the temperature is increased, the density of the supercritical gas is reduced, consequently lowering its solubility parameter; and thus the equilibrium solubility of the solute is reduced. Conversely, if the temperature is reduced, the density increases, raising the solubility parameter; and the equilibrium solubility in the gas phase is increased. The pressure effect is the opposite from that of temperature. The solubility increases with pressure increase and decreases with pressure decrease. These effects are maximized near the critical point, and thus it is usually advisable to operate in the near-critical to supercritical region but close to the critical temperature, i.e., within 10°F or so. This minimizes the pressure required to achieve the desired gas phase solubility. The optimum operating conditions for near-critical and supercritical extraction are pressures from one to six times the critical pressure at a reduced temperature of 0.95 to 1.4. (The reduced temperature is the actual temperature divided by the critical temperature, both on the absolute temperature scale.)

Just as with conventional liquid-liquid extraction, the supercritical fluid solvent can be a blend of two or more components. These materials are chosen to optimize the interaction energy between the solute and solvent and thus control solubility. Also, modifiers called *entraining agents* are often used to enhance the solubility of solutes in the supercritical phase. These work by having a high solubility in the supercritical phase on their own and a reasonably strong molecular affinity for the solute it is desired to solubilize. In other words, they work through their effect on the solubility parameter of the dense gas mixture.

15.3.2 Specific Applications

A near-critical or supercritical fluid can be used to extract liquids in a way entirely analogous to liquid-liquid extraction, or to extract solutes from solids in a way analogous to leaching. For the near-critical or supercritical fluids, higher pressures with the attendant higher cost of equipment are required. Some of the applications analogous to liquid-liquid extraction which have reached commercialization are:

- Critical Fluid Systems (CFS; Woburn, MA) CFS process for recovery of alcohols, ketones, and organic acids from aqueous process streams using carbon dioxide as solvent
- Residuum oil supercritical extraction (ROSE) process (Kerr-McGee) for deasphalting and fractionating heavy oils and residua from petroleum refining using supercritical pentane and butane solvents
- Krupp process for re-refining lubricating oils using supercritical ethane as solvent
- CFS process for recovery of organics from wastewaters using near-critical carbon dioxide

Critical Fluid Systems (CFS) Process

CFS has developed and commercialized a variety of processes using near-critical or supercritical carbon dioxide as extraction solvent. The liquid-liquid extraction processes for recovering oxygenated chemicals from aqueous process streams have an economic advantage over distillation because much less energy is required per gallon of product. It is expected that this advantage will apply to many products made by fermentation. Here, the product is produced as a dilute aqueous solution due to the limited product concentrations that the fermentation organisms can tolerate.

ROSE Process

The residuum oil supercritical extraction (ROSE) process was developed by the Kerr-McGee Refining Corporation in the late 1950s for recovery of high-quality oils from petroleum residua. This process is very similar to the subcritical propane deasphalting process that has been in use by petroleum refiners for many years for removing asphalt from heavy oils and residua. However, the solvent for the ROSE process is supercritical pentane, and the process operates at higher pressure and temperature. Its main advantages are less expensive solvent recovery and the ability to make different grades of oil by stepwise depressuring of the extract phase to precipitate oils of different quality.

Krupp Process

The Krupp process was developed by the Krupp Research Institute in West Germany and uses supercritical ethane to extract lube oil base stocks from waste oils. The waste oils are contaminated with metals, water, naphtha, PCBs, oxidation products, and additives. A preliminary distillation is used to remove naphtha and water, and the rest of the contaminants are rejected by the supercritical extraction step. The extraction is carried out at about 1500 psig and 108°F in a packed column fed at the bottom with supercritical ethane and at the top with the waste oil. The ethane dissolves the quality lube oil components from the feed as the feed flows down the column countercurrent to the rising ethane. The undissolved impurities flow out the bottom of the column and the oil extract flows out the top dissolved in the solvent. Just as with the ROSE process, different grades of lube base stock can be obtained from the extract by reducing the pressure in stages to precipitate the least soluble components first, then those of intermediate solubility, and finally the most soluble components. This final step removes the last of the dissolved oil from the ethane and purifies it for recycling via a compressor.

CFS Process for Wastewater Treating

In view of the successful development of subcritical liquid-liquid extraction processes for recovery of organic solutes from wastewaters and the use of supercritical carbon dioxide to recover organics from aqueous process streams, it would seem obvious that supercritical fluid extraction should be applicable to recovery of organics from wastewaters. Critical Fluid Systems has developed such a process and designed and constructed a commercial unit for using carbon dioxide to recover organics from wastewaters. However, as of mid-1991 it has not yet been put into service. Assuming that its operation will be an economic success, we can expect a lot more applications of this type. Until commercial experience is obtained with supercritical extraction of organics from wastewaters, we will not know whether conventional solvent extraction or supercritical fluid extraction holds the most promise for this application.

15.3.3 Supercritical Extraction Equipment

The equipment used for supercritical fluid extraction is essentially the same as for liquid-liquid extraction except that it must be built to operate at higher pressure. It also usually needs a recycle compressor in the solvent recovery loop if solvent recovery is achieved by flashing the supercritical solvent from the product streams.

15.3.4 Common Problems and Troubleshooting

The problems common to supercritical extraction are the same as those encountered with subcritical extractions, that is, inadequate mixing intensity, settling, entrainment, flooding, and solubility control. However, the lower viscosity and greater density difference between the phases provided by supercritical solvents make the operational problems less severe. The major additional problem is the need for a compressor in the solvent recycle loop.

15.3.5 Economics

The capital cost of the extraction section of a supercritical extraction plant is normally higher than that of a conventional extraction plant due to the higher pressure ratings for the vessels and the need for a compressor. However, simplified solvent recovery equipment and lower energy costs for the overall process frequently lead to an economic advantage for the supercritical plant. Some of these benefits can also be achieved by using subcritical but volatile solvents such as liquid propane and liquid carbon dioxide as extraction solvents. Increased use of such processes can be expected.

15.4 SOLVENT PRECIPITATION

Solvent precipitation uses the same solubility theory discussed in Section 15.2.2. However, solvent precipitation differs from conventional liquid-liquid extraction in that the added solvent is not selectively dissolving the impurities. It is altering the solvent properties of the feed so that the impurities can no longer stay in solution. The added solvent dissolves in the carrier of the waste contaminants and precipitates the impurities by generating a new carrier solvent which cannot hold the impurities in solution. The impurities then precipitate out of solution to form a second liquid phase or a solid phase. An example of this type of operation is the separation of impurities such as spent additives and asphaltenes from used motor oils by extraction with light hydrocarbons, ethers, or esters. This is often done in the re-refining of used motor oils. The precipitation solvents usually have lower solubility parameters than the bulk of the used oil, which in turn has a lower solubility parameter than the used oil impurities. The added solvents dissolve in the used oil and generate an oil-solvent mixture which cannot hold the impurities in solution. The impurities then precipitate out of solution as a second liquid phase. This second phase can be removed from the oil-solvent phase by gravity settling, centrifuging, etc. The added precipitation solvent can be recovered from the purified oil and asphalt phases by distillation.

15.4.1 Fundamentals

Solvent precipitation techniques are based on the concept that high molecular weight and/or highly polar materials such as asphaltenes would normally not dissolve in oil. However, they are held in solution as a colloid by intermediate *peptizing agents*. The peptizing agents are soluble in the oil but also interact strongly with the high molecular weight polar compounds that constitute asphaltenes, thus holding them in solution. In petroleum the peptizing agents are very similar in molecular structure to asphaltenes, but of much lower molecular weight so that they dissolve in hydrocarbons. In solution theory terms, the peptizing agents raise the solubility parameter of the oil to a level near enough to that of the asphaltenes so that they can be held in solution. However, if a low solubility parameter diluent such as a light hydrocarbon (from propane to heptane) is added to the oil, the solubility parameter of the oil phase is lowered. When enough light hydrocarbon has been added, the peptizing agent and thus the asphaltenes cannot be held in solution. Fractionation of the petroleum fraction can be achieved by gradually increasing the light hydrocarbon to oil ratio. At first the diluent just dissolves in the oil. Then, as the quantity of diluent gets high enough, a second phase of asphaltenes begins to precipitate out of the oil phase. The most polar, highest molecular weight fractions precipitate first, followed by less polar and lower molecular weight fractions as the quantity of precipitating agent is increased. The precipitate can be either a solid or a liquid depending on the melting and solution properties of the precipitate. Asphaltenes are usually a solid precipitate while asphalt is usually precipitated as a liquid and includes more of the peptizing agents.

15.4.2 Specific Applications

The major application of solvent precipitation is the removal of undesirable contaminants such as asphaltenes and polar aromatic compounds from heavy oil fractions in petroleum refining. This allows recovery of high-quality feedstocks for further processing into lubricating oils. This process has been adapted to the re-refining of waste lubricating oils to remove decomposition products of the oil along with additives, polymers, etc. The solvents which have been tried include light hydrocarbons and low molecular weight alcohols, ethers, esters, and ketones. Processes which have been developed to use solvent precipitation in the re-refining of waste oils are (1) the DOE solvent process, developed at the Bartlesville Energy Technology Center, which uses a mixture of methyl ethyl ketone, isopropanol, and *n*-butanol as solvent, (2) the Snamprogetti process which uses light paraffins, typically propane, as solvent, and (3) the IFP process, developed by the Institute Français du Pétrole, which uses subcritical propane as solvent. The Snamprogetti and IFP processes are described by Short et al. (1987) and the DOE process is discussed by Cotton et al. (1980).

Table 15.4 Vendors of Leaching Equipment

Critical Fluid Systems 500 W. Cummings Park Suite 6600 Woburn, MA 01801 (617) 937-0800	Crown Iron Works P.O. Drawer 1364 Minneapolis, MN 55440 (612) 331-6400
Dorr-Oliver Incorporated P.O. Box 3819 Milford, CT 06460 (203) 876-5432	Eimco Process Equipment Co. 669 West Second, 200 South Salt Lake City, Utah 84110 (801) 526-2000
FMC Corporation 200 East Randolph Drive Chicago, IL 60601 (800) 621-4500	French Oil Mill Machinery Co. 1088 Greene St., P.O. Box 920 Piqua, OH 45356 (513) 773-3420
Humboldt Decanter, Inc. Dept. S-1 3200 Pointe Pky., (Norcross) Atlanta, GA 30092 (404) 448-4748	Mixing Equipment Co. 138 Mt. Read Blvd. Rochester, NY 14603 (716) 436-5550

15.4.3 Solvent Precipitation Equipment

The equipment used for solvent precipitation is just like that used for solvent extraction if the precipitate is a liquid phase. The same manufacturers listed in Table 15.2 for liquid-liquid extraction also provide suitable equipment for solvent precipitation. If the precipitate is a solid phase, the required equipment will be more like that needed for leaching, and the equipment manufacturers listed in Table 15.4 should be contacted to acquire appropriate hardware.

15.4.4 Common Problems and Troubleshooting

Most of the problems encountered in solvent precipitation are the same as those common to liquid-liquid extraction, i.e., adequate mixing without flooding, sufficient settling to prevent entrainment, and generation and maintenance of two flowing phases. This is made particularly difficult by the high viscosity of precipitated asphalt phases and the problems encountered when the precipitates are solids rather than liquids. Both of these types of problems can be reduced by operating at a higher temperature to reduce viscosities and liquefy solids. It may be necessary to change solvents to do this and still maintain the desired solubilities. However, solvent mixtures can often be used in place of single solvents to adjust the relationship between temperature and solubility. For example, mixtures of propane and butanes are often used by petroleum refiners to adjust the operating temperature of the deasphalter and the physical properties of the asphalt that is precipitated from heavy oils. If it is expected that the precipitate encountered will be a solid, as in solvent crystallization, then suitable solids-handling equipment can be provided in the original design of the solvent precipitation unit.

15.5 LEACHING

When solvent extraction is applied to solid materials, it is called *leaching*. It can be either a potential hazard, as when rainwater passes through a hazardous waste dump and dissolves toxic materials, or a useful tool for removing hazardous wastes from contaminated solids. It is commonly used in the food industry for recovery of seed oils and decaffeination of coffee. In the mineral and mining industries, it is used for recovery of metal values. In the waste minimization area, the primary applications are for regenerating adsorbents, de-oiling sludges, removing contaminants from soil, and recovering precious metals from auto exhaust converters or other spent catalysts.

The solvent used for leaching processes can be a subcritical liquid or a supercritical fluid. It must compete with forces binding the solute to the solid in order to solubilize it. These forces can be very weak, as with oil contamination on mill scale, of intermediate strength, as with adsorbed contaminants on clay or activated carbon, or very strong, as with chemical binding of metals in ores. Correspondingly, the solvents used to dissolve the solute can vary from those having purely physical interactions through those having weak chemical interactions (such as polar forces), through those having intermediate strength chemical forces (such as chemical complexing), to those exhibiting the strong chemical forces of ionic bonding. Once the solute has been solubilized in the solvent by the leaching process, it can be treated further by other processes, including liquid-liquid extraction, to purify and recover the solute, i.e., the leachate.

15.5.1 Fundamentals

The basic difference between leaching and the other extraction processes is that a *solid* rather than a fluid must be intimately mixed with and then separated from the solvent. This may require longer residence times for the solvent-solid mixture to come to equilibrium in each contacting stage. This is especially true if the solid is porous and solvent must diffuse into the pores to dissolve the desired solute. The solvent needs to fulfill two major functions to be a good leaching solvent. First, it must be able to desorb the desired solute from the surface of the solid and, second, it should be able to dissolve the released solute to keep it from being mechanically remixed with the feed solids and trapped in the matrix during phase separation. Thus, choosing a suitable solvent for a leaching operation requires some knowledge of the nature of the binding forces between the feed solids and the components to be recovered from them. For example, as was discussed in Chapter 14 on adsorption, clays tend to adsorb polar molecules more strongly than non-polar molecules. Thus, while clays are effective for adsorbing polar components from non-polar solutions such as hydrocarbons, they are not effective for recovering organics from aqueous solutions. This is because the strongly polar water molecules occupy all the adsorption sites and cannot be displaced by the less polar organic molecules. However, to recover organics from spent clay, a small amount of water added to a good solvent for the organics will displace the organics from the surface of the clay very effectively. This then permits the organics to dissolve in the solvent and be recovered from the solids.

In contrast to the clays, the carbon adsorbents tend to adsorb non-polar molecules more strongly than polar ones. This permits carbon to be used effectively to adsorb organics from aqueous wastewater solutions. However, to regenerate the spent carbon adsorbent requires a solvent of low solubility parameter to compete effectively for the weakly interactive adsorption sites on the carbon. The solvent can then dissolve the relatively non-polar solute released from the adsorbent. This, along with ease of solvent recovery from solids, is what made supercritical fluids look like attractive solvents for regenerating spent carbon from adsorption processes.

All sorts of binding forces and surface areas are encountered in the solids generated either by nature or by man. For example, solids (1) can exhibit acidic or basic functionality due to either Lewis or Bronsted sites on the solid surface, (2) can have ion exchange properties, (3) can participate in chemical reactions, and (4) can act as adsorbents for a wide range of organic chemicals. This makes the choice of the optimum leaching solvent for a given contaminant on a given solid very dependent on the properties of the solid and the contaminant.

15.5.2 Specific Applications

A variety of processes for extracting solutes from solids have been developed for use in hazardous waste minimization applications. Examples of these are now described.

Regeneration of Adsorbents

Spent adsorbents such as activated carbons, clays, silicas, and aluminas are usually regenerated and sent back to the adsorption process. Regeneration techniques involve removing adsorbed organic material from the surface of the adsorbent and then reactivating the surface, if necessary. These techniques include stripping with steam or hot inert gas, washing with a solvent for the adsorbed material, thermal desorption, and even combustion for non-combustible adsorbents. The regeneration can be done in place or in a separate facility and, with the exception of combustion, the adsorbed material can be recovered for further use or disposal.

The new addition to this regeneration arsenal is extraction of the adsorbed organic material from the adsorbent using supercritical carbon dioxide. This is done at 250°F and 3000 psi. If the adsorbent is regenerated in situ, the adsorption vessel must be built to withstand these conditions. As with other solvent extraction processes, the benefit from using supercritical gas rather than a liquid solvent is that the process is less energy intensive. It also allows much simpler and more complete solvent recovery from both the solids and the extract. The process has been developed, but no commercial installations are known to have been made.

Cleaning Soils

The use of supercritical carbon dioxide as an extraction solvent for cleaning up contaminated soil has been studied extensively by Knopf et al. (1990) at Louisiana State University. These researchers prepared topsoils and subsoils intentionally contaminated with DDT, PCBs, and toxaphene. They then studied the removal of these contaminants by leaching the soil with supercritical carbon dioxide, both with and without 5% methanol as a desorber and solubility enhancer. The presence of some water in the soil seems to assist the leaching of organics in much the same way as methanol added to the solvent when leaching dry soil. Methanol losses to the leached soil were apparently not measured. Their results show that leaching with supercritical carbon dioxide can effectively clean contaminated soil containing organic materials. The main advantage of supercritical gas over liquid solvents is the ease of recovering solvent from cleaned soil. This results in savings from reduced energy consumption and lower solvent losses.

Breaking Sludges and Emulsions

Sludges and emulsions are frequently encountered in industrial processing and petroleum refining. They usually involve the intimate mixing of water, oil, and solids to form a relatively stable semisolid material much like mayonnaise. This material is frequently stabilized by the presence of either natural or contaminating surfactants and finely divided solids. Historically, this material has usually been disposed of in landfills or by incineration. However two new processes have appeared to break these sludges and emulsions to recover their oil content and prepare clean solids and water for disposal. These are the B.E.S.T. process and the CFS liquid propane extraction process for treating sludges.

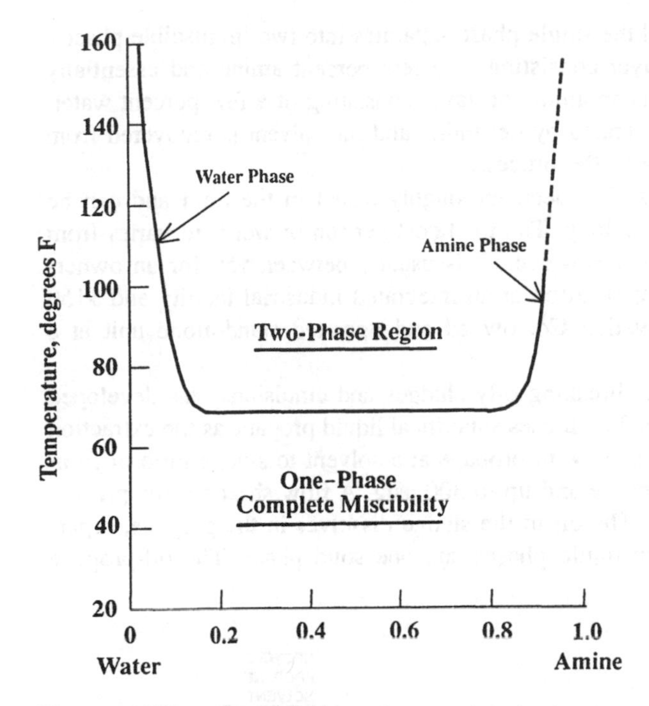

Figure 15.9 Phase behavior of water-triethylamine system.

The *B.E.S.T. (basic extraction sludge treatment) process* was developed by the Resources Conservation Company (Seattle, WA). It takes advantage of the unusual miscibility properties of aliphatic amines such as triethylamine in water. Below 68°F triethylamine and water are completely miscible, while above 68°F the solution splits into two phases. The mutual solubility of the amines and water decreases with rising temperature in the manner illustrated in Figure 15.9. This is called a lower-crtical-solution-temperature phase diagram. To operate the B.E.S.T. process, the feed sludge and the amine are mixed at a temperature of about 40°F and a solvent to sludge ratio of about 2. This dissolves all the liquid components and breaks any emulsions or stable suspensions. The released solids can then be removed by any solid-liquid separation process, such as filtration or centrifugation. Because the solution is alkaline, heavy metals in the sludge precipitate as hydrated oxides and are removed along with the solids fraction. The solids cake is dried in an indirectly heated dryer where the solvent vapors are driven off, condensed, and reused. Typically the dried solids contain more than 99% solid matter.

After removal of the solids fraction, the single-phase amine-oil-water fraction is heated to about 130°F. This is above the lower-critical-solution temperature for the system, and the single phase separates into two immiscible phases. One phase is a water layer consisting of a few percent amine and essentially no oil. The other phase is an amine-oil layer consisting of a few percent water. These layers are then separated by decanting and the solvent is recovered from both layers for recycling to the process.

The B.E.S.T. process has been thoroughly tested in the field and can be considered a proven technology. The total cost per ton of wet feed varies from one application to another. However, it is usually between $50 for an owner-operated, large-volume application at an integrated industrial facility and $150 for a Resources Conservation Co. owned and operated stand-alone unit at a Superfund site.

The CFS process for breaking oily sludges and emulsions was developed by Critical Fluid Systems, Inc. It uses subcritical liquid propane as the extraction solvent. The sludge is treated with propane at a solvent to sludge ratio of from 1 to 3 at ambient temperature and up to 300 psig. A flow sheet for the process is given in Figure 15.10. The oil in the sludge dissolves in the propane, separating the sludge into two liquid phases and one solid phase. The oil-propane

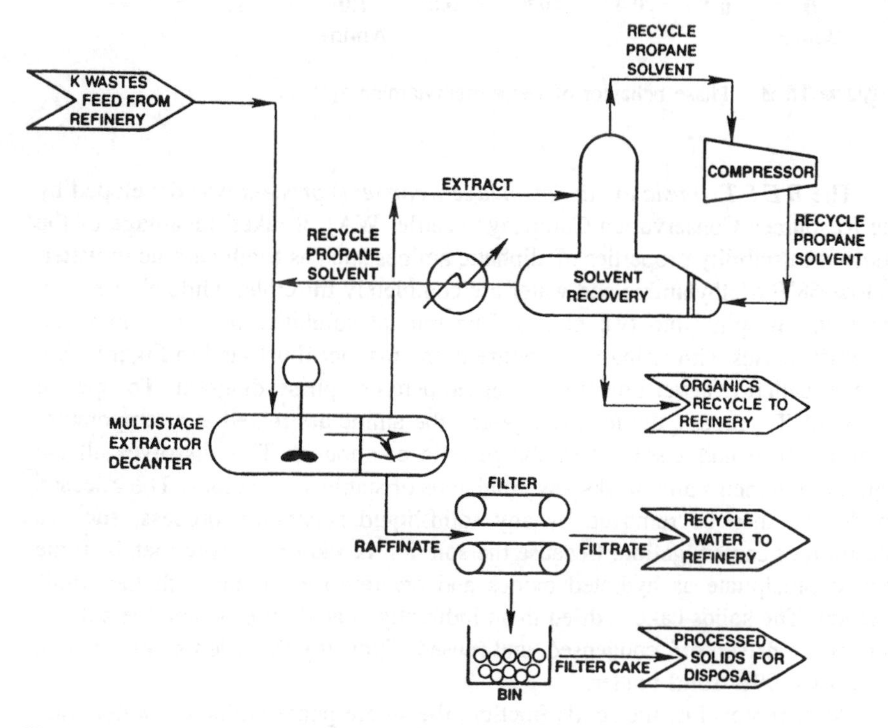

Figure 15.10 Flow diagram for CFS sludge process.

phase is separated from the water-solids slurry by decanting, and the solids are removed from the water layer by filtration. Propane is recovered for recycling to the process by distillation using vapor compression of the distillate vapor to minimize the heat requirement. The CFS system has been designated by the EPA as a best demonstrated available technology (BDAT) for treating refinery wastes.

Recovery of Oil from Mill Scale

The CFS process for extraction with near-critical liquid propane has also been applied in bench scale tests by de Fillippi and Chung (1985) to leaching of mill scale from steel production. This removes lubricating oil so that the mill scale can be recycled to the blast furnace. The tests were made at room temperature and about 300 psig and showed that about 80 to 100% of the oil could be easily removed. The recovered oil can be used as a fuel component. Near-critical carbon dioxide was much less effective than propane.

Recovery of Active Metals from Spent Catalysts

Precious metals and precious metal alloys supported on ceramic materials have been used for a long time as catalysts in the petroleum refining industry and more recently as the active component in automotive exhaust converters. When such catalysts have been deactivated past the point of simple regeneration, the precious metals can be recovered to be used to manufacture new catalysts. The metals recovery is usually performed by the catalyst manufacturers and uses leaching techniques to dissolve either the metal support or the metals themselves. After separation of the solids and solution by filtration or centrifuging, the separated streams can be used in the process for manufacturing new catalyst.

15.5.3 Leaching Equipment

Leaching equipment is much different from that used for extracting fluid feeds due to the fact that a solid phase must be transported to, mixed with, and separated from a solvent in the extraction apparatus. This is enough of a problem with liquid solvents and is further exaggerated when fluids near to or above their critical temperatures are used as extraction solvents. In general, mixing is provided either by mechanical agitation or by pumping the solvent through a captive fixed or fluid bed of the solids being extracted. The choice of leaching equipment depends on the type and volume of solids to be processed. Usually small amounts of solids are processed batchwise and large volumes are processed in continuous countercurrent fashion. The various types of equipment are now discussed and major suppliers of leaching equipment are listed in Table 15.4.

Fixed and Moving Bed Percolation Systems

Percolation systems are normally used with coarse solids when the flow of solvent through the bed of solids is relatively uniform. The simplest form of a percolation system is the pile of solids placed on a solvent-impermeable base. The solvent is sprayed onto the upper surface of the pile and percolates through the solid. It collects on the base and runs off to a suitable storage tank or pond. As it percolates through the pile of solids, the solvent dissolves the desired material from the solids and carries it out of the pile. A pumparound system is usually used to recycle the solvent back to the top of the pile. When the solvent is rich enough in solute to make solute recovery economic, it is regenerated. This type of operation is common for the recovery of metal values from low grade ores and waste piles in the mineral and mining industries. It is also effective when exposure to oxygen from the air is helpful in the solution process. The solvents must be relatively available, cheap, non-toxic, and sufficiently selective for the desired solute to get good removal from the solids. Furthermore, they must recover the desired materials in reasonable purity. This usually requires some compromise among the properties of potential solvents. In the mineral industries, the solvents are usually aqueous solutions of acids, bases, and oxidizing or reducing agents. These reagents convert insoluble metal values to soluble forms which then dissolve in the solvent.

Fixed bed percolation systems more commonly use either open or closed tanks equipped with a false bottom support for the fixed bed of solids. Open tanks can be used with innocuous or non-volatile solvents, while closed tanks must be used with volatile or toxic solvents. The solvent can be fed either continuously or intermittently to either the top or bottom of the bed of solids. This permits the bed to be operated (1) in a fully submerged state, (2) in a drained state where the solvent is sprayed onto the top of the bed of solids, which are thus exposed to air as the solvent percolates through the bed, or (3) in an intermittent fill-and-dump mode where the holding time in the submerged part of the cycle can be varied to give time for solution of difficult to dissolve materials.

Fixed bed percolation systems can be operated batchwise in a single vessel with successive additions of fresh solvent until the solids are adequately depleted of the material to be removed. Alternatively, they can also be arranged as multi-bed systems with the solvent passing through successive beds in series in such a way as to simulate counterflow of solvent and solids. These operations are illustrated in Figure 15.11. In actuality, when using a battery of fixed beds in series, as in 15.11b, the fresh solvent inlet is moved as the beds of solids become depleted of solute. For example, when bed 3 is spent, the fresh solvent feed is moved to bed 2, a new charge of fresh solids is put into bed 3, and bed 3 now takes the place of bed 1.

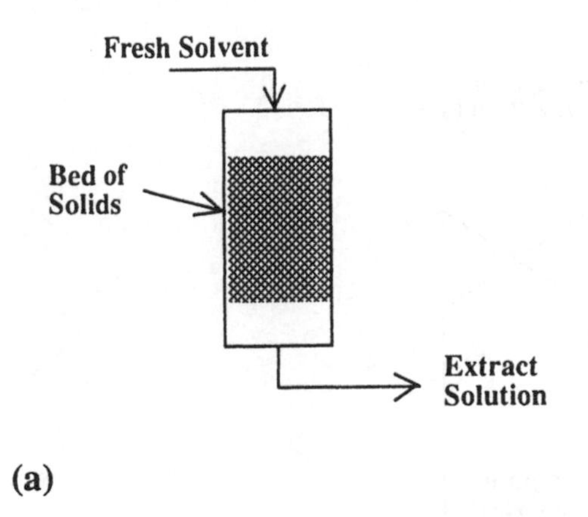

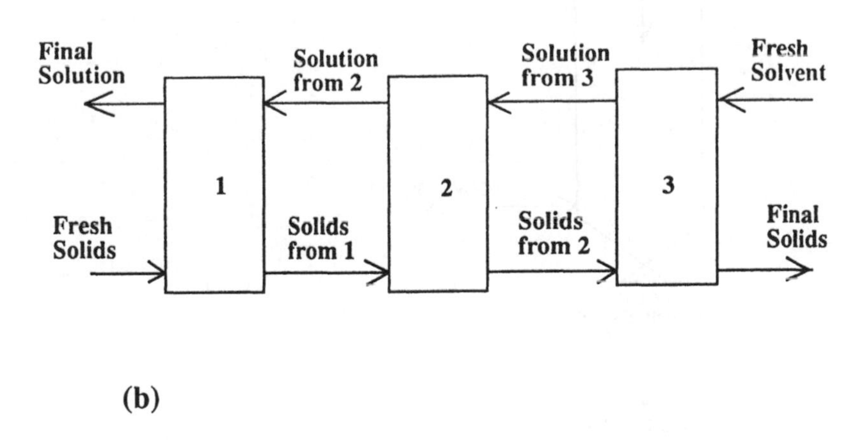

Figure 15.11 Fixed bed percolation systems. (a) Batch operation in a single vessel, (b) countercurrent operation in three vessels.

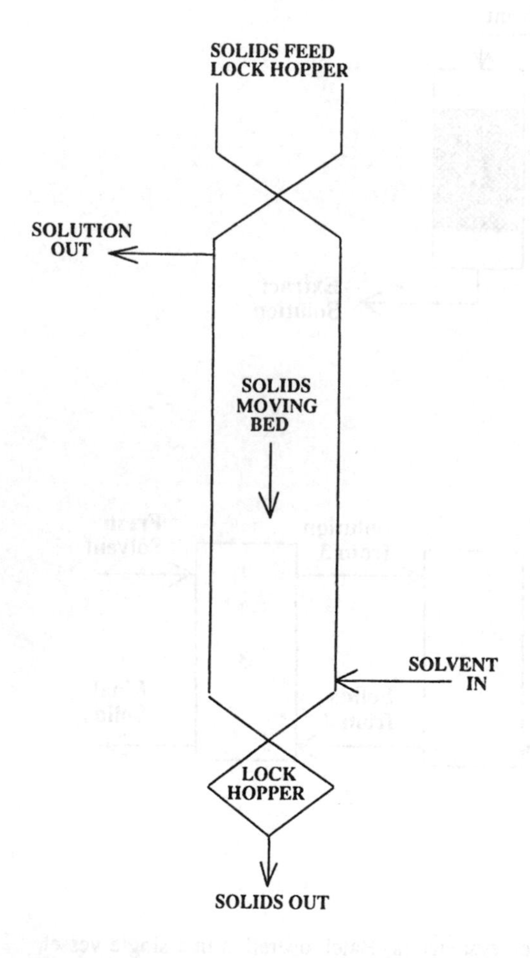

Figure 15.12 Moving bed countercurrent leaching.

In addition to changing the point of introduction of fresh solvent to get countercurrent flow through a series of solid beds, some mechanical devices have been invented in which the solid beds are actually moved through counter-flowing solvent. An example of such a device is illustrated in Figure 15.12. Here the solid bed moves by gravity against a counterflow of solvent. Lock hoppers are used to feed fresh solids at the top of the column and remove spent solids at the bottom. Careful control of solvent flow rate is required to get satisfactory operation.

Agitated Extraction Vessels

In contrast to leaching by percolation, where the solvent passes through the interstices of a fixed bed of solids, agitated extraction vessels disperse the solids in the solvent. This increases the rate of solution and permits operation with smaller particle size solids. After the solids are held in the dispersed state for an adequate length of time to approach equilibrium, they are allowed to settle from the solvent. The solids and solvent are then moved countercurrently to the next dispersion stage until they exit from the leaching apparatus. The agitation and dispersion may be provided by mechanical devices such as mixers and rakes. They can also be achieved using fluid-driven dispersion devices such as air lifts, fluid beds, and jet eductors.

A typical agitated extraction vessel is shown in Figure 15.13. The solvent and solids are fed to the batch vessel, which is provided with mixing rakes on a stirring shaft and a bed support to hold settled solids. The solids and solvent are mixed for the desired leaching time and then allowed to settle. The solution is drained through the bed support and out of the bottom of the vessel. The solids are removed from on top of the bed support through the side manhole. Such vessels can be used either singly or in series.

Continuous Countercurrent Systems

Continuous countercurrent multistage leaching is carried out either in:

1. Extraction batteries wherein the solids do not move from stage to stage but the fresh solvent injection point is moved in steps through the battery of solid beds to get simulation of true countercurrent flow
2. Conveyer systems wherein the solids are moved by mechanical means through a counterflowing pool of solvent, often using screw conveyers
3. Decanting systems wherein the solids are mechanically mixed with solvent, settled, or centrifuged to separate the solid and solvent phases, with the settled solids and clarified solvent sent in opposite directions through the battery of decanters

An example of the first type of operation was given in Figure 15.11b and the third type is illustrated in Figure 15.14. In Figure 15.14 a series of decanting

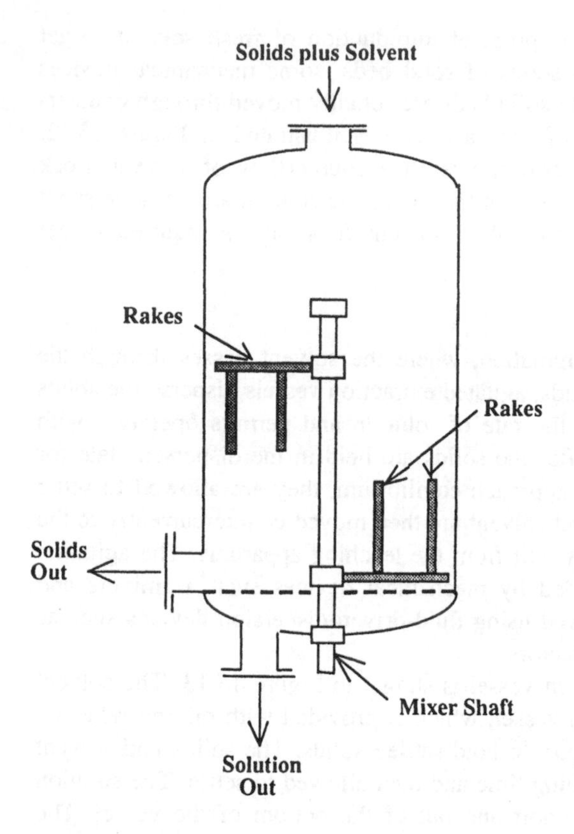

Figure 15.13 Mechanically agitated leaching vessel.

thickeners is used to obtain counterflow contacting of solids and solvent. Fresh solvent enters and spent solids leave unit No. 4 while fresh solids enter and concentrated solution leaves unit No. 1. Contacting is provided by mechanical devices, and solids and liquid are separated in the clarifier region in the upper part of the vessels. The clarified solution overflow and the concentrated solids underflow pass countercurrently between vessels.

15.5.4 Common Problems and Troubleshooting

The problems most frequently encountered in leaching of solids have to do with the effective contacting of liquids and solids and the effective recovery of liquids from solids. For example, in percolation operations the problem is to get satisfactory distribution of flow through the bed. This is required to promote good contacting of the entire solid bed with the solvent and avoid channeling, i.e.,

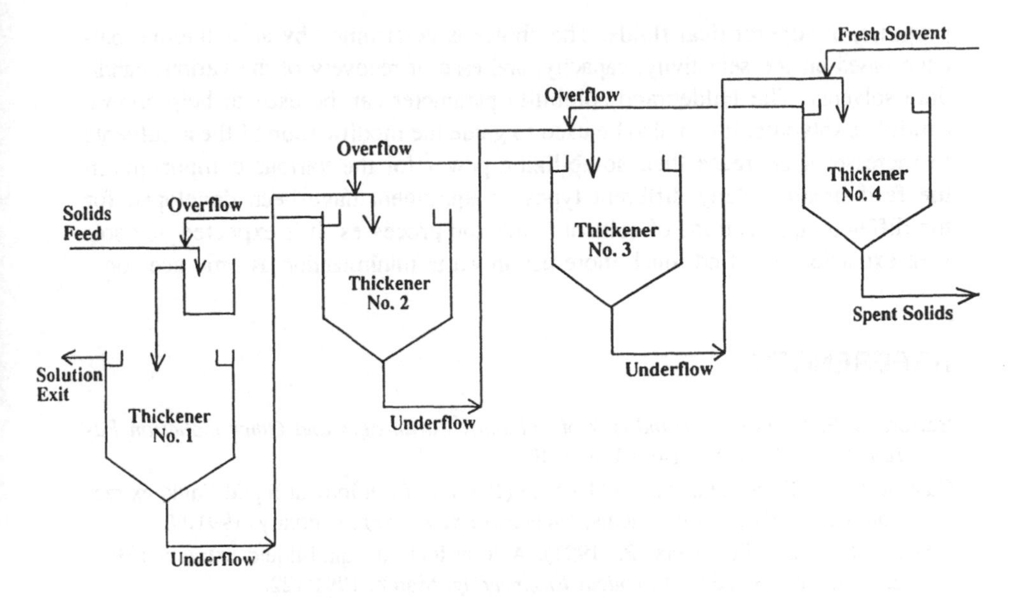

Figure 15.14 Flow plan for a countercurrent decantation system.

streaming of the solvent through only a portion of the solid bed. Good contacting is achieved by proper sizing of solids in the bed and adequate design of the solvent distribution devices. Coarser solids give better solvent flow distribution through the bed but slow down the rate of solution of the solids in the solvent.

Recovery of solvent from the wet solids is usually done by a drying or stripping operation which requires volatilizing the solvent from the solid particles. Hot air or steam can be used as the stripping agent and the solvent can be recovered by condensation, adsorption, or scrubbing from the stripping agent. To make drying of the solids and removal of the solvent easier, it is possible to use a volatile liquid or supercritical fluid as solvent. However, the price to be paid for this easier solids drying and solvent recovery is operation under pressure with increased equipment costs. Lock hoppers to introduce solids into pressurized systems and pressure vessels to hold the solid beds may be needed. An economic balance must be made to decide if the easier solvent recovery will offset these increased costs.

15.6 SUMMARY

In summary, solvent extraction processes are very useful for the recovery of valuable materials for reuse or sale and the removal of banned materials from waste streams. The solvents can be conventional liquids, condensed vapors under pressure, or supercritical fluids. The choice is determined by an economic balance based on the selectivity, capacity, and ease of recovery of the various candidate solvents. The Hildebrand solubility parameter can be used to help choose candidate solvents. It can also be used to guide the modification of these solvents to increase or decrease their solubilizing power for the various components in the feed stream. Many different types of equipment have been developed for the different applications of solvent extraction processes. It is expected that solvent extraction will find much more use in waste minimization as time goes on.

REFERENCES

Barton, A. F. M. (1983). *Handbook of Solubility Parameters and Other Cohesion Parameters*, CRC Press, Boca Raton, FL.

Cusack, R. W., Fremeaux, P., and Glatz, D (1991). A fresh look at liquid-liquid extraction, Part 1, Extraction systems, *Chemical Engineering*, February, 1991:66.

Cusack, R. W. and Fremeaux, P. (1991). A fresh look at liquid-liquid extraction, Part 2, Inside the extractor, *Chemical Engineering*, March, 1991:122.

Cusack, R. W. and Karr, A. (1991). A fresh look at liquid-liquid extraction, Part 3, Extractor design and specification, *Chemical Engineering*, April, 1991:112.

Cotton, F. O., Brinkman, D.,W., Reynolds, J. W., et al. (1980). *Pilot-Scale Used Oil Re-refining Using a Solvent Treatment Distillation Process*, BETC/RI-79/14, Bartlesville, OK, January, 1980.

de Fillippi, R. P. and Chung, M. E. (1985). *Laboratory Evaluation of Critical Fluid Extractions for Environmental Applications*, EPA/600/2-85/045(NTIS No. PB 85-189843), April, 1985.

Diepen, G. A. M. and Scheffer, F. E. C. (1948). The Solubility of Naphthalene in Supercritical Ethylene, *Journal of the American Chemical Society*, *70*:4085.

Hansen, C. M. (1967). The three-dimensional solubility parameter—Key to paint component affinities, *Journal of Paint Technology*, *39*:104, 505.

Hildebrand, J. H. and Scott, R. L. (1964). *Solubility of Non-Electrolytes*, 3rd. ed., Dover, New York.

Knopf, F. C., Dooley, K. M., Ghonasgi, D., et al. (1990). Supercritical CO_2-Cosolvent Extraction of Contaminated Soils and Sediments, *Environmental Progress*, *9*(4):197.

Martin, A., Wu, P., Adjei, A., et al. (1981). Extended Hansen Solubility Approach: Naphthalene in Individual Solvents, *Journal of Pharmaceutical Sciences*, *70*(11):1260.

Seymour, R. B. (1984). *Handbook of Chemistry and Physics* (R. C. Weast, M. J. Astle, W. H. Beyer, eds.), CRC Press, Boca Raton, FL, p. C-699.

Short, H., Hunter, D., and Parkinson, G. (1987). Three re-refining methods for waste lube oils, *Chemical Engineering*, July 20:21.

Wallin, B. K., Condren, A. J., and Waldan, R. L. (1981). *Removal of Phenolic Compounds from Wood Preserving Wastewaters*. EPA-600/52-81-043, April, 1981.

Index

Milton Keynes UK
Ingram Content Group UK Ltd.
UKHW021905071024
449327UK00021B/1621

9 780367 401719